Lecture Notes in Networks and Systems

Volume 85

The series “Lecture Notes in Networks and Systems” publishes the latest developments in Networks and Systems—quickly, informally and with high quality. Original research reported in proceedings and post-proceedings represents the core of LNNS.

Volumes published in LNNS embrace all aspects and subfields of, as well as new challenges in, Networks and Systems.

The series contains proceedings and edited volumes in systems and networks, spanning the areas of Cyber-Physical Systems, Autonomous Systems, Sensor Networks, Control Systems, Energy Systems, Automotive Systems, Biological Systems, Vehicular Networking and Connected Vehicles, Aerospace Systems, Automation, Manufacturing, Smart Grids, Nonlinear Systems, Power Systems, Robotics, Social Systems, Economic Systems and other. Of particular value to both the contributors and the readership are the short publication timeframe and the world-wide distribution and exposure which enable both a wide and rapid dissemination of research output.

The series covers the theory, applications, and perspectives on the state of the art and future developments relevant to systems and networks, decision making, control, complex processes and related areas, as embedded in the fields of interdisciplinary and applied sciences, engineering, computer science, physics, economics, social, and life sciences, as well as the paradigms and methodologies behind them.

**** Indexing: The books of this series are submitted to ISI Proceedings, SCOPUS, Google Scholar and Springerlink ****

More information about this series at http://www.springer.com/series/15179

Gheorghe Ion Gheorghe
Editor

Proceedings of the International Conference of Mechatronics and Cyber-MixMechatronics – 2019

Springer

Editor
Gheorghe Ion Gheorghe
Bucharest, Romania

ISSN 2367-3370 ISSN 2367-3389 (electronic)
Lecture Notes in Networks and Systems
ISBN 978-3-030-26990-6 ISBN 978-3-030-26991-3 (eBook)
https://doi.org/10.1007/978-3-030-26991-3

This Springer imprint is published by the registered company Springer Nature Switzerland AG
The registered company address is: Gewerbestrasse 11, 6330 Cham, Switzerland

Preface

The third International Conference of Mechatronics and Cyber-MixMecatronics (ICOMECYME) was held in Bucharest, Romania, on 5–6 September, 2019. This conference is envisioned as a forum and an opportunity to researchers, engineers, professors, PhD students and graduate students as well as business representatives from all over the world to present their research results and development activities. It comes as a consequence of the expansion of the field of mechatronics, which has come to step into the world of newer transdisciplinary fields of adaptronics, integronics and cyber-mix-mechatronics.

Originally entitled "International Conference on Innovations, Recent Trends and Challenges in Mechatronics, Mechanical Engineering and New High-Tech Products Development" (MECAHITECH), the conference was held for the first time in 2009.

With a new name and derived from an event that deeply penetrated into the academic community, gathering specialists from all over the world—including North America, South America and Asia, the conference facilitates reflection on the current state of the addressed fields and discussions about potential future directions for research.

The papers presented at the conference will be a ramp for PhD, PhD Students, researchers and engineers, who will have to be prepared to cross-traditional boundaries in order to accommodate the new technologies. This volume will thus be a valuable addition to the literature, it will examine the intersection between mechatronics, cyber-mechatronics and cyber-mixmechatronics, as well as other related disciplines, and it will assess the implications for industry throughout the world.

I am particularly grateful to the authors for their contributions and all the participating experts for their valuable advice. Furthermore, I thank the staff for their cooperation and support, and especially, all members of the international programme committee and the organizing committee for their work in preparing and organizing the conference.

On behalf of the organizing committee, I would like to thank Springer for its professional assistance and particularly to Ms. Varsha Prabakaran and Mr. Holger Schäpe, who supported this publication.

Gh. Gheorghe
Editor in Chief and Conference Chairman

Contents

A Simple Loading System Applied for Needle Bending Measurement

Jan Hošek(✉) and Šárka Němcová(✉)

Faculty of Mechanical Engineering, Czech Technical University in Prague, Technická 4, Prague 6, Czech Republic
{jan.hosek, sarka.nemcova}@fs.cvut.cz

Abstract. The paper presents a simple system for load and deformation measurement of a needle-like tubular structure to simulate mechanical conditions and loads on miniature endoscope during eye surgery operation. The system allows for loading force measurement with a strain gauge load cell, the deformation measurement with an optical triangulation sensor and needle deformation measurement with a remote camera. The paper presents the system structural design, the system calibration and its application for the needle bending test measurement.

Keywords: Load cell · Needle · Bending · Deformation · Endoscope · Eye surgery

1 Introduction

Stress and strain measurement is a common technique for testing mechanical structures. There exist a large variety of bend loading machines for testing standard samples. Such universal systems are typically large to achieve high enough stiffness to allow for reliable data measurement. Small or portable systems are usually custom designed to meet demands for special sample measurement. In our case we aim for a small portable loading system capable to produce a desired level of loading force (0–10)N in a small contact point and able to measure an induced sample deformation caused by the load. The system needs to be stiff enough and clearance-free to give reliable data measurement.

We want to use such a system for loading tests of various samples, but firstly, we want to use it for stiffness measurement of a needle-like tubular structure to simulate mechanical conditions and loads on a miniature endoscope under eye surgery operation. A standard commercial testing systems are designed to follow the requirements of ISO 80369 standards. We also found few examples of needle loading systems designed for different aims, usually to test a needle bending during its insertion to a tissue [1–3]. Eventually, we designed a new system which differs from the standard design of loading systems, but which fulfilled our needs.

This paper presents the system structural design, the system calibration and its application for the needle bending test measurement.

G. I. Gheorghe (Ed.): ICOMECYME 2019, LNNS 85, pp. 1–10, 2020.
https://doi.org/10.1007/978-3-030-26991-3_1

2 The Loading System Design

The system design must solve few possible problems. While the needle is a relatively soft structure we have expected a relatively large deformation under a small force load. For this reason, there is needed to solve how to separate the sample deformation measurement from the loading structure motion position measurement.

The loading structure needs to be provided with an exact contact point of the sample. The applied force needs to be measured together with deformation of a commercial load sell provided with strain gauges. While the load cell is deformed by applied force, the sample deformation needs to be measured as a position change of a mechanical part in front of the load cell to eliminate the influence of the load cell deformation to the measured data. The needle deformation can be measured e.g. with a camera. To achieve even higher resolution, we decided to measure the position of a stiff part of the loading cell directly, with a precise optical gauge. In such a case the actual position of the basement part of the load cell is not important. The only demand on the basement part of the load cell is that its motion needs to be clearance-free. Such motion is commonly provided with linear guides and a driving lead screw, at least two or three precise and expensive parts. As we expect the sample deformations in a range of few millimeters we replaced the common linear motion with a rotary motion around a remote center of motion. An approximate motion of this kind can be easily realized as clearance-free with a flexure rotary joint or a self-aligning ball support.

A measured sample is fixed in a XYZ movable support where no fine motions are needed. The fine loading motion is performed just by the loading system rotation. Keeping in mind the Abbe principle [4], we needed to hold the loading axis free for placement of a distance measurement gauge. This condition leads to using a bar cell gauge allowing for an off axis design of the system. We used a commercially available 1 kg bar load cell with the HX711 balance module, processed by an Arduino Uno minicomputer. To achieve a point-like load contact to a tubular needle we provided the load cell with a contact edge perpendicular to the sample axis. The edge was realized by a right angle optical prism. The contact between the prism and the needle surface occurs at the prism bevel surface or its edge during the loading. The prism fixed to the load cell is always loaded with compression force only. The basement of the load cell is an aluminum beam allowing for a rotary motion around the axis at the distance of L = 150 mm from the applied load force contact position. The rotary axis is formed by centers two balls in self-aligning ball supports of diameter D = 6 mm. A small axial preload eliminates any clearance. The mass and the corresponding torque of the aluminum beam is fully compensated for with a soft compression spring. From this zero position the aluminum beam with the load cell can be adjusted with a micrometer screw in contact with a ball fixed on the top of the beam. The real position of the load cell part in contact with the sample is measured with the Micro-Epsilon optoNCDT 1402-5 gauge having 1 μm distance resolution in 5 mm wide position measurement working range, and under 1.5 kHz reading frequency. The overall view to the physical realization of the loading system with manual load adjustment and electronic loading force and contact position reading is shown in Fig. 1.

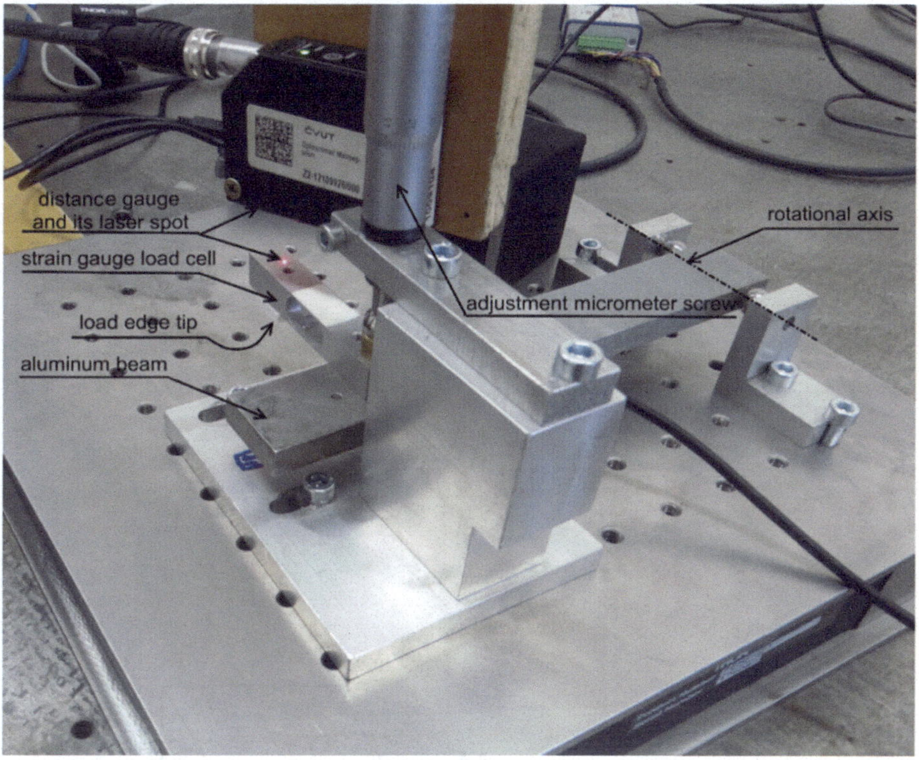

Fig. 1. A photo of the physical realization of the loading system with manual load adjustment and electronic loading force and sample contact position reading.

3 The Loading System Analysis

The system load cell is permanently fixed to the aluminum beam so they move together by rotation around rotational axis made by two balls in contact with precisely reamed holes and lifted up with a soft spring so as to be in a point contact with the micrometer screw contact surface. The load cell can be assumed as three independently movable parts connected together with clearance-free flexure rotary joints. The Part 3 of the load cell is permanently fixed to the aluminum beam and performs the same rotary motion around the axis as the aluminum beam. The Part 2 consists of a parallelogram composed of two levers rotating in four flexure rotary joints. The Part 1 represents a rigid metal beam of the load cell capable to perform a linear motion with regard to the Part 3 when a non-zero force F is applied to the loading edge of the optical prism. This position change of the Part 3 is registered with the non-contact optical sensor magnetically fixed to the loading system basement. It provides the Part 3 position information independently on whether it was moved by micrometer screw adjustment or the load cell response to the contact force F. A detail view of the load cell and a schematic view of its Parts 1, 2, and 3 is shown in Fig. 2.

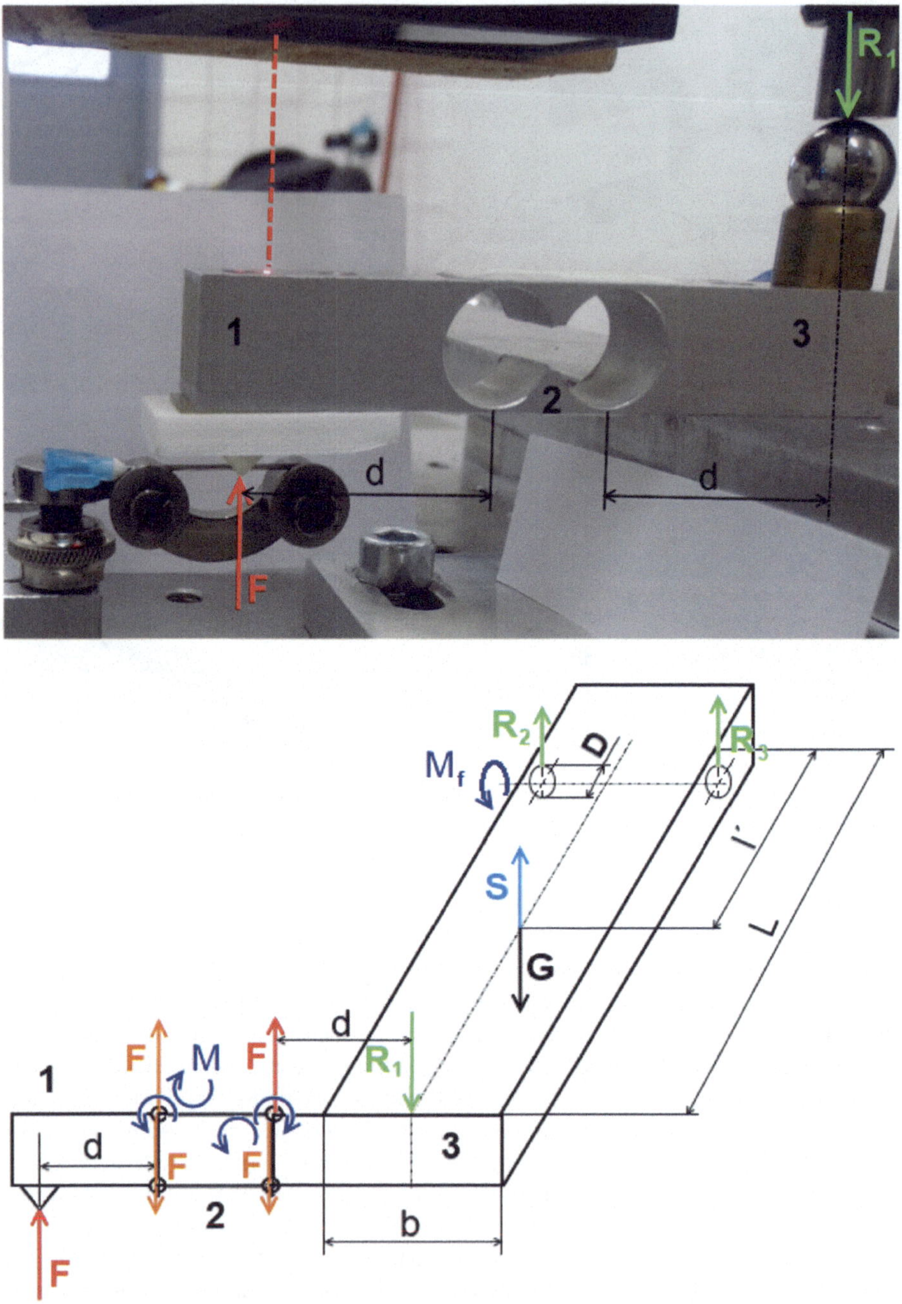

Fig. 2. A detail view to the load cell and the schematic view to its Parts 1, 2 and 3 with all forces and torques emerging within the system.

The force F applied to the Part 1 is compensated for with the inverse reaction F and torque M at the contact with the Part 2. The Part 2 transfers the reaction and toque to the contact with the Part 3, where the force F emerges as the same as the contact force F. The torque M is given by relation

$$M = Fd. \tag{1}$$

The Part 3 must reach a static balance described by equations

$$M + Fd + R_2\frac{b}{2} - R_3\frac{b}{2} = 0 \tag{2}$$

$$F - R_1 + S - G + R_3 + R_2 = 0 \tag{3}$$

$$FL - R_1L + (S - G)l' - M_f = 0, \tag{4}$$

where b = 30 mm is the ball support width, d = 26.5 m is the distance from the flexure joint to the force axis. G is the gravity force compensated by the spring force $S = G + k\varphi l'$, where k it the spring stiffness, φ is the rotation angle around the axis measured from the balance position where $S = G$, l' = 100 mm and L = 150 mm are the distances of the corresponding forces from the rotational axis, and M_f is the friction torque given by relation

$$M_f = f\frac{D}{2}(R_3 + R_2). \tag{5}$$

The reactions R_2 and R_3 compensate the torque produced by the contact force F. The reactions can be expressed by relations

$$R_3 = R_0 + R' \tag{6}$$

$$R_2 = R_0 - R', \tag{7}$$

what simplifies the solution while

$$R_3 + R_2 = 2R_0 \tag{8}$$

$$R_3 - R_2 = 2R'. \tag{9}$$

Solving for the individual reactions we get

$$R' = 2\frac{d}{b}F = 1.766F \tag{10}$$

$$R_1 = F\left(1 + \frac{2Dfd}{b}\right) + k\varphi l'\frac{l' + \frac{D}{2}f}{L + \frac{D}{2}f} \tag{11}$$

$$R_0 = F\left(\frac{1}{2} + \frac{Dfd}{b}\right) + \frac{k}{2}\varphi l'\left(\frac{l' + \frac{D}{2}f}{L + \frac{D}{2}f} - 1\right). \tag{12}$$

The loading system design dimensions were selected to reach a high ratio of the friction torque arm $D/2$ and the actuation torque arm L, where $2L/D = 50$. This assures that the system does not possess any motion backlash due to the friction. We also analyzed the influence of the approximate motion of the contact edge to the applied force F. The force F is changed by 10^{-4} times for the z travel distance up to ± 2.12 mm from the horizontal position. If there is acceptable force F uncertainty of $10^{-3}F$, the total travel in z axis around the horizontal position can reach up to ± 6.71 mm, which is much more than the used optical distance sensor working range.

4 The Needle Deformation Measurement

The aim of the system is to use it for various loading reasons. Actually we are interested in bending characteristics of an eye endoscope under development. We decided to simulate the eye endoscope bending with a bending of a syringe needle of corresponding dimension 23G.

4.1 Needle Three-Point Bending Test

The 3-point bending test is a standardized test method for determination of flexural properties of beam-like specimens. For this reason, we used it as the first insight to the experimental task of the needle bending. We put the needle on top of two 9 mm diameter rollers with center's distance of 19 mm. We adjusted the contact loading to the center of the rollers and performed the loading in 100 μm steps set with the micrometer screw. The needle deformation response is shown in Fig. 3.

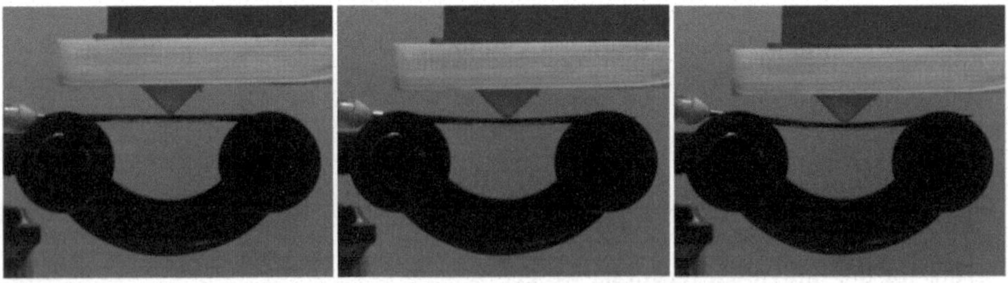

Fig. 3. A detail view to the tested needle under load from zero (left image) to maximum used load (right image).

We took simultaneous reading of the load cell signal with 3.5 Hz and distance gauge signal with 50 Hz frequency. We synchronized the data to the maximum load time point and interpolated them to receive a limited number of data for further processing. The recorded and processed data are shown in Fig. 4.

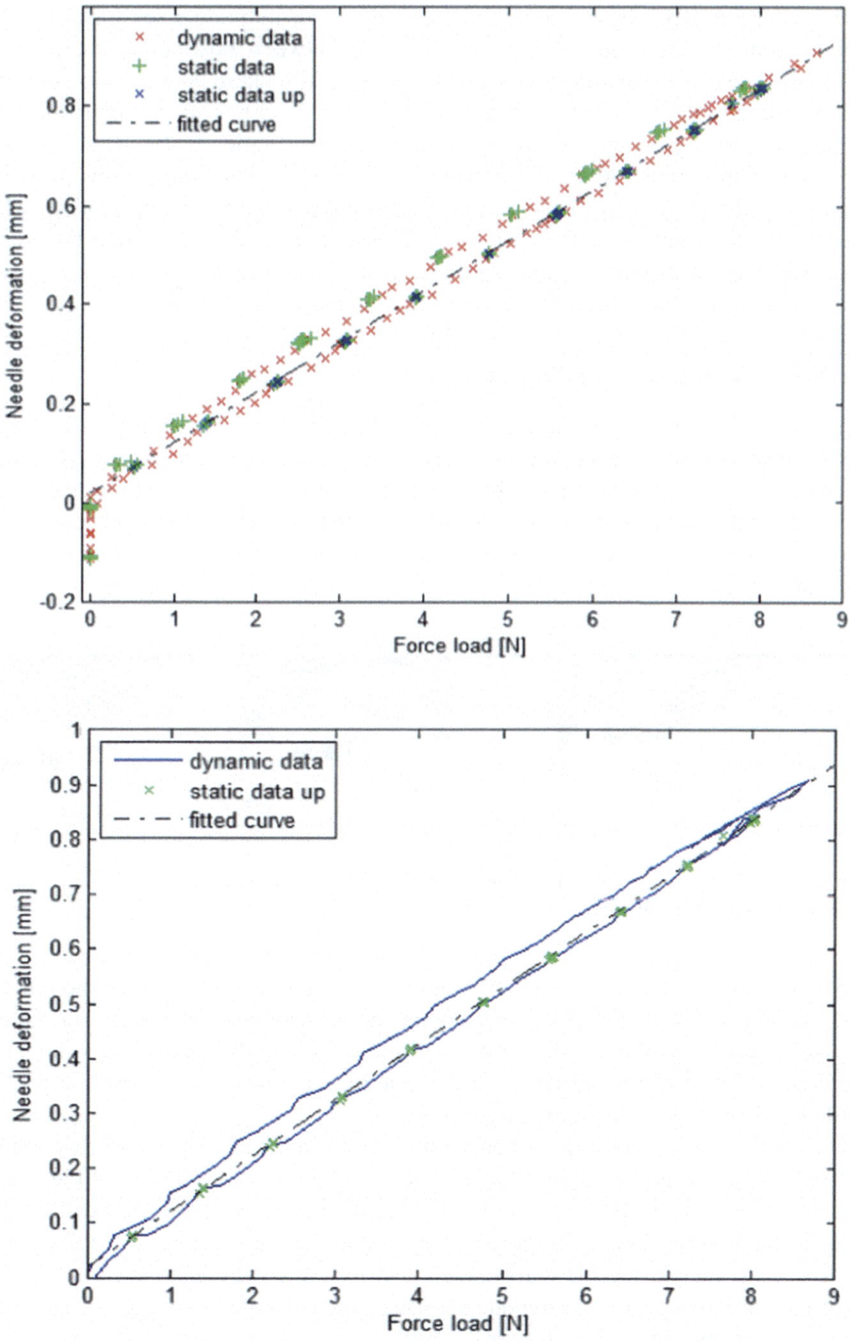

Fig. 4. The needle deformation measurement data under 3-point bending test.

The measured data show different characteristics for individual set position values by the micrometer screw and for dynamic motion in between those static set points, and also for the loading force's increases and decreases. The static data points show very good reproducibility (marked with green or blue crosses). The dynamic data - in between the static positions where needle deformation varies - show always lower needle deformation under given force load compared to the static data points. It may be explained by the presence of friction forces and torques during needle motion, which decreases the bending needle reaction force. The static data show excellent linearity of $R = 0.9999$ for the increase of the loading force. Unloading data points show distinguished hysteresis and higher needle deformation caused by the applied force.

4.2 Needle Two-Point Bending Test

We prepared also a system for a needle 2-point bending test, as such load corresponds better to the expected load scheme of the eye endoscope, where the tube is fixed at the tool end and is loaded with a force by eye cornea structure. We fixed the needle to a 23G trocar and loaded it at given distance from the port by the loading cell optical prism bevel edge. The needle deformation response under 2-point bending test is shown in Fig. 5.

Fig. 5. A detail view to the tested needle under 2-point bending load from zero (left image) to maximum used load (right image).

We kept the same measurement conditions for 2-point bend test as for the previous 3-point bend test just with a changed value of the individual step distance set with the micrometer screw. As the needle has lower stiffness and higher deformation under the same load level, we performed loading data measurement in 250 μm steps. The recorded and processed data are shown in Fig. 6.

The measured data show similar characteristics as in the case of the 3-point bending test. The static data show excellent linearity $R = 0.9998$ for the increases of the loading force. Unloading data points show distinguished hysteresis of higher needle deformation to the applied force. To the contrary to the 3-point bending test, the dynamic data give always higher needle deformation than for the static data load case. It may be caused by a different accumulation and release of the deformation energy of the needle during loading and unloading due to the different level and opposite direction of the friction force between the steel needle surface and glass loading prism in its contact point. The correctness of this explanation needs to be proved with exact deformation profile measurement of the needle during its loading and unloading.

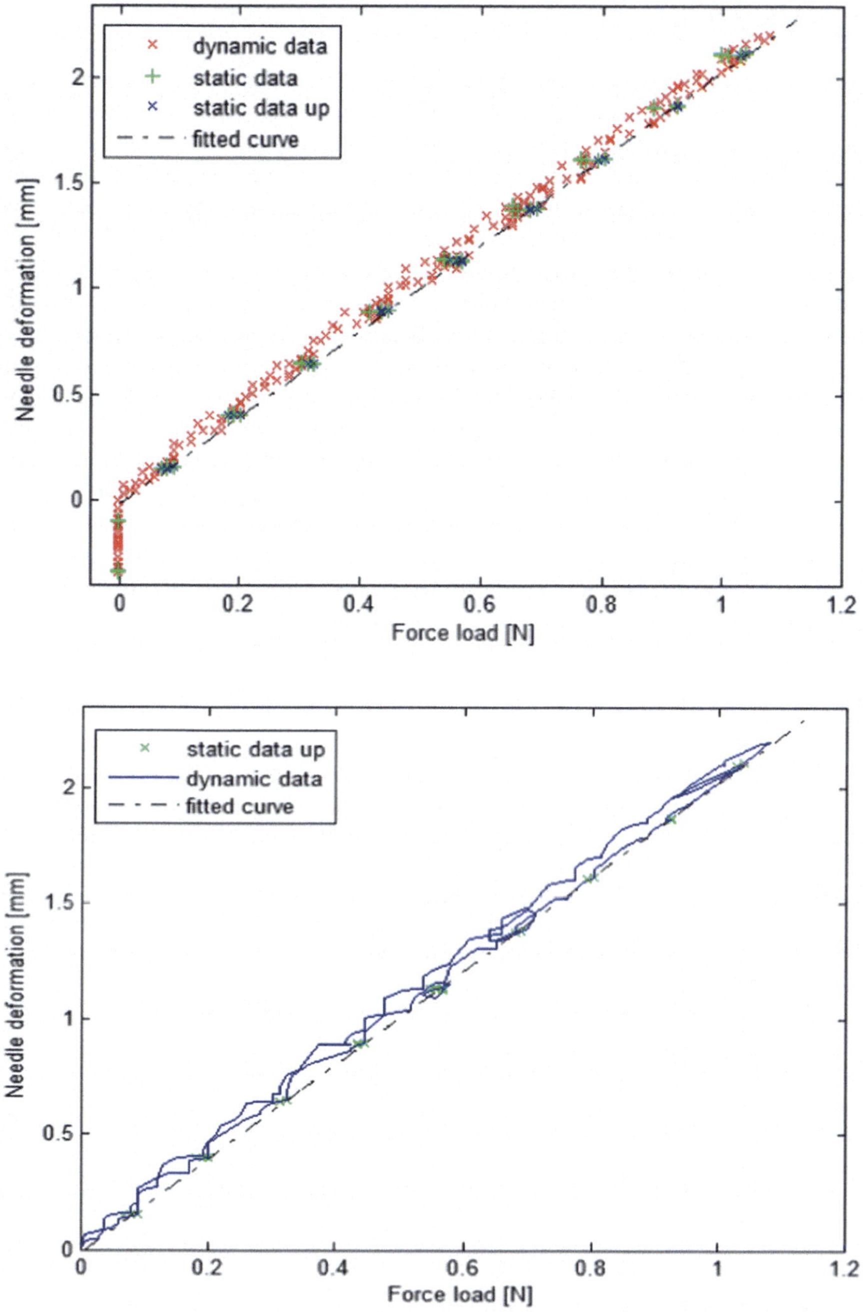

Fig. 6. The needle deformation data under 2-point bending test.

5 Conclusion

The paper presented a new design of a simple loading system based on replacing the standard linear motion of the loading tip with approximate circular motion. We performed a mechanical analysis of the proposed system and proved a high reliability of about 10^{-4} within assumed working range. We performed the system calibration and load/deformation measurement in 3-point and 2-point bending test of a needle specimen. Obtained data show excellent linearity for load increases. Notable hysteresis observed in the case of the specimen unloading needs to be correctly explained. The system sensitivity was able to distinguish a discrepancy between linear characteristics of the static data measurement and dynamic data points. In the case of 3-point bending load we found that the needle deformation is always lower for the dynamic data points than for the static data points. The 2-point bending load shows completely opposite behavior where the needle deformation is always higher for the dynamic data points than for the static data points. We tried to explain the described data difference by different accumulation and release of the deformation energy of the needle during its loading and unloading due to the different level and opposite direction of the friction force between the steel needle surface and glass loading prism in its contact point. The correctness of this explanation needs to be proved with exact deformation profile measurement of the needle during its loading and unloading.

References

1. Van de Berg, N.J., de Jong, T.L., van Gerwen, D.J., Dankelman, J., van den Dob-belsteen, J.J.: The influence of tip shape on bending force during needle insertion. Sci. Rep. **7**, 40477 (2017)
2. Mahvash, M., Dupont, P.E.: Mechanics of dynamic needle insertion into a biological material. IEEE Trans. Biomed. Eng. **57**(4), 934–943 (2010)
3. Misra, S., Reed, K.B., Schafer, B.W., Ramesh, K.T., Okamura, A.M.: Mechanics of flexible needles robotically steered through soft tissue. Int. J. Robot. Res. **29**(13), 1640–1660 (2010)
4. Abbe, E.: Messapparate für Physiker (in German). Zeitschrift fur Instrumentenkunde **10**, 446–448 (1890)

Researches on the Use of Ozone Generators for Wastewater Treatment

Roza Albertino Giovani[1(✉)], Albu Nicoleta Dorina[1], Donțu Octavian[1], Moga Corina[2], Băran Nicolae[1], and Constantin Mihaela[1,2]

[1] Politehnica University of Bucharest, Splaiul Independenței no. 313, sector 6, Bucharest, Romania
bertu_san@yahoo.com
[2] DFR Systems SRL, Bucharest, Romania

Abstract. The paper presents the need for water aeration, increasing the dissolved oxygen content in water. The formation of ozone and its use is exposed. The authors chose a TCB - 30003 ozone generator to be used in the experimental installation. Thus, atmospheric air and subsequently atmospheric air with ozone generated by CORONA type electric discharge are introduced in a water tank. At the end of the paper the experimental obtained data are compared.

Keywords: Ozone generator · Water depollution · Wastewater treatment

1 Introduction

Increasing the dissolved oxygen concentration in water is accomplished by introducing into the water an oxygen containing gas: this process is defined as aeration or oxygenation of water.

The literature [1–3] uses the term aeration or oxygenation; it is proposed to make the following distinction:

- Water aeration means the introduction of atmospheric air into the water (21% O_2 + 79% N_2);
- Water oxygenation means introducing a gaseous mixture as follows:
  - Atmospheric air + oxygen from a cylinder in volumetric ratio (25%, 50%, 75%, 100%),
  - Air with low nitrogen content (r_{O2} = 95%; r_{N2} = 5%) supplied by oxygen concetrators,
  - Air containing ozone.

Water aeration is carried out in the following fields [4–6]:

- In sewage treatment plants;
- In water treatment processes, removal of dissolved inorganic substances or chemical elements such as iron, manganese, etc., by oxidation and formation of sedimentable compounds or which can be retained by boiling;

G. I. Gheorghe (Ed.): ICOMECYME 2019, LNNS 85, pp. 11–21, 2020.
https://doi.org/10.1007/978-3-030-26991-3_2

- On biological treatment of waste water, either through the activated sludge process or with bio filters;
- In disinfection processes, by ozonation of raw water captured from a source in the purpose of its drinking;
- In separating and collecting emulsified fats from wastewater.

Water oxygenation is a mass transfer process with wide application in water treatment. Oxygenation equipment's are based on the dispersion of one phase into the other, for example gas in the liquid, energy consuming process.

Water oxygenation is carried out in the following fields [7–9]:

- In sewage treatment techniques;
- Maintaining an oxygen concentration in aquariums;
- The operation of swimming pools;
- In the medical field in the supply of oxygen to some lung patients.

Aeration or oxygenation of water leads to increased dissolved oxygen in water.

Aeration is necessary to improve water quality to avoid oxygen deficiency in systems where there is biochemical oxygen demand over water self-aeration capacity to remove toxic gases that can be found in water and in the wastewater treatment process [10, 11].

The main purpose of water aeration, irrespective of the industry and the reason of is use, is to increase or maintain an optimal level of dissolved oxygen in a mass of water.

The oxygen required for the aeration process is taken from the atmospheric air and introduced into the water. For this aeration to be effective, uniform air dispersal must be ensured throughout the mass of water in a tank or basin; the air must be uniformly distributed so as to ensure the oxygen demand.

Dissolved oxygen content is the most important indicator of water quality.

Fishes, for example, need to survive up to 5 mg/dm^3 of dissolved oxygen [12].

The amount of oxygen in the water is consumed by different biological or chemical processes. The amount left in the water as a result of these processes depends on the rate of de oxygenation and the oxygenation rate (aeration), which can occur naturally or artificially [12].

By aerating water is meant the transfer of oxygen from atmospheric air into water, which is actually a phenomenon of transferring a gas into a liquid.

The most common method of removing impurities of organic nature under the action of a biomass of aerobic bacteria is the introduction of gaseous oxygen into the wastewater.

Oxygen originates most frequently from atmospheric air, in this case the process is known as water aeration (Fig. 1).

Dissolved oxygen in water is known as dissolved oxygen (Fig. 1) and is measured in mg O_2/dm^3.

From the above figure one can see that each molecule of water consists of an oxygen molecule connected to two hydrogen molecules (the blue sphere coupled to two pink spheres). The oxygen molecules (the blue spheres) constituting the dissolved oxygen can be found among water molecules The maximum amount of oxygen that can be dissolved in water depends on a number of physical and chemical parameters

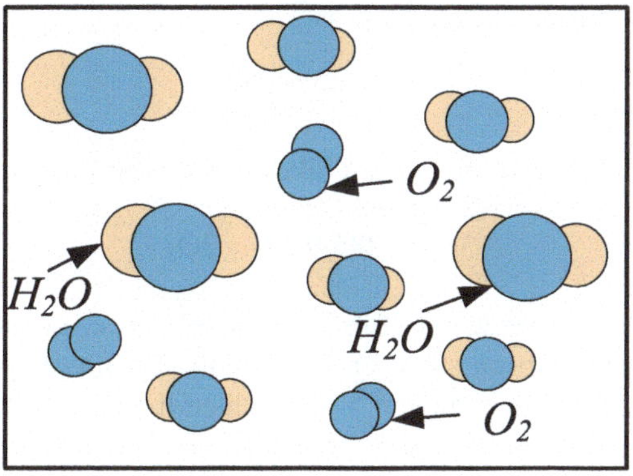

Fig. 1. View of the molecular structure: dissolved oxygen [12].

such as: atmospheric pressure, water temperature, salinity, the degree of water turbulence [12].

Water temperature is an important factor, so the warmer the water, the lower the dissolved oxygen concentration. So [4]:

- at t = 10 °C, in clean fresh water, an amount of 11.3 mgO_2/dm^3 can be absorbed;
- at t = 25 °C, in clean water, only 8.3 mgO_2/dm^3 can be absorbed.

2 Ozone Formation and Its Use

(a) Ozone formation mechanism

Ozone (O_3) is a variety of oxygen (allotropic form of oxygen), the molecule of which consists of three oxygen atoms (the triatomic molecule). Ozone is free from atmospheric oxygen, resulting from ultraviolet radiation or from electrical discharge [1]. Under these factors, some of the atmospheric oxygen decomposes into free oxygen atoms. The ozone formation mechanism is the following: absorbing a radiant energy quantum hν, the oxygen molecule (O_2) passes into the activated molecule (O_2^*):

$$O_2 + h\nu = O_2^* \tag{1}$$

it then decomposes into two oxygen atoms:

$$O_2^* \rightarrow 2O \tag{2}$$

and free oxygen atoms act on molecular oxygen O_2 by turning it into ozone:

$$2O + O_2 \rightarrow 2O_3 \tag{3}$$

The global ozone formation equation is:

$$3O_2 + hv \rightarrow 2O_3 \tag{4}$$

The formation of ozone from oxygen takes place with energy consumption:

$$3O_2 + 285.9\ \text{kJ} = 2O_3 \tag{5}$$

Therefore, when the reverse transformation takes place, ozone frees up the same amount of energy that has been spent on its formation from oxygen.

Lighting and electric discharge can also produce the reverse transformation:

$$2O_3 \leftrightarrow 3O_2 \tag{6}$$

Ozone can be prepared by physical or chemical means. Due to the fact that ozone decomposes and turns into oxygen again, according to the previous reversible reaction, it cannot be a pure ozone preparation, but only an ozonisation of oxygen.

By physical means, ozone can be prepared by means of electric effluents (electric discharge without sparks) on oxygen in a device called ozoniser.

(b) The properties of ozone

Ozone is a smell of garlic gas, easily recognizable even in a concentration of 1 vol. Ozone at 500,000 vol. air. In the gaseous state is blue; in liquefied state is dark blue, and violet, almost black in solid state. It is poorly soluble in water.

Ozone has a much stronger oxidative action than oxygen, because in the reactions it takes part, the molecule decomposes into an oxygen atom, which is very reactive, and an oxygen molecule:

$$O_3 = O + O_2 \tag{7}$$

Disinfection is a necessary step to destroy or inactivate micro-organisms and prevent the spread of dangerous diseases. The properties of ozone, especially its ability to oxidize, have led to its use in water treatment. Ozone is an “ideal” reagent because it does not introduce adverse effects in water or in the atmosphere.

(c) Use of ozone

Ozone, having a much more prominent effect than oxygen, can be used to deodorize and disinfect air from showrooms or hospitals as well as sterilizing drinking water.

An ozone generator for water and air is useful for [15]:

- Strong disinfectant effect due to 500 mg O_3/h;
- Destroys microbes in air or water: Staphylococcus, Bacillus coli, Bacillus pyocyaneus, Pseudomonas fluorescens, Salmonella, Typhimorium, Shigella flexneri, Vibrio cholerae, hepatitis A virus;
- Inhibition of fungus, Aspergillums, Pecilocin and VUQ protozoa;
- Deplete spores in air or water;
- Removes unpleasant smells and toxins;
- Disintegrate pesticides, chemical fertilizers and hormones;
- Oxide ions of heavy metals and organic substances.

3 Ozone Generator

Experimental researches will use an ozone generator [15] called ozoniser, type TCB - 30003 (Figs. 2 and 3).

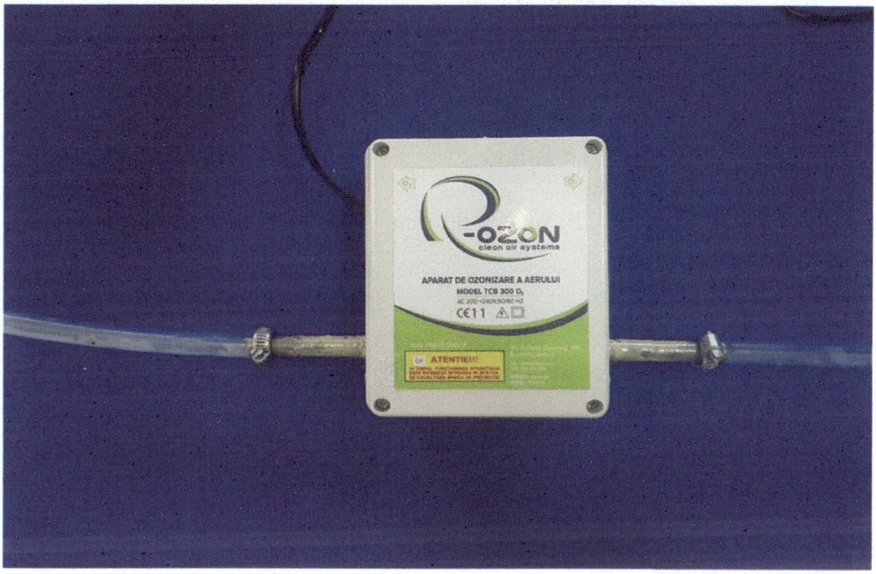

Fig. 2. Plan view of the ozone generator connected to the pipes

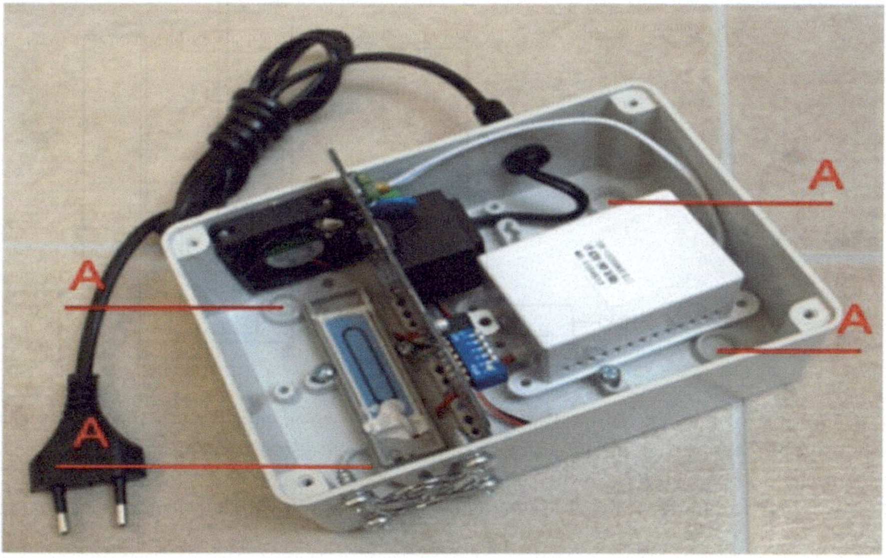

Fig. 3. Components of the ozone generator [15]

The ozone generator consists of a programmable timer module and an ozone system. Depending on the consumer's needs (interior space and other parameters), the ozone generator can be delivered with the FQM-P300 type ozonizing module with a capacity of 300 mg/h m^3.
Basic technical features:

- Supply voltage: 210 V–250 V, 50–60 Hz;
- Ozone flow rate: $\dot{m} = 300 \frac{mg}{hm^3} = 0.3 \frac{10^{-3} kg}{hm^3}$ [15];
- Maximum power: 100 mA;
- Equipped with fuse of 2A;
- Fan: 12 V, which creates an airflow;
- Air flow: up to 13.5 m^3/h;
- Dimensions: 150 mm × 120 mm × 80 mm;
- Timer accuracy: 7%;
- Weight, g: up to 600 g;
- Conditions of use: from −20 °C to +50 °C;
- Humidity: up to 75%;
- Ozone generator power: P = 20 W.

4 Installation Scheme

The scheme of the laboratory installation is shown in Fig. 4.

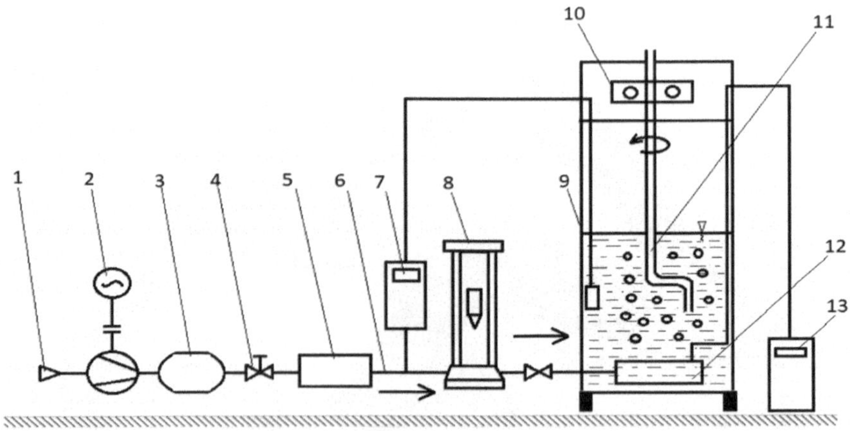

Fig. 4. Scheme of the experimental installation for water ozonation. 1 - air filter; 2 - electro compressor; 3 - compressed air tank; 4 - pressure reducer; 5 - ozone generator; 6 - pipe; Digital thermometer; 8 - rotameter; 9 - water tank; 10 - mechanism for rotating the oxygen probe in water; 11 - Oxygenometer probe; 12 - fine bubble generator; 13 - digital manometer

Data on the devices used can be found in [16–18].

The experimental installation was designed and built to allow:

(a) Measurement of a gas flow rate (air + ozone) from 10 dm^3 to 5 dm^3/min;
(b) Measurement of the pressure and temperature of the gas entering the fine bubble generator (FBG).

Figure 5 shows the fine bubble generator.

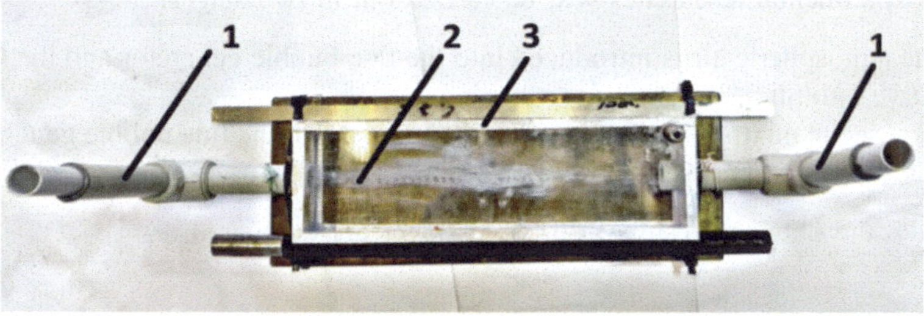

Fig. 5. Plan view of the fine bubble generator. 1 - connections for air entry; 2 - plate with orifices 0.5 mm; 3 - metal frame supporting the orifices plate

The orifices ϕ 0.5 mm were executed in the orifice plate by modern technologies (spark-erosion).

5 The Purpose and Researches Methodology

(a) The purpose of experimental researches
Experimental research aims to:

- validating the operation of a FBG with 37 orifices of ϕ 0.5 mm using air;
- validating the operation of the same FBG using a gaseous mixture of air + ozone.

Experimental researches were carried out within the laboratory of the Department of Thermotechnics, Engines, Heat and Refrigeration Equipment's and aimed to experimentally determination of the variation of dissolved oxygen in water in time.

(b) Researches methodology
Each measurement involves the following steps:

1. Check that the 37 orifices with ϕ 0.5 mm works, i.e. air is introduced into the fine bubble generator;
2. Fill the tank with water up to H = 0.5 m;
3. Measure C_O, t_{H2O}, t_{air};
4. Insert the fine bubble generator and record the start time of the experiment (τ);
5. Every 15 min, remove the bubble generator out of the tank, measure the dissolved oxygen concentration (CO_2);

6. When a horizontal plane of the function C = f(τ) is reached, the measurements stops with the following condition: C ≈ Cs;
7. From previous researches [16–18, 19], the concentration of dissolved oxygen in water tends to saturation after a two-hour period. So, the measurements of the oxygen concentration will be at the times: 15 min; 30 min; 45 min; 60 min; 75 min; 90 min; 105 min; 120 min.
8. At the end of the measurements clean the oxygen sensor and empty the water from the tank.

The experimental researches will be carried out in two stages:

I - The atmospheric air is introduced into the fine bubble generator and the CO_2 = f (τ) curve is built;
II - A gaseous mixture (air + ozone) is introduced into the fine bubble generator and the CO_2 = f(τ) curve is built.

Finally, the obtained results in the two stages are compared.

6 Experimental Obtained Results

I - The atmospheric air is introduced into the fine bubble generator and the CO_2 = f(τ) curve is built;

The results of the measurements are shown in Table 1.

Table 1. Experimental obtained data.

Nr.crt	0	1	2	3	4	5	6	7	8
τ [min]	0	15	30	45	60	75	90	105	120
t_{H2O} [°C]	21	21	21	21	21	21	21	21	21
t_{air} [°C]	19.4	19.4	19.4	19.4	19.4	19.4	19.4	19.4	19.4
C [mg/dm^3]	3.13	5.85	7.15	7.75	8.17	8.35	8.55	8.80	8.98

For the water temperature of 21 °C the value of Cs = 8.90 mg/dm^3.

Based on the data in Table 1, the function CO_2 = f(τ) is graphically represented in Fig. 6.

Figure 7 shows a photograph of an FBG before being inserted into the water tank.

Figure 8 shows the bubble curtain issued by FBG.

II - A gaseous mixture (air + ozone) is introduced into the bubble generator and the CO_2 = f(τ) curve is built;

The results of the measurements are shown in Table 2.

For water temperature of 21 °C the value of Cs = 8.90 mg/dm^3.

Based on the data in Table 2, the function CO_2 = f(τ) was graphically represented.

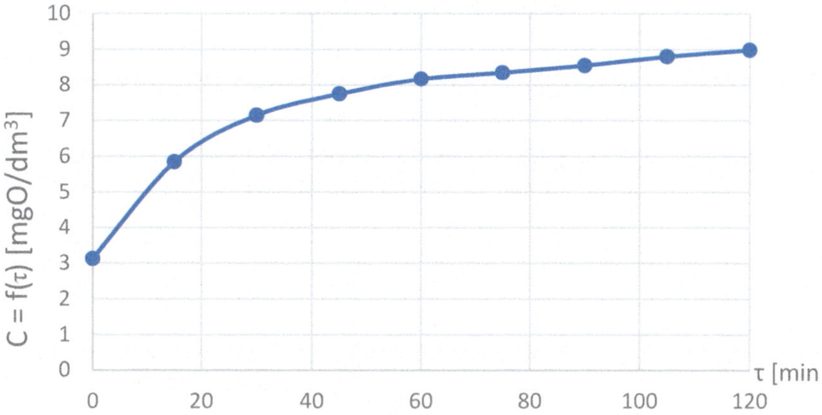

Fig. 6. The function $CO_2 = f(\tau)$

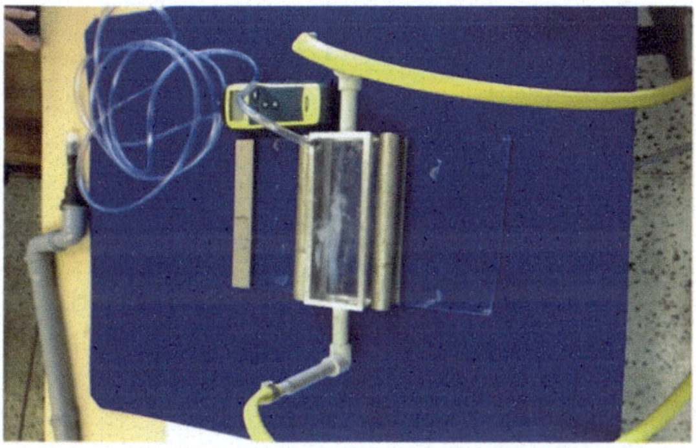

Fig. 7. FBG with pressure measuring connection with differential manometer

Fig. 8. A fine bubble generator in operation

Table 2. Experimental data.

Nr.crt	0	1	2	3
τ [min]	0	15	30	45
t_{H2O} [°C]	21	21	21	21
t_{air} [°C]	19.4	19.4	19.4	19.4
C [mg/dm^3]	3.13	6.92	8.04	8.64

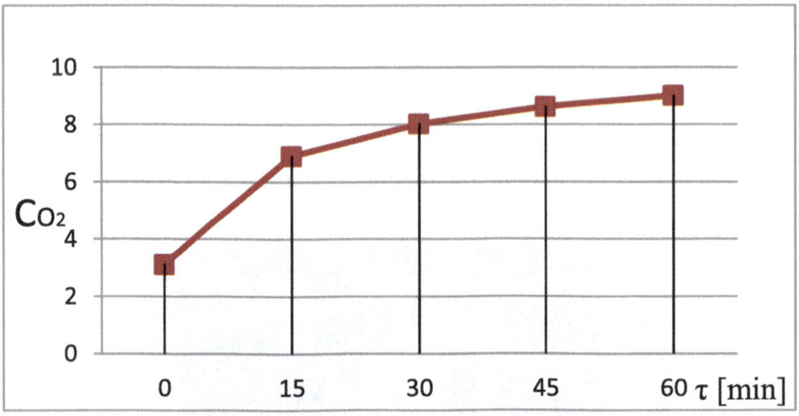

Fig. 9. The dependence of $CO_2 = f(\tau)$ on the operation of an FBG on the introduction of a gaseous mixture (air + ozone)

7 Conclusions

Following the two-stage experimental researches:

I - Introduction of atmospheric air into the water tank;
II - Introduction of a gaseous mixture consisting of atmospheric air and ozone, starting from common data,

- Water volume: $V = 0.125\ m^3$;
- Initial concentration of dissolved oxygen in water: $CO_2 = 3.65$ mg/dm^3;
- $t_{H2O} = 21$ °C;
- $t_{air} = 19{,}4$ °C;
- p = 760 mmHg.

the following were found: Fig. 9 shows that when a gaseous mixture (air + ozone) is introduced into the water, the time at which Cs is reached is reduced by half, which makes this process much more efficient.

References

1. Tănase, E.B.: Influenţa compoziţiei gazului insuflat în apă asupra conţinutului de oxigen dizolvat. Teză de doctorat, Universitatea POLITEHNICA din Bucureşti, Facultatea de Inginerie Mecanică şi Mecatronică (2017)
2. Căluşaru, I.: Influenţa proprietăţilor fizice ale lichidului asupra eficienţei proceselor de aerare. Teză de doctorat, Universitatea POLITEHNICA din Bucureşti, Facultatea de Inginerie Mecanică şi Mecatronică (2014)
3. Căluşaru, I.M., Băran, N., Pătulea, A.: Researches regarding the transfer of oxygen in water. In: The 3rd International Conference on Mechanic Automation and Control Engineering, 27–29 July, pp. 2617–2620. IEEE Computer Society CPS, and then Submitted to be Indexed by Ei Compendex, Baotou, China (2012)
4. Hand, D.W., Hokanson, D.R., Crittenden, J.C.: Air stripping and aeration, Chap. 4. In: MWH's Water Treatment: Principles and Design, 3rd edn. Wiley (2012)
5. Oprina, G., Pincovschi, I., Băran, G.: Hidro-Gazo-Dinamica Sistemelor de aerare echipate cu generatoare de bule. Ed. POLITEHNICA PRES, Bucureşti (2009)
6. Robescu, D., Robescu, D.L.: Procedee, instalaţii şi echipamente pentru epurarea apelor. Litografia UPB, Bucureşti (1996)
7. Stoianovic, S., Robescu, D., Stamatoiu, D.: Calculul şi construcţia echipamentelor de oxigenare a apelor. Editura CERES, Bucureşti (1985)
8. Mattock, G.: New Process of Waste Water Treatment and Recovery. Ellis Harwood Ltd., Publishers, Chichester (1978)
9. Robescu, D.L., Stroe, F., Presura, A., Robescu, D.: Tehnici de epurare a apelor uzate. Editura Tehnică, Bucureşti (2011)
10. Oprina, G.: Contribuţii la hidro-gazo-dinamica difuzoarelor poroase. Teză de doctorat, Universitatea Politehnica din Bucureşti, Facultatea de Energetică (2007)
11. Droste, L.R.: Theory and Practice of Water and Wastewater Treatment. Wiley, Hoboken (1996)
12. Rasha, R.M.: Influenţa arhitecturii generatoarelor de bule fine asupra concentraţiei de oxigen dizolvat în apă. Teză de doctorat, Universitatea POLITEHNICA din Bucureşti, Facultatea de Inginerie Mecanică şi Mecatronică (2017)
13. https://ro.wikipedia.org/wiki/Oxigen
14. http://www.ozonfix.ro/intrebari-frecvente-ozon
15. http://www.ozonfix.ro
16. Căluşaru, I.M., Băran, N., Pătulea, A.: Determination of dissolved oxygen concentration in stationary water. Revista de Chimie, **12**(63), 1312–1315 (2012)
17. Băran, N., Pătulea, A., Căluşaru, I.: The determination of the oxygen transfer speed in water in nonstationary conditions. In: International Proceedings of Computer Science and Information Technology, Mechanical Engineering, Robotics and Aerospace, pp. 267–272 (2011)
18. Pătulea, A., Căluşaru, I., Băran, N.: Researches regarding the measurements of the dissolved concentration in water. In: Advanced Materials Research, vol. 550–555, pp. 3388–3394. Trans Tech Publications, Switzerland (2012)

Study the Mechanical Behavior of Contact Lenses

Dana Rizescu(✉), Lucian Bogatu, and Ciprian Ion Rizescu

University Politehnica of Bucharest,
Splaiul Independentei no. 313, Bucharest 060042, Romania
dana.rizescu@upb.ro

Abstract. The paper presents a study concerning mechanical behavior of contact lenses. There were considered different materials for contact lenses like: conventional Hydrogel, Silicone-Hydrogel with different lens powers. The lenses were tested using a vertical test stand from Hans Schmidt, equipped with Imada force transducer. The lenses were "pushed" on a plane surface using a certain transducer accessory. Also, a FEM simulation of contact lenses was developed in Solidworks in order to study the lenses behavior for different peripheral curves, widths, junctions and lens diameter. There were considered the lens deflection, lens deformation and lens stresses. The diameter of a soft contact lens should be a specific amount larger than the actual corneal diameter. One conclusion of this study is: contact lens diameter may be a better predictor than base curve is of lens behaviour on eye.

Keywords: Contact lenses · Lens FEM simulation · Lens behaviour

1 Materials for Contact Lenses

1.1 Silicone Hydrogel

There is known that silicone hydrogel contact lenses are advanced soft lenses that allow more oxygen to pass through the lens to the cornea than regular soft ("hydro-gel") contacts. Another advantage consists of that silicone hydrogel lenses enable up to five times more oxygen to reach the cornea than regular hydrogel lenses. Silicone hydrogel contact lenses sometimes are erroneously called silicon hydrogel lenses. Where is the confusion? Silicon is a very common mineral. In fact, ordinary sand is composed primarily of silicon dioxide (silica).

The high oxygen transmissibility (Dk/t) of silicone hydrogel (SiHy) allows more oxygen to pass through the lens to the cornea than hydrogel (Hy) lenses, which helps maintain optimal ocular health.

1.2 Hydroxyethyl Methacrylate (HEMA)

The development of hydrogels started in the beginning of the sixties, with the production of hydroxyethyl methacrylate (HEMA) based polymers [1], which presented a swelling capacity of 40–50%. These hydrogels were applied mainly in the development

G. I. Gheorghe (Ed.): ICOMECYME 2019, LNNS 85, pp. 22–28, 2020.
https://doi.org/10.1007/978-3-030-26991-3_3

of contact lenses that completely changed the course of Ophthalmology evolution. The need of optimization of these lenses and to solve other problems in the medical field led to the development of a second family of hydrogels. HEMA-based hydrogel lenses are still the most popular type of contact lens. These lens materials are copolymers of HEMA and other hydrophilic or lipophilic monomers such as Nvi-nyl pyrrolidone (NVP) and methacrylates that contribute to the wide range of water contents of both ionic and non-ionic materials.

2 Experimental Tests and Results

2.1 Experimental Setup

The experimental setup (see Fig. 1) consists of: 1 - deflection measuring system, 2 - Imada force transducer, 3 - plane end for pushing the contact lens, 4 - the tested contact lens. Maximum testing force is 5 N with 0.001 N accuracy. The Imada transducer is connected to PC for record the pushing force and time. The testing end is of a cylindrical shape, with a plane surface pushing the contact lens. The pressing process of the contact lens is presented above (see Figs. 2 and 3).

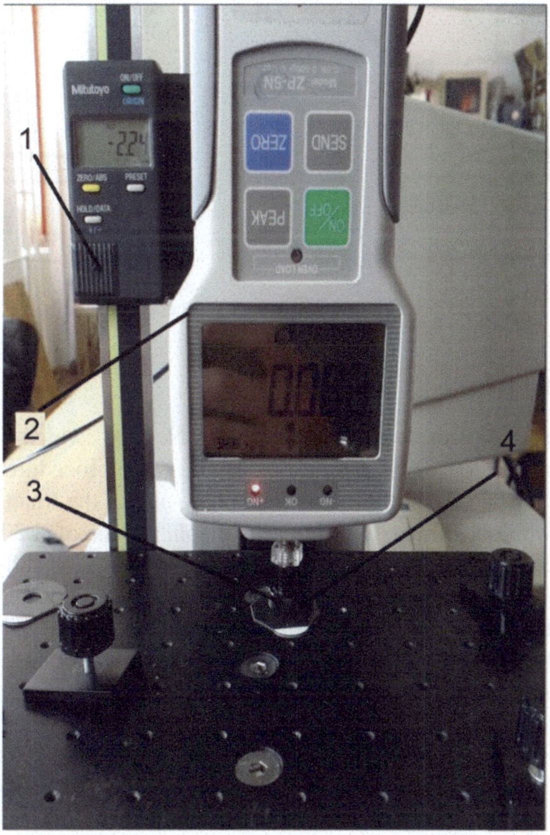

Fig. 1. The experimental setup HV 5 N

Fig. 2. Vertical test stand from Hans Schmidt with Imada Transducer 5 N.

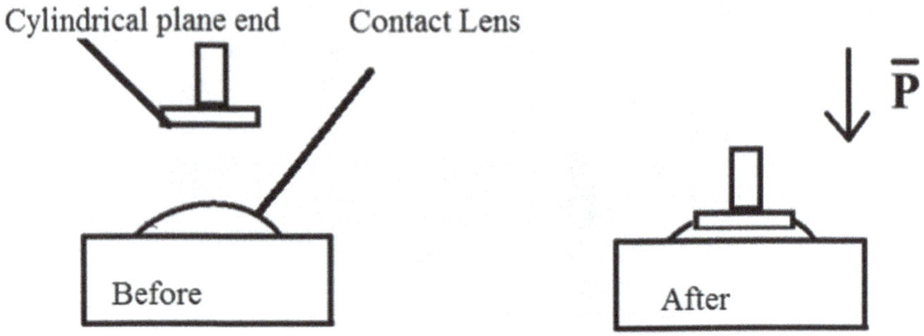

Fig. 3. The pressing process of the contact lens

2.2 Experimental Results

Compression forces are of the order of thousandth Newton, which corresponds to the real adaptive forces. All lenses were tested using the same procedure: placed and fixed on the table, then "pushed" with the cylindrical plane end of the Imada Transducer. The pressing process was performed for a maxim displacement of 2.20 mm using a 0.2 step. There were considered various powers contact lenses, with different edge shapes [2, 3]. All experimental results were analyzed and then represented on a graph, developed in Matlab environment (see Figs. 4 and 5).

The pressing forces ranges belong to the [0.05 … 0.09] N and there are no substantial differences between different powers of the contact lenses. For example, in Fig. 4 is represented the contact lens with +5dpt power. Considering the lens displacement of 0.836 mm the pressing force is 0.0042 N. as it is marked in Fig. 5. For the same displacement the simulation reveals a pressing force of 0.005 N. Modeling both contact lens materials there were considered several parameters as: elastic modulus, Poisson's ratio, mass density, tensile strength, Yield strength, thermal expansion coefficient, thermal conductivity, specific heat. On one face(s) was applied normal force 0.0005 N using uniform distribution.

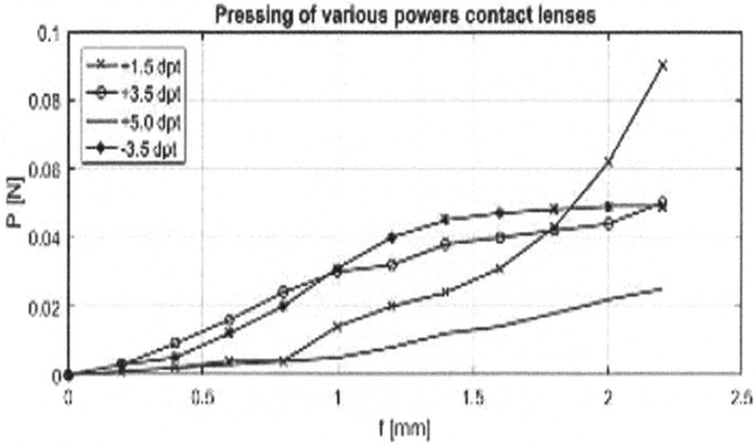

Fig. 4. Pressing of various powers contact lens: +1.5, +3.5, + 5.0, −3.5

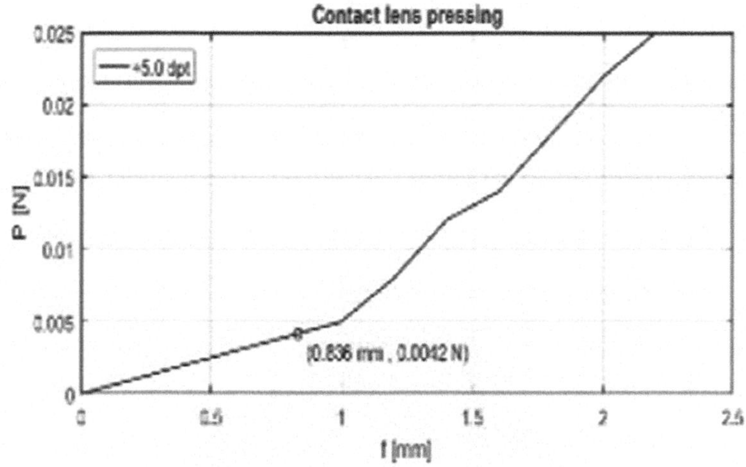

Fig. 5. Contact lens (+5.0) pressing

3 FEM Simulation Results

One aim of this study is to determine the influence of the base curve on the movement, on the corneal surface mechanical alterations and on the subjective comfort of different contact lenses for corneas having central curve radius flatter than 7.80 mm [4, 5]. A simulation model of the contact lens and the cylindrical plane end were developed (see Figs. 5 and 6). There were considered different rays for the edge region of the contact lens. If the thickness of the lens at the edge is to large there are major difference between the experimental results and simulated results for both materials: hydroxyethyl methacrylate (HEMA) and silicone hydrogel.

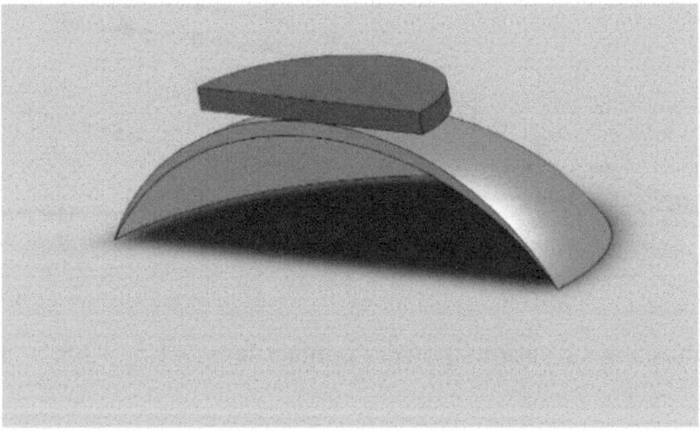

Fig. 6. A section through the contact lens and the plane end

After applying normal force of 0.0005 N, using uniform distribution, there were computed the stresses, displacements, strains as it is presents in Figs. 7, 8, 9, 10 and 11. The maximum displacement, 0.836 mm, is reached for normal force of 0.0005 N, as it was mentioned above. The tested lenses have a base curve diameter equal to 8.3 mm and a lens diameter of 14 mm.

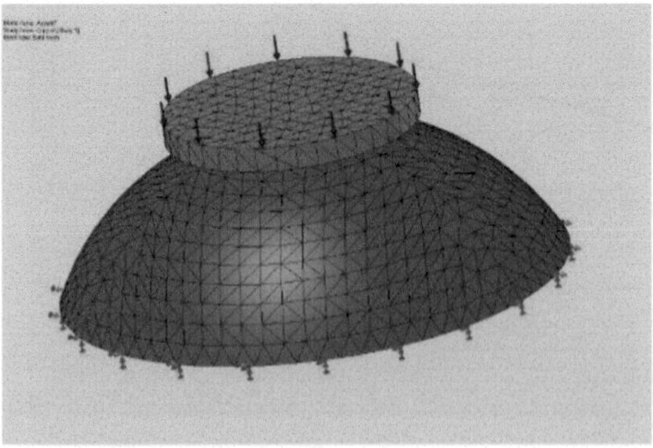

Fig. 7. Meshing and fixing the contact lens and the plane end

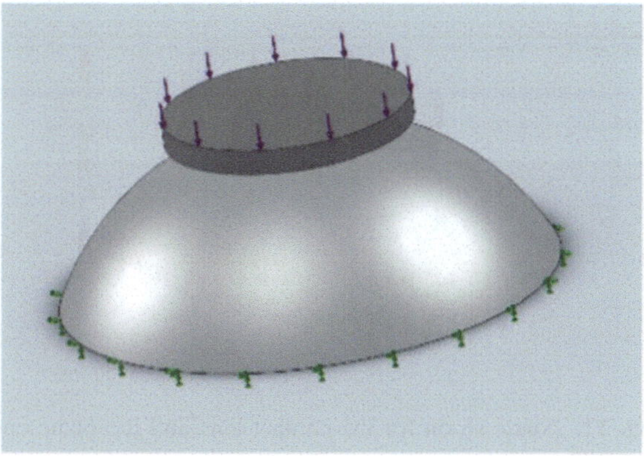

Fig. 8. Fixing the contact lens and the plane end

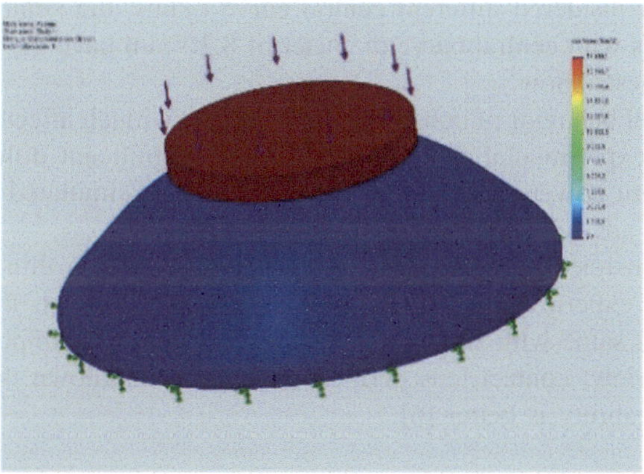

Fig. 9. Static stresses (von Misses) for the contact lens and the plane end

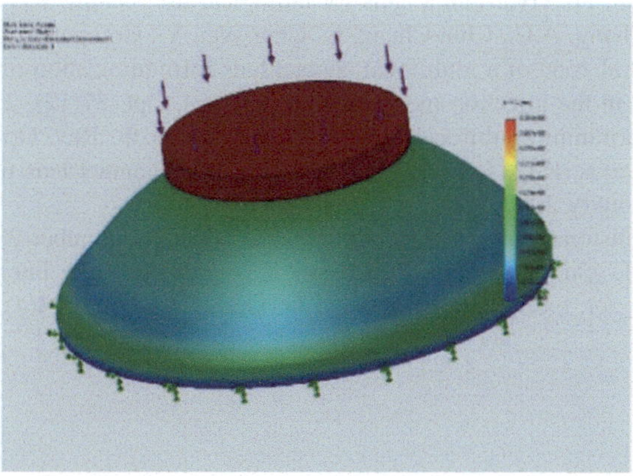

Fig. 10. Static displacement for the contact lens and the plane end

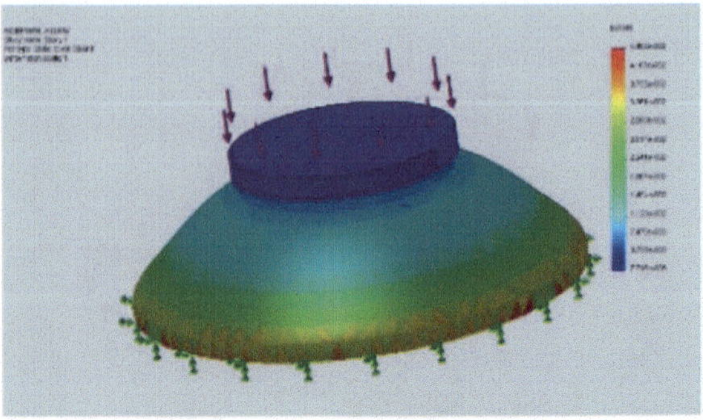

Fig. 11. Static strain for the contact lens and the plane end

4 Conclusions

Even there were considered different central curve radius, the simulation shows that only contact lenses with central curve diameter of 8.30 mm have the same behavior as the experimental tests show.

The mechanical behavior of contact lenses is not very much affected by the con-tact lens power. The experimental tests show that are insignificant differences be-tween lenses with different powers. The experimental results and simulated ones for pressing the contact lenses are very closed.

Another conclusion of this work is that contact lenses with thin edges are on the same behavior on experimental test and simulation. As the lens is thinner the experimental test are the same with simulation test. This statement is important considering another point of view: contact lens permeability. There is known that: as the lens is thinner, the permeability is better [6].

References

1. Wichterle, O., Lim, D.: Hydrophilic gels for biological use. Nature **185**, 117–118 (1960)
2. Vu, L.T., Chao-Chang, A.C., Chia-Cheng, L., Chia-Wei, Y.: Compensating additional optical power in the central zone of a multifocal contact lens forminimization of the shrinkage error of the shell mold in the injection molding process. Appl. Opt. **57**(12), 2981–2991 (2018)
3. Davis, R.L.: Determining multifocal parameters for a better fit. Rev. Optom., August 2016
4. Becherer, P.D., Davis, R.L., Kempf, J.A.: A personalized contact lens prescription. Contact lens spectrum, January 2007
5. Becherer, P.D.: Customizing with care. Optometric Office, September 2013
6. Young, G., Hall, L., Sulley, A., Osborn-Lorenz, K., Wolffsohn, J.S.: Inter-relationship of soft contact lens diameter, base curve radius, and fit. Optom. Vis. Sci. **94**(4), 458–465 (2017)

Mechanical Strength of Stripped Optical Fiber

R. El Abdi[1(✉)], R. Leite Pinto[1,2], P. Lallinec[2], and M. Poulain[3]

[1] Université de Rennes - CNRS, Institut de Physique de Rennes. UMR 6251, 35000 Rennes, France
relabdi@univ-rennes1.fr

[2] Entreprise Acome - Usines de Mortain, 50140 Mortain, France

[3] Institut des Sciences Chimiques de Rennes-UR1, UMR CNRS 6226, 35042 Rennes Cedex, France

Abstract. A novel removal technique of silica optical fiber coating with chemical stripping gel was used. We observed that the fiber strength decreases from 5.48 to 4.96 GPa in the case of tensile test and from 0.06 to 0.05 GPa in the case of bending test when the coating was removed from the cladding surface. But the stress corrosion parameter varies little considering the measurement accuracies.

Keywords: Optical fiber · Coating · Stripping gel · Tensile test · Bending test · Stress corrosion parameter

1 Introduction

The commercial single-mode fibers are typically composed of silica glass for the core and cladding and acrylate protective coating. Sakaguchi and Hibino [1] were interested by the fatigue in the low-strength of silica optical fibers. The fatigue behavior is mainly characterized by crack growth parameter n_d (stress corrosion parameter). The value of n_d for low-strength fibers which contain macroscopic flaws has not been sufficiently clarified, because only a few examinations have been made on low strength fibers [2, 3], although many studies have been made on high-strength fibers which have no macroscopic flaws [4, 5]. The allowable loading condition needed to prevent any growth of the macroscopic flaw was discussed in order to assure high reliability for the fiber.

Chen and Chang [6] have focused on the fracture mechanics of silica optical fibers to evaluate the strength and fracture characteristics of single and multi-mode fibers subjected to uniaxial tensile testing and two-point bending. The fracture strength data of both single and multi-mode under either testing were found to be very similar. The fracture stresses at 50% fracture probability for tensile testing and two-point bending were 4.5 and 5.1 GPa, respectively. The fracture characteristics of each tested optical fiber specimen were evaluated by using an SEM. A critical flaw on the surface of glass fiber was found to be the fracture origin for specimens under either tension or bending as expected.

But, the influence of the coating on the fiber strength has not been plainly studied when the fiber is subjected to bending or tensile tests. The protective coating that

G. I. Gheorghe (Ed.): ICOMECYME 2019, LNNS 85, pp. 29–35, 2020.
https://doi.org/10.1007/978-3-030-26991-3_4

surrounds the glass fiber plays an important role in protecting the glass cladding surface of the optical fiber from mechanical and chemical damages. However, the knowledge of the mechanical coating strength is one of the factors that determine the lifetime of the fabricated fiber-based devices. In this paper, silica optical fibers with and without polymer coatings were submitted to tensile and bending tests using a new stripping gel to remove the optical fiber coating materials that leaves the surface of the glass cladding intact.

2 Fiber, Stripping Gel Test Benches Used

2.1 Fiber Used

The used monomode silica fiber has an acrylate coating. This fiber was manufactured using the Plasma activated Chemical Vapor Deposition (PCVD) process which produces a totally synthetic, ultra-pure fiber. The combined coating diameter is 242 ± 5 µm, the clad diameter is 125 ± 0.7 µm and the coating thickness is 58.5 ± 0.5 µm.

2.2 Stripping Gel Used

The used gel is a stripping gel based on methylene chloride and methanol. We have spread a gel layer on the surface of a glass plate specimen. Glass sample was exposed to gel during 20 h and cleaned using cellulose paper, rinsed with soap and ethanol. Figure 1 gives the surface details when AFM of microscope-grade was used for a glass surface not exposed to the stripping gel (Fig. 1a) and that exposed to the stripping gel (Fig. 1b). One can note that there is no chemical reaction when stripping gel was spread. Gel used doesn't attack SiO_2 molecules for a short term application such as coating stripping.

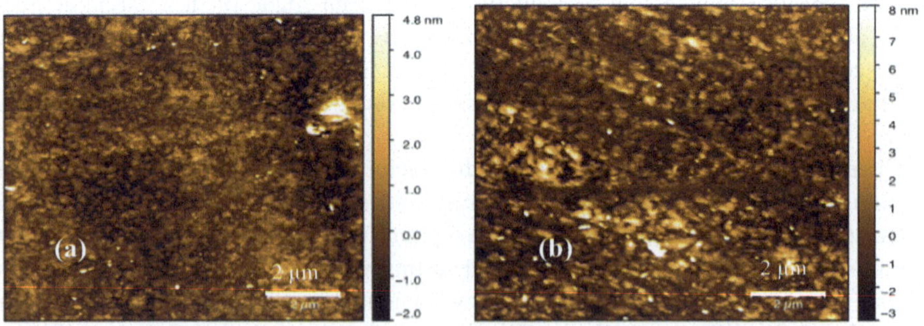

Fig. 1. (a) Glass surface without stripping gel and (b) exposed to stripping gel

The surface roughness S_q corresponds to the RMS (Root Mean Square gradient) of the altitude of the statistical distribution of heights in the sample. S_q varied from 0.64 up to 2.20 nm for unexposed glass surface and from 0.7 up to 2.00 nm for exposed glass surface. Both measured S_q correspond to the RMS of bulk microscope glass and indicate that there is no action between gel and SiO_2 molecules during almost a day.

Figure 2 gives details of stripped fiber.

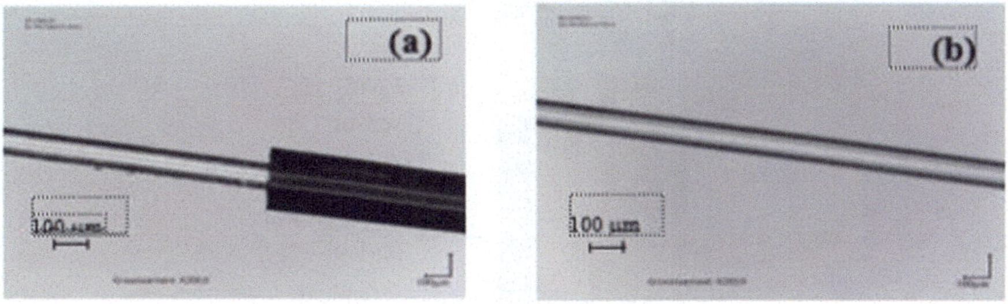

Fig. 2. Stripped optical fiber: (a) gel was applied only on a fiber part; (b) all fiber was stripped

2.3 Test Benches Used

Two experimental test benches were used:

a/. A two points bending bench made up of a displacement plate which is mounted on an aluminum plate. The first thrust block is movable and mounted on the displacement plate, while the second thrust block is fixed on a force sensor. The optical fiber is positioned between the two thrust blocks in such a way that it forms a "U". To avoid slipping, the fiber is positioned in the grooves of the thrust blocks. During the test, load and displacement are recorded, allowing the load/displacement curve to be obtained (Fig. 3). Bending speeds were from 0.1 to 8.0 mm/s and fiber length was 10 cm.

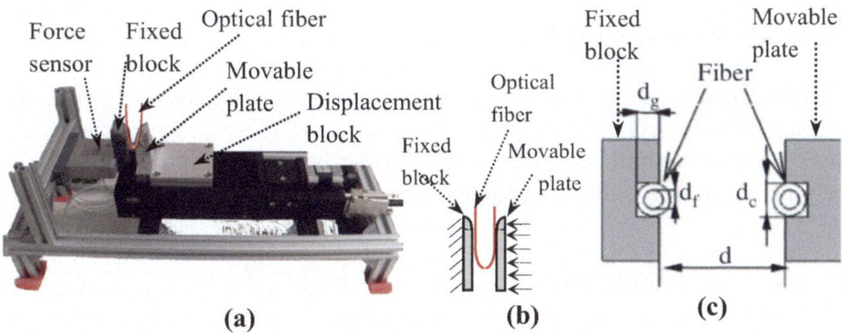

Fig. 3. (a) Bending bench used; (b) and (c) fiber between thrust blocks.

b/. The dynamic tensile test consists of subjecting fibers to a deformation under a constant velocity until rupture. The used tensile bench was a Lloyd 50 K bench with a 100 N sensor (Fig. 4). The fiber is rolled three times around two pulleys; the lower pulley is fixed and the upper pulley is movable with different velocities (0.8, 2.6, 4.4,

6.2 and 8 mm/s). These velocities, expressed as a percentage of the initial sample length (500 mm), correspond to 1.6×10^{-3} s^{-1}, 5.2×10^{-3} s^{-1}, 8.8×10^{-3} s^{-1}, 12.4×10^{-3} s^{-1} and 16×10^{-3} s^{-1}.

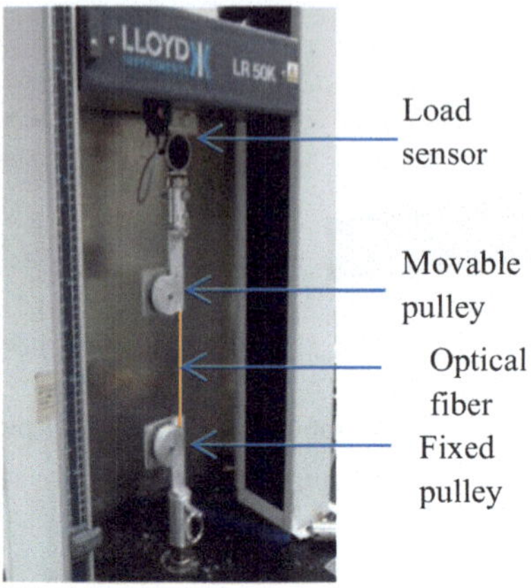

Fig. 4. Tensile test bench used

Tensile testing was performed in a controlled environment with 46–52% relative humidity with a maximum of 5% humidity variation for each series of the tensile tests.

During the test, the tensile load was measured using a dynamometric cell (load sensor) while the fiber deformation was deduced from the displacement between the fixed lower pulley and the mobile higher pulley (Fig. 4). The testing procedure used 20 samples for each velocity.

3 Theoretical Approach

The stress corrosion parameter n_d refers to the crack growth parameter which is typically 20 or more for silica fiber.

The statistical Weibull law gives a relationship between the probability ***F*** of fiber rupture with a length L and the applied stress σ:

$$Ln\left[\frac{1}{L}\left\{Ln\left(\frac{1}{(1-F)}\right)\right\}\right] = m[Ln(\sigma) - Ln(\sigma_o)] \tag{1}$$

where m is a size parameter and σ_o is a scale parameter.

The evolution of $Ln\left[\frac{1}{L}\left\{Ln\left(\frac{1}{(1-F)}\right)\right\}\right]$ (Failure cumulative probability in (%)) according to $Ln(\sigma_F)$ is called Weibull diagram (σ_F is the failure stress). On the other hand, one can give the stress change according to the stress rate $[\dot{\sigma}_F]$ as follow:

$$Ln[\sigma_F] = \left(\frac{1}{1+n_d}\right) Ln[\dot{\sigma}_F] + b \quad (2)$$

The main parameter which defines the fiber mechanical strength was the stress corrosion parameter under dynamic loadings $n_{d.}$

Matthewson [7] gives for different applied strength the crack propagation type and the n_d values (Table 1).

Table 1. Types of defect and strength ranges for silica [7]

Defect type	n_d	Strength	Comments
Pristine	20	>7 GPa	Surface phenomena
Subthreshold	10–20	0.3–7 GPa	Crack initiation dominates by residual stresses
Postthreshold	30	1–300 GPa	Crack propagation dominates by residual stresses
Macroscopic crack	40	<1 MPa	Sharp, residual stress free cracks

4 Results and Discussion

Figures 5 and 6 give the Weibull distribution of coated and stripped fibers for tensile and bending tests for different faceplate velocities. For each velocity, the median fiber strengths of the coated or stripped fibers were close one to another. As a result, the presence of the coating modifies very little the mechanical strength of the fiber.

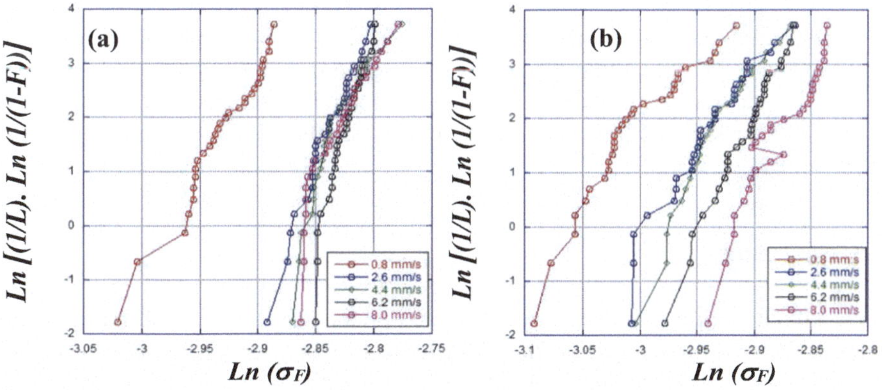

Fig. 5. Weibull curves for two point bending test: (a) coated fibers; (b) stripped fibers

Table 2 gives n_d values and the median strength for coated fibers and stripped fibers obtained from bending and tensile tests. The stress corrosion parameter n_d is related to the Weibull curve slope. This parameter varies little since all the failure stresses of stripped fiber have decreased by the same amount compared with those of the coated fiber.

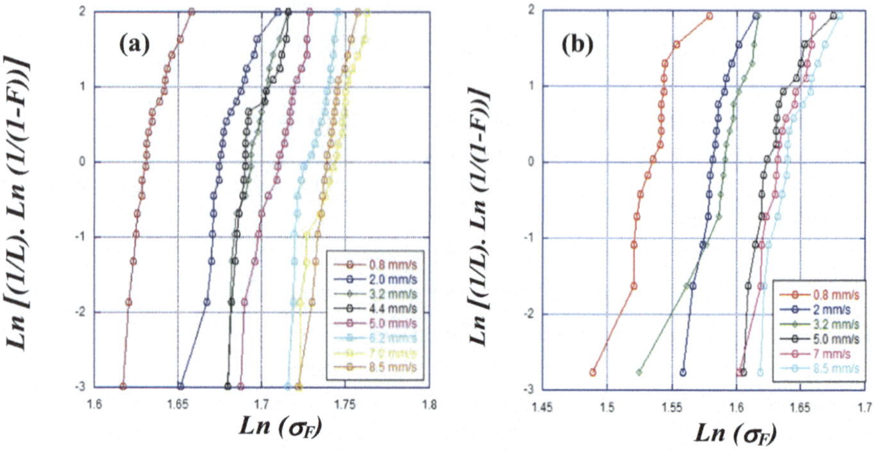

Fig. 6. Weibull curves for tensile test: (a) coated fibers; (b) stripped fibers

Table 2. Mechanical fiber parameters

	Coated fibers		Stripped fibers	
	Stress corrosion parameter n_d	Median fiber strength (GPa)	Stress corrosion parameter n_d	Median fiber strength (GPa)
Tensile test	20.28	5.48 ± 0.1	18.61	4.96 ± 0.13
Bending test	20.73	0.058 ± 0.003	21.72	0.052 ± 0.003

If the coating provides protection to optical fiber against detrimental environmental effects and leads to a great fiber flexibility, it can also increase the fiber strength which is a significant parameter in the case of fibers used in telecommunication networks. The strength loss for stripped fibers can be close to 10% when fiber is submitted to tensile test or to bending test.

5 Conclusion

A new tripping gel used allows for a clean and undamaged surface and AFM analysis confirms SiO_2 isn't damaged on the short term. Stripping using gel allows for good measurements that were unreliable using other methods. In addition, only a few minutes after the coat of gel were sufficient to remove the fiber coating.

Only 10% mechanical resistance loss was due to coating stripping. The stress corrosion parameter n_d changes little. The high value of n_d obtained during the bending test for stripped fibers was certainly due to the difficulties during the experiment, because after the stripping of fibers, it is necessary to quickly test them otherwise they will be damaged by the environmental conditions (dust, oxygen, humidity…).

References

1. Sakaguchi, S., Hibino, Y.: Fatigue in low-strength silica optical fibres. J. Mater. Sci. **19**, 3416–3420 (1984)
2. Gulati, S.T., Helfinstine, J.D., Justice, B., McCartney, J.S., Runyan, M.A.: Measurement of stress corrosion constant n for optical fibres. Am. Ceram. Soc. Bull. **58**, 1115–1117 (1979)
3. Donaghy, F.A., Nicol, D.R.: Evaluation of the fatigue constant n in optical fibers with surface particle damage. J. Am. Ceram. Soc. **66**, 601–604 (1983)
4. Kalish, D., Tariyal, B.K.: Static and dynamic fatigue of a polymer-coated fused silica optical fiber. J. Am. Ceram. Soc. **61**, 518–523 (1978)
5. Sakaguchi, S., Kimura, T.: Influence of temperature and humidity on dynamic fatigue of optical fibers. J. Am. Ceram. Soc. **64**, 259–262 (1981)
6. Chen, C.P., Chang, T.H.: Fracture mechanics evaluation of optical fibers. Mater. Chem. Phys. **77**(11), 110–116 (2002)
7. Matthewson, M.J.: Optical fiber reliability models. SPIE Crit. Rev. Opt. Sci. Technol. **50**, 3–31 (1993). Fiber optics reliability and testing

The First and Second-Order Theory of Shearing and Compression in Case of the Beam Fixed at One End and Supported to Helical Spring at the Other End

Cornel Marin(✉)

FIMM, Valahia University of Târgovişte,
Aleea Sinaia Nr. 13, Targoviste, Romania
marin_cor@yahoo.com

Abstract. The purpose of this paper is to present the results of two methods of calculating beams subjected simultaneously to bending and compression: no taking into account the influence of the beam displacements on momentum of compressive force (first order theory) and taking into account (second order theory). The relations obtained for one particular case in the MATHCAD program and we obtained the diagrams of the sectional shear and bending efforts, and the diagrams of deformation beam (displacements and slopes). These diagrams allow the comparative study of the results obtained by the two mentioned theory. We also simulate the position of the transverse force on the beam so that the maximum displacement is obtained, by the first theory and the second theory of calculus.

Keywords: Beam · Bending · Compression · Second-order theory

1 Introduction

The stability calculus of elastic beams subjected to compressive and bending loads are solving using two methods: *the first order theory* - without taking into account the effect of compression force - *and the second order theory* - taking into account the effect of compression force [3, 7, 8].

In this chapter we consider the particular case, a simple straight beam with profiled standard section, having the section area A and axial moments of inertia I_y know, which is fixed at one end and supported elastic at the other.

The beam is subject to bending and compression simultaneous loads by the transverse compressive force Q and the axial force P (smaller than buckling critical force P_{cr}). We obtain the functions of bending moments, rotation and displacements for the curved fiber beams, according to *the first order theory of calculus* and also to *the second order theory of calculus* - i.e. taking into account the influence of compressive force. These analytical expressions are introduced into MATHCAD professional program and we obtain the diagrams of bending moment $M_1(x)$, rotation $\varphi_1(x)$ and displacements $w_1(x)$ for first order theory. Also, we obtain the diagrams of bending moment $M_2(x)$, rotation $\varphi_2(x)$ and the displacements $w_2(x)$ for second order theory.

G. I. Gheorghe (Ed.): ICOMECYME 2019, LNNS 85, pp. 36–50, 2020.
https://doi.org/10.1007/978-3-030-26991-3_5

2 The First Order Theory for Calculus of Beam Subjected to Compressive and Bending Loads

In Fig. 1 it consider a particular case of straight beam of length L, having the section area A and axial moments of inertia I_y known. For this case, we wrote the functions of bending moments, of rotation and displacements for the beams, according to the first-order theory – i.e. without taking into account the influence of compressive force on bending moments.

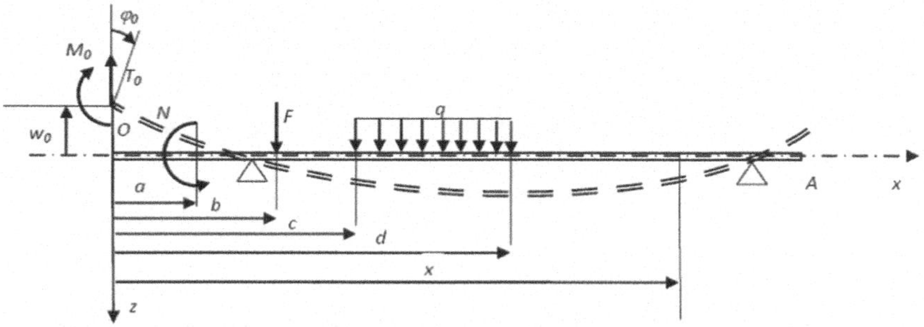

Fig. 1.

The beam is subjected to bending and compression loads by the external moment N, the transverse force F, the uniformly distributed force q and one axial compression force P - whose effect on bending moments can be neglected [2, 3].

In Fig. 1 we note:

- w_0, φ_0 - the displacement and rotating of the left end of the beam;
- T_0, M_0 - the corresponding shear forces and bending moment for the left end of the beam.

We obtain the displacement $w_1(x)$ and rotating $\varphi_1(x)$ functions as solution of the differential equation of curved fiber of beam [1, 4–6]:

$$\frac{d^2w}{dx^2} = -\frac{M_{iy}(x)}{E \cdot I_y} \tag{1}$$

Where the bending moment $M_{iy}(x)$ function is written using step function $\Phi(x - a)$ of MATHCAD program, as particular relationship for this case:

$$\begin{aligned} M_{iy}(x) &= M_0 \cdot \Phi(x) + N \cdot \Phi(x-a) + T_0 \cdot x \cdot \Phi(x) - F \cdot (x-b) \cdot \Phi(x-b) \\ &\quad - \frac{q_0}{2} \cdot \Phi(x-c) \cdot (x-c)^2 + \frac{q}{2} \cdot \Phi(x-d) \cdot (x-d)^2 \end{aligned} \tag{2}$$

By integrating the differential Eq. (1) we obtain the rotation function $\varphi 1(x)$ of the section beam using step function $\Phi(x - a)$; for this purpose we approximate the first derivative function of displacements with the slope:

$$\frac{dw_1(x)}{dx} = tan[\varphi_1(x)] \cong \varphi_1(x)$$

$$\varphi_1(x) = \frac{dw_1(x)}{dx} = \varphi_0 - \frac{1}{EI_y}[M_0 \cdot x \cdot \Phi(x) + N \cdot (x-a) \cdot \Phi(x-a) + \frac{T_0}{2} \cdot x^2 \cdot \Phi(x) - \frac{F}{2}(x-b)^2 \cdot \Phi(x-b) - \frac{q_0}{6} \cdot (x-c)^3 \cdot \Phi(x-c) + \frac{q}{6} \cdot (x-d)^3 \cdot \Phi(x-d)] \tag{3}$$

By integrating the differential relation (3) we obtain the displacement function $w_1(x)$ of the curved beam:

$$w1(x) = w_0 + \varphi_0 \cdot x - \frac{1}{EI_y}\left[\frac{M_0}{2} \cdot x^2 \cdot \Phi(x) + \frac{N}{2} \cdot (x-a)^2 \cdot \Phi(x-a) + \frac{T_0}{6} \cdot x^3 \cdot \Phi(x) - \frac{F}{6}(x-b)^3 \cdot \Phi(x-b) - \frac{q_0}{24} \cdot (x-c)^4 \cdot \Phi(x-c) + \frac{q}{24} \cdot (x-d)^4 \cdot \Phi(x-d)\right]; \tag{4}$$

The expressions (2), (3) and (4) vas introduced into the MATHCAD professional program to draw the bending, rotation and displacement diagrams.

3 The Second Order Theory for Calculus of Beam Subjected to Compressive and Bending Loads

In Fig. 2 it consider a same particular case of straight beam of length L, having the section area A and axial moments of inertia I_y known. For this particular case, we wrote the functions of bending moments, of rotation and displacements for the curved fiber beams, according to the second-order theory - i.e. taking into account the influence of compressive force on bending moments.

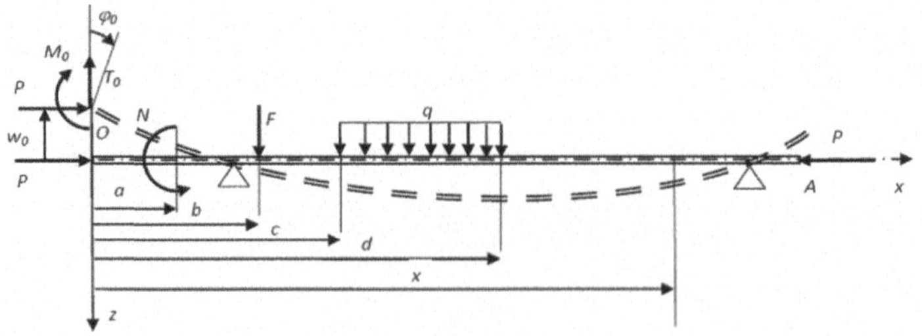

Fig. 2.

The beam is subjected to bending and compression loads by the external moment *N*, the transverse force *F*, the uniformly distributed force q and one axial compression force P - whose effect on bending moments cannot be neglected [3, 7, 8].

For this particular case we note also:

- w_0, φ_0 are the displacement and rotating of the left end of the beam;
- T_0, M_0 - the corresponding forces and moment of the left end of the beam.

We obtain the displacement $w_2(x)$ and rotating $\varphi_2(x)$ functions as solution of the differential equation of curved fiber of beam [1, 3]:

$$\frac{d^2M_i}{dx^2} = -q + P \cdot \frac{d^2w}{dx^2} \tag{5}$$

If we derivates twice the differential equation of curved fiber (1) and taking into account the relationship (5) we obtained the differential equation:

$$\frac{d^4w}{dx^4} + \frac{P}{E \cdot I_y} \cdot \frac{d^2w}{dx^2} = q \tag{6}$$

The solution of the differential Eq. (6) is written as follow:

$$\begin{aligned} w2(x) &= C_1 sin(kx) + C_2 cos(kx) + C_3 x + C_4 + \bar{w} \\ \varphi 2(x) &= kC_1 cos(kx) - kC_2 sin(kx) + C_3 + \frac{d\bar{w}}{dx} \end{aligned} \tag{7}$$

If we derivates twice the differential Eq. (7) we obtained the bending moment function $M_i(x)$ respectively the shear functions $T(x)$:

$$\begin{aligned} M2(x) &= -EI_y \left(-k^2 C_1 sin(kx) - k^2 C_2 coskx + \frac{d^2\bar{w}}{dx^2}\right) \\ T2(x) &= EI_y \left(k^3 C_1 cos(kx) - k^3 C_2 sinkx - \frac{d^3\bar{w}}{dx^3}\right) \end{aligned} \tag{8}$$

where: *k* is one parameter depending of the compressive force *P*:

$$k^2 = \frac{P}{E \cdot I_y} \tag{9}$$

- C_1, C_2, C_3, C_4 are the constants determined from the boundary conditions of curved beam (displacements and rotations of the beam ends);

$\bar{w}(x)$, $\bar{\varphi}(x) = \frac{d\bar{w}}{dx}$ are the particular solutions suitable for external loads *N, F, q*. This particular solution for external loads N, F and q is written [1, 3, 7, 8]:

$$\begin{aligned} &\bar{w}_N(x) = -\tfrac{N}{P}(1 - cos(k \cdot (x - a))) \quad for \ x \geq a \\ &\bar{w}_F(x) = \tfrac{F}{kP}(k \cdot (x - b) - sin(k \cdot (x - b))) \quad for \ x \geq b \\ &\bar{w}_q(x) = \tfrac{q}{kP}\int_c^d (k \cdot (x - t) - sin(k \cdot (x - t)))dt \quad for \ x \in (c, d) \end{aligned} \tag{10}$$

If the origin parameter values: w_0, φ_0, M_0 *and* T_0 are know, it can written the boundary conditions take into account the relationship (7) and (8):

$$\begin{aligned} &w_2(0) = w_0 = C_2 + C_4 \\ &\varphi_2(x) = \varphi_0 = kC_1 + C_3 \\ &M_2(0) = M_0 = E \cdot I_y(k^2 \cdot C_2) \\ &T_2(0) = T_0 + P \cdot \varphi_0 = E \cdot I_y(k^3 \cdot C_1) \end{aligned} \tag{11}$$

The expressions of constants C_1, C_2, C_3, C_4 are obtained as followed:

$$C_1 = \frac{T_0}{kP} + \frac{\varphi_0}{k}; \quad C_2 = \frac{M_0}{EI_y \cdot k^2} = \frac{M_0}{P}; \quad C_3 = -\frac{T_0}{P}; \quad C_4 = w_0 - \frac{M_0}{P} \tag{12}$$

Substituting expressions functions (12) into relations (7) and (8) we obtained the relationship of displacement $w_2(x)$, rotating $\varphi_2(x)$, bending moment $M_2\ (x)$ and shear functions $T_2(x)$ functions:

$$\begin{aligned} &w_2(x) = w_0 + \tfrac{\varphi_0}{k} sin(k \cdot x) - \tfrac{M_0}{P}(1 - cos(k \cdot x)) - \tfrac{T_0}{k \cdot P}(k \cdot x - sin(k \cdot x)) + \bar{w} \\ &\varphi_2(x) = \varphi_0 \cdot cos(k \cdot x) - \tfrac{M_0}{P} k \cdot sin(k \cdot x) - \tfrac{T_0}{P}(1 - cos(k \cdot x)) + \tfrac{d\bar{w}}{dx} \\ &M_2(x) = EI_y\left(\varphi_0 \cdot k \cdot sin(k \cdot x) + \tfrac{M_0}{P} k^2 \cdot cos(k \cdot x) + \tfrac{T_0}{P} k \cdot sin(k \cdot x) - \tfrac{d^2\bar{w}}{dx^2}\right) \\ &T_2(x) = EI_y\left(\varphi_0 \cdot k^2 \cdot cos(k \cdot x) - \tfrac{M_0}{P} k^3 \cdot sin(k \cdot x) + \tfrac{T_0}{P} k^2 \cdot cos(k \cdot x) - \tfrac{d^3\bar{w}}{dx^3}\right) \end{aligned} \tag{13}$$

4 Particular Case Study of First and Second Order Theory of Calculus

For particular case study we consider a straight beam from Fig. 3 with *length* L = 5 m, profiled standard section *I18* (STAS 565-80) having section area A = 27.9 cm^2 and axial moments of inertia I_y = 81.3 cm^4 (I_z = 145 cm^4). The beam is fixed at one end and supported to helical spring at the other end. The beam is subject to bending and compression loads by one transverse force Q = 2,5 kN applied to embedding distance αL (α = 0,5) and one axial force $P = 0{,}25 \cdot P_{cr}$ (P_{cr} is the buckling critical force), at the other end.

The material beam has the following mechanical characteristics: longitudinal elastic modulus $E = 2{,}1 \cdot 10^5$ MPa; yield stress σ_c = 150 MPa, coefficient of thinness to buckling λ_0 = 105. Cylindrical helical spring has the following characteristics: d = 18 mm, D = 100 mm, elastic shear modulus $G = 85 \cdot 10^3$ MPa, n = 9 *rings*.

The study was done for the following cases:

CASE 1: For first-order theory of calculus we calculate the reaction moment/forces M_O, T_O, V_A and drawing the diagrams of moments, displacements and rotations, with (2), (3), (4) relationship and MATHCAD program facilities.
CASE 2: For second-order theory of calculus we calculate the reaction moment/forces M_O, T_O, V_A and drawing the diagrams of moments, displacements and rotations, with (13) relationship and MATHCAD program facilities.

For first-order theory and second-order theory of calculus we drawing the comparative diagrams of bending moment, displacements and rotations.

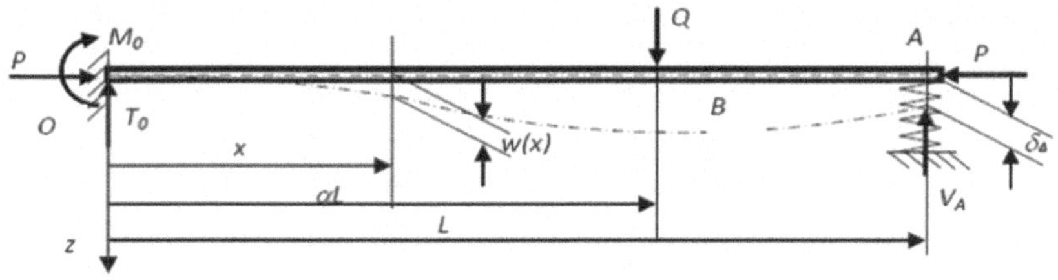

Fig. 3.

CASE 1: The first-order theory of calculus
The unknown reaction moment/forces M_O, T_O, V_A are obtained from equilibrium equations of Mechanics, and the condition of proportionality of reaction V_A with displacement w_A [11, 12] (Fig. 3):

$$w_A = \delta_A \cdot V_A; \qquad \delta_A = \frac{8 \cdot D^3 \cdot n}{G \cdot d^4} \tag{14}$$

The equilibrium equations of Mechanics are written:

$$\begin{aligned} &\sum F_z = 0 \qquad \Rightarrow \quad T_0 + V_A - Q = 0 \\ &\sum M_{yO} = 0 \qquad \Rightarrow \quad M_0 + Q \cdot \alpha L - V_A \cdot L = 0 \end{aligned} \tag{15}$$

The condition of proportionality (14) lead to expression of reaction T_O:

$$T_0 = \frac{1 - \alpha - \frac{(1-\alpha)^3}{3} + \frac{2 \cdot E \cdot I_y}{L^3} \cdot \delta_A}{\frac{2}{3} + \frac{2 \cdot E \cdot I_y}{L^3} \cdot \delta_A} \cdot Q \tag{16}$$

The bending moment $M_1(x)$ function is written using step function $\Phi(x-a)$ of MATHCAD program, for this case as particular relationship:

$$M1(x) = M_0 \cdot \Phi(x) + T_0 \cdot x \cdot \Phi(x) - Q \cdot (x - \alpha L) \cdot \Phi(x - \alpha L) + V_A \cdot (x - L) \cdot \Phi(x - L) \tag{17}$$

The relationship (3) and (4) of rotation function $\varphi_1(x)$ and displacements function $w_1(x)$ are becoming:

$$\begin{aligned} \varphi 1(x) &= -\frac{1}{EI_y}\left[M_0 \cdot x \cdot \Phi(x) + T_0 \cdot \frac{x^2}{2} \cdot \Phi(x) - Q\frac{(x-\alpha L)^2 \cdot}{2}\Phi(x-\alpha L) + V_A\frac{(x-L)^2}{2} \cdot \Phi(x-L)\right] \\ w1(x) &= -\frac{1}{EI_y}\left[M_0 \cdot \frac{x^2}{2} \cdot \Phi(x) + T_0 \cdot \frac{x^3}{6} \cdot \Phi(x) - Q\frac{(x-\alpha L)^3 \cdot}{6}\Phi(x-\alpha L) + V_A\frac{(x-L)^3}{6} \cdot \Phi(x-L)\right] \end{aligned} \tag{18}$$

The diagrams of moments, displacements and rotations for this particular case are obtained with relationship (17), (18) and MATHCAD program facilities; they are drawing in Fig. 4.

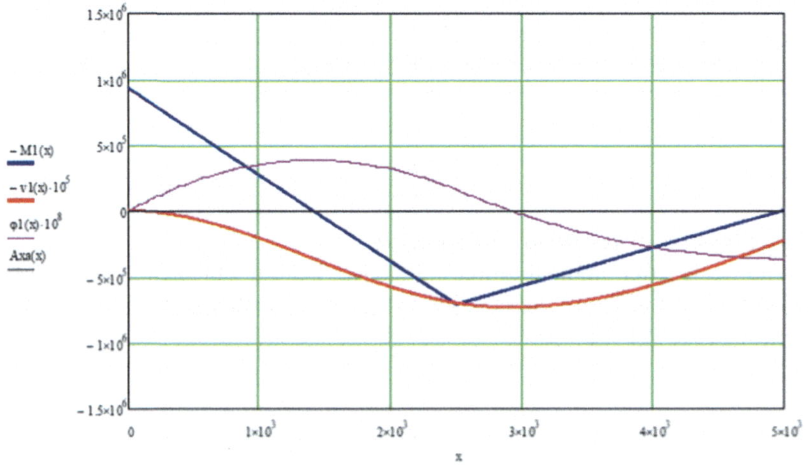

Fig. 4.

We can seen from the Fig. 4 that the maximum bending effort is obtained for $x = \alpha L$ and the maximum displacement is obtained from condition:

$$\frac{d\varphi 1(x)}{dx} = 0$$

Using MATHCAD program facilities (root function) we obtain:

$$\begin{aligned} &M_1(\alpha L) = 7.6466 \times 10^5 \ \text{Nmm} \\ &root(\varphi_1(x), x, 2000, 4000) = 2.8631 \times 10^3 \ \text{mm} \\ &w_{1\max} = 7.446 \ \text{mm} \end{aligned} \tag{19}$$

CASE 2: The second-order theory of calculus

The reaction moment/forces M_0, T_0, V_A are obtained from equilibrium equations of Mechanics (15), and boundary condition of displacement [1, 3]:

$$\begin{aligned} w(L) &= V_A \cdot \delta_A \\ M_{iy}(L) &= 0 \end{aligned} \tag{20}$$

The relationships (13) of displacements $w_2(x)$, rotation $\varphi_2(x)$, and bending moment $M_2(x)$, are writing for this particular case as follow:

$$\begin{aligned} w_2(x) &= -\tfrac{T_0}{k\cdot P}(k\cdot x - sin(k\cdot x)) - \tfrac{M_0}{P}(1 - cos(k\cdot x)) + \tfrac{Q}{k\cdot P}(k(x-\alpha L) - sin(k\cdot(x-\alpha L)))\cdot \Phi(x-\alpha L) \\ \varphi_2(x) &= -\tfrac{T_0}{P}(1 - cos(k\cdot x)) - \tfrac{M_0}{P}\cdot k\cdot sin(k\cdot x) + \tfrac{Q}{P}(1 - cos(k\cdot(x-\alpha L)))\cdot \Phi(x-\alpha L) \\ M_2(x) &= \tfrac{T_0}{k}\cdot sin(k\cdot x) + M_0\cdot cos(k\cdot x) + \tfrac{V_B}{k}\cdot sin(k\cdot(x-\alpha L))\cdot \Phi(x-\alpha L) - \tfrac{Q}{k}\cdot cos(k\cdot(x-\alpha L))\cdot \Phi(x-\alpha L) \end{aligned} \tag{21}$$

Taking into account the relationships is achieved the expression of T_0, M_0, V_A

$$\begin{aligned} T_0 &= -Q\frac{(1-cos(kL))\cdot sin(kL\cdot(1-\alpha)) - cos(kL)\cdot\left((1-\alpha)kL - sin((1-\alpha)kL) - \left(\frac{kL}{L}\right)^3 E\cdot I_y\cdot\delta_A\right)}{(1-cos(\alpha\cdot kL))\cdot sin(kL) - cos(kL)\cdot\left(kL - sin(kL) - \left(\frac{kL}{L}\right)^3 E\cdot I_y\cdot\delta_A\right)} \\ M_0 &= -Q\cdot L\frac{sin(kL)\cdot\left((1-\alpha)kL - sin((1-\alpha)kL) - \left(\frac{kL}{L}\right)^3 E\cdot I_y\cdot\delta_A\right) - sin((1-\alpha)kL)\cdot\left(kL - sin(kL) - \left(\frac{kL}{L}\right)^3 E\cdot I_y\cdot\delta_A\right)}{(1-cos(\alpha\cdot kL))\cdot sin(kL) - \cos(kL)\cdot\left(kL - sin(kL) - \left(\frac{kL}{L}\right)^3 E\cdot I_y\cdot\delta_A\right)} \\ V_A &= Q - T_0 \end{aligned} \tag{22}$$

The diagrams of moments, displacements and rotations for this particular case (the compressive force $P = 0{,}25\cdot P_{cr}$) are obtained with relationship (21) and MATHCAD program facilities; they are drawing in Fig. 5.

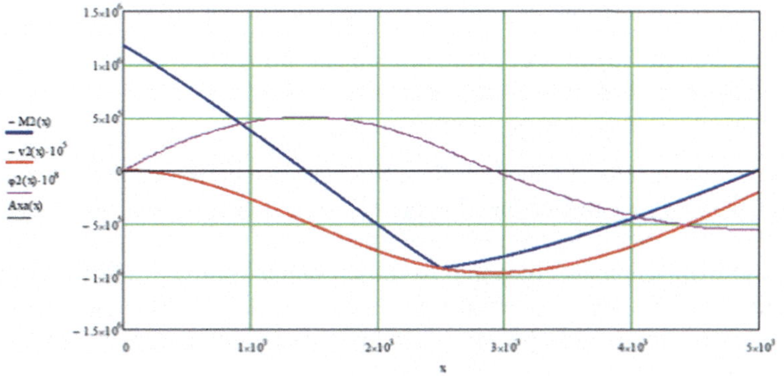

Fig. 5.

We can seen from the Fig. 5 that the maximum bending effort is obtained for $x = \alpha L$ and the maximum displacement is obtained from condition: $\frac{d\varphi 1(x)}{dx} = 0$

$$M_2(\alpha L) = 9.9185 \times 10^5 \text{ Nmm}; \quad root(\varphi_2(x), x, 2000, 4000) = 2.8776 \times 10^3 \text{ mm}$$
$$w_{2\max} = 10.0195 \text{ mm} \tag{23}$$

The critical buckling forces P_{cr} are obtained for this case from denominator zero condition of relationship (22) of reaction T_0, M_0.

We note the denominator as function *F(kL):*

$$F(kL) = (1 - cos(\alpha \cdot kL)) \cdot sin(kL) - cos(kL) \cdot \left(kL - sin(kL) - \left(\frac{kL}{L}\right)^3 E \cdot I_y \cdot \delta_A \right) = 0 \tag{24}$$

The first solution is obtained with root function of MATHCAD program facilities:

$$\text{root}\left[\cos(\text{kL}) \cdot \left[\text{kL} - \sin(\text{kL}) - (\text{kL})^3 \cdot \frac{\text{E} \cdot \text{Iy}}{\text{L}^3} \cdot \delta\text{A}\right] - (1 - \cos(\text{kL})) \cdot \sin(\text{kL}), \text{kL}, 4, 6\right] = 4,4561$$

We obtain in MATHCAD the numerical value of buckling forces for this case:

$$P_{cr} = \frac{(kL)^2}{L^2} \cdot E \cdot I_y \qquad \Rightarrow P_{cr} = 1.3561 \times 10^5 \text{ N} \tag{25}$$

The value of axial force used in calculus is:

$$P = 0,25 \cdot P_{cr} = 33,9025 \times 10^3 \text{ N} \tag{26}$$

In order to represent the *comparative diagrams* of bending moment, displacements and rotations obtained with first-order theory (Fig. 4) and second-order theory of calculus (Fig. 5) are drawing with MATHCAD program facilities:

- The comparative diagrams of bending moment in Fig. 6.
- The comparative diagrams of displacements in Fig. 7.
- The comparative diagrams of rotations in Fig. 8.

The numerical results obtained for the particular case presented in this section let to conclusion that the **second-order theory** is an accurate method and the diagrams obtained show that increasing compressive force lead to increasing values of moments, rotations and displacements compared to those obtained through the **first-order theory**. Figures 7 and 8 show also that the position of section corresponding to maximum bending moment (the middle beam) is not located at same distance with the section corresponding to the maximum displacement.

Also, can be observed small difference for the position of section corresponding to the maximum displacement for first order theory (x_{1max} = 2,8631 mm) and second order theory (x_{2max} = 2,8776 mm) we obtained large differences between the displacements value (v_{1max} = 7,446 mm, v_{2max} = 10,0195 mm).

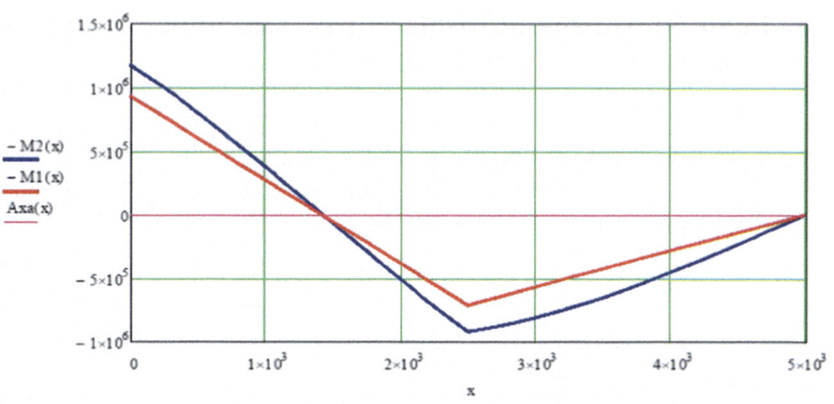

Fig. 6. The comparative bending moment diagrams

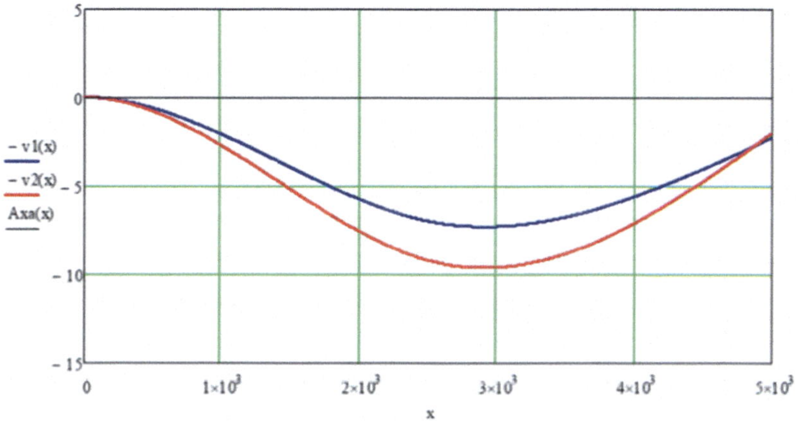

Fig. 7. The comparative displacement diagrams

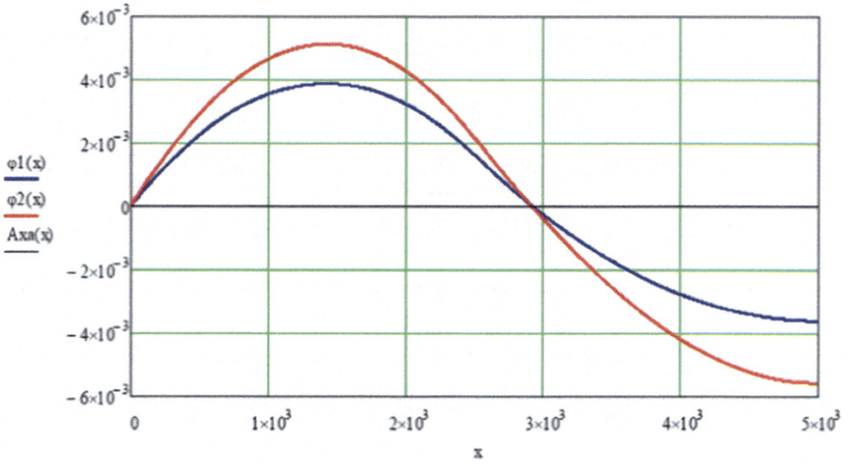

Fig. 8. The comparative rotations diagrams

5 Simulation of Maximum Displacement Variation with Distance αL of External Force Q

Figures 9, 10, 11, 12 and 13 shows the variation of bending moment, displacement and rotation corresponding to five different parameter values $\alpha = 0{,}1$; $\alpha = 0{,}3$; $\alpha = 0{,}5$; $\alpha = 0{,}7$; $\alpha = 0{,}9$. So we found the position of the transverse force on the beam Q so that is obtained the maximum displacement, both using the first-order theory and the second-order theory of calculus.

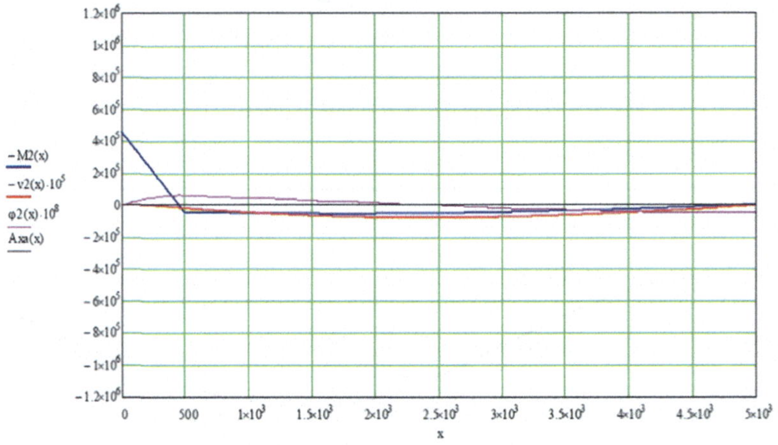

Fig. 9. Variation of bending moment, displacements and rotations for $\alpha = 0{,}1$ $x_{max} = 2435{,}9$ mm; $V_{2max} = 0{,}8165$ mm.

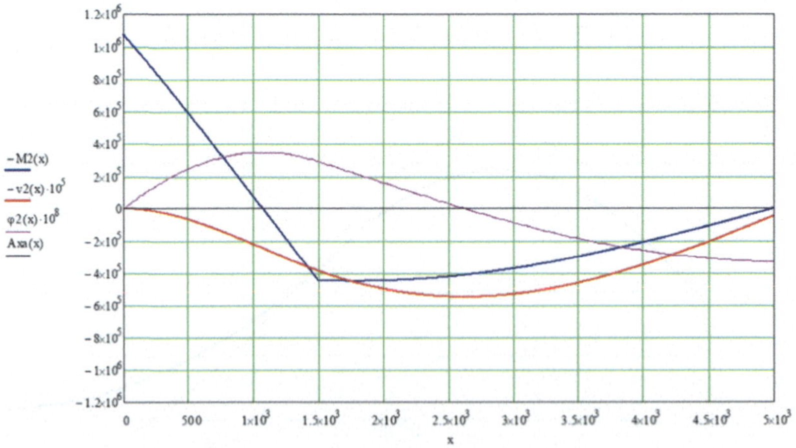

Fig. 10. Variation of bending moment, displacements and rotations for $\alpha = 0{,}3$ $x_{max} = 2625{,}1$ mm; $V_{2max} = 5{,}4856$ mm.

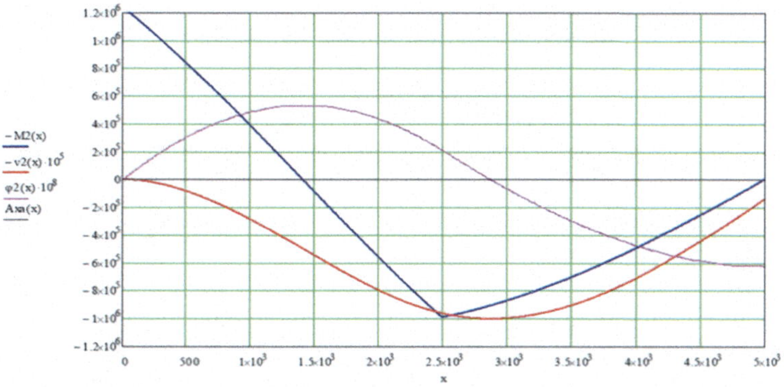

Fig. 11. Variation of bending moment, displacements and rotations for $\alpha = 0{,}5$ $x_{max} = 2877{,}6$ mm; $V_{2max} = 10{,}0195$ mm.

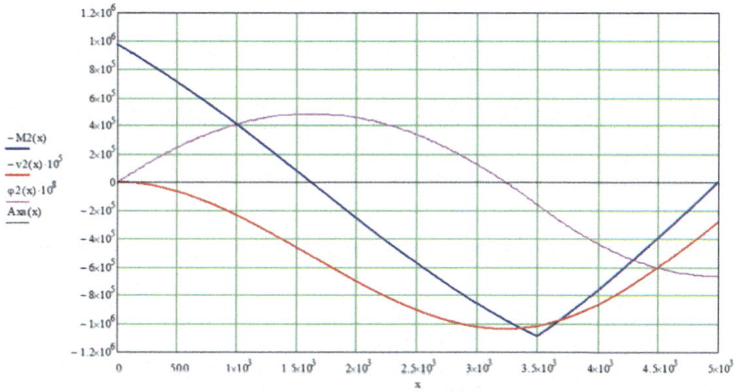

Fig. 12. Variation of bending moment, displacements and rotations for $\alpha = 0{,}7$ $x_{max} = 3234{,}3$ mm; $V_{2max} = 10{,}4003$ mm.

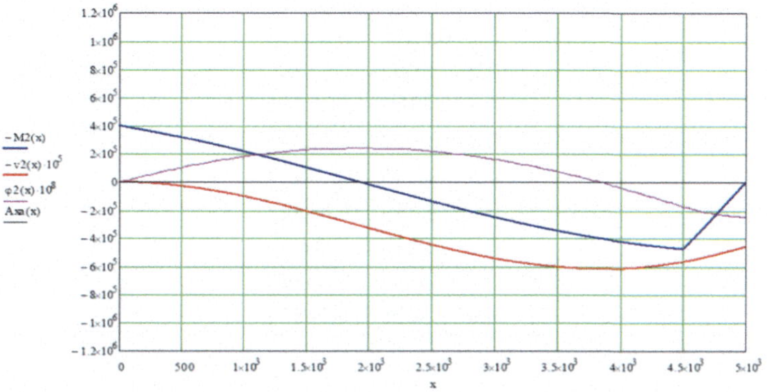

Fig. 13. Variation of bending moment, displacements and rotations for $\alpha = 0{,}9$ $x_{max} = 3845{,}3$ mm; $V_{2max} = 6{,}1485$ mm.

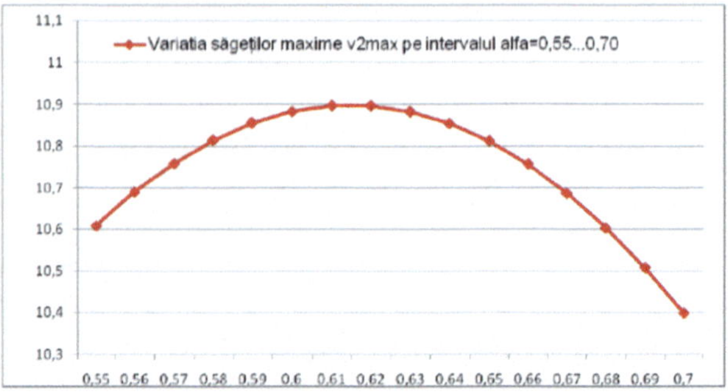

Fig. 14. The variation of Maximum displacement v2max depending to α

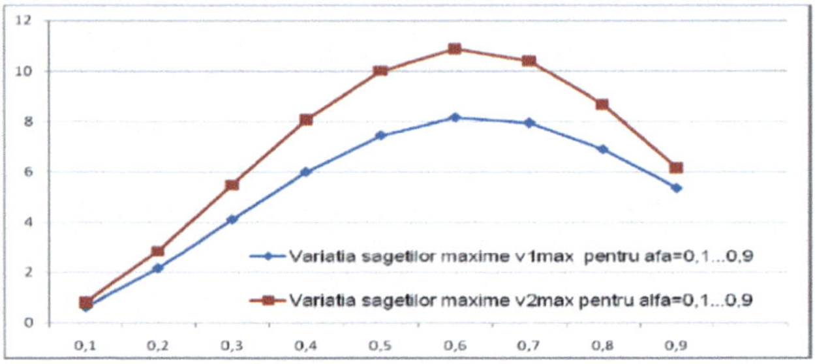

Fig. 15. Maximum displacement obtained by first and second theory

Table 1.

Variation of maximum displacement v_{1max} *(first theory) si* v_{2max} *(second theory)*		
α	v_{1max}	v_{2max}
0,1	0,6312	0,8165
0,2	2,1692	2,8578
0,3	4,1102	5,4856
0,4	5,9993	8,0657
0,5	7,446	10,0195
0,6	**8,149**	**10,8839**
0,7	7,9418	10,4003
0,8	6,8942	8,6689
0,9	5,3605	6,1485

Table 2.

Variation of maximum displacement (frame 0.55 … 0.7) v_{1max} *(first theory) si* v_{2max} *(second theory)* $\alpha = 0.55$ … $\alpha = 0.70$

α	v_{1max}	v_{2max}
0,55	7,9049	10,6105
0,56	7,9716	10,6917
0,57	8,0295	10,7599
0,58	8,0785	10,8148
0,59	8,1183	10,8562
0,6	8,149	10,8839
0,61	8,1703	**10,8978**
0,62	8,1823	10,8976
0,63	**8,1847**	10,8834
0,64	8,1777	10,8552
0,65	8,1613	10,8131
0,66	8,1355	10,7573
0,67	8,1006	10,688
0,68	8,0565	10,6052
0,69	8,0035	10,5093
0,7	7,9418	10,4003

6 Conclusions

From the numerical results obtained for the particular case presented in this chapter it can be concluded that the method of second-order theory is an accurate compared to those obtained through the first-order theory and the simulation by increasing of α parameter indicate increasing and decreasing of deformations as shown in Figs. 9, 10, 11, 12 and 13; this result can allow to find the distance αL of transversal force Q which displacement is the maximum (Table 1 corresponding to Fig. 15 and Table 2 corresponding to Fig. 14). By simulating the displacement variation for the two theories, the position of force Q was obtained so that the displacement was maxim; the position of section corresponding to maximum displacement is not same for the two theories:

1. *The first theory:* $\alpha L(v_{1max})$ = 0,63 L; v_{1max} = 8,1847 mm
2. *The second Theory:* $\alpha L(v_{2max})$ = 0,61 L; v_{2max} = 10,8978 mm

References

1. Simulescu, I.: Buckling and instability - Lecture 17, Ed. U.T.C.B Bucuresti (2007). Chap 6
2. Anghel, A.: Rezistenţa materialelor. Partea I. Ed. Universităţii din Ploiesti, Ploiesti (2004)
3. Anghel, A.: Rezistenţa materialelor. Partea a II-a. Ed. Universităţii din Ploiesti (2005)
4. Marin, C.: Aplicaţii ale teoriei elasticităţii şi plasticităţii în inginerie. Ed. Biblioth (2007)
5. Marin, C.: Probleme de rezistenţa materialelor rezolvate in MATHCAD. Ed. Bibliotheca Targoviste, 2012
6. Marin, C.: Rezistenţa materialelor. Partea I- Solictări simple. Ed. Bibliotheca (2013)

Control of an Autonomous Mobile Waste Collection Robot

Mihai Mărgăritescu[1(✉)], Paul-Nicolae Ancuţa[1], Eduard Valentin Canale[1], Dănuţ Iulian Stanciu[1], Dan Dumitriu[2], and Cornel Mircea Brişan[3]

[1] National Institute of Research and Development in Mechatronics and Measurement Technique, Şos. Pantelimon 6-8, Bucharest, Romania
mihai.margaritescu@gmail.com
[2] Institute of Solid Mechanics of the Romanian Academy, Constantin Mille 15, 10141 Bucharest, Romania
dan.dumitriu@imsar.ro
[3] Technical University of Cluj-Napoca, Cluj-Napoca, Romania
cmbrisan@gmail.com

Abstract. The SIRAMAND project, deployed in the period 2018–2020, aim to develop an autonomous mobile robot capable to recognize different types of solid waste such as plastic bottles or aluminum doses placed on alleys of parks, streets, etc., pick up and storage them in a container behind the transporter. The robot must track a programmable trajectory and stop when the vision system detects an object of interest, the principle of detection being described. Adjusting the angular velocities of the two front drive wheels, the system follows a theoretical trajectory. The relations between the external and internal coordinates of the robot are presented, the guidance on the desired trajectory based on the GPS system being the main control option, with possibilities of using of supplementary sensors. To reduce the deviations due to the expected noisy GPS signal, a Kalman filter was simulated. An alternative solution consists in robot guidance using curbs, as a wall follower vehicle.

Keywords: Waste collection · Autonomous robots · Mobile robots

1 Introduction

In the context of the anticipated developments of the smart city concept, there are different attempts to replace the routine work of street and green spaces caretakers through the creation of autonomous mobile robotic systems. The trend of robotizing transport, delivery services using drones, elder care services is a reality, and the field of "cleaning robots" has grown in recent years. Even though efforts are being made to selectively collect waste, major cities are affected by their presence on streets, parks, other types of public spaces, by waste meaning mainly plastic bottles, metallic doses, bags, papers, etc. For the transporter of the mobile system, there are specific characteristics: Assuming that the mobile robot is assigned a particular cleaning area (a pavement, an alley), defined by a series of GNSS/GPS coordinates, it must be possible

G. I. Gheorghe (Ed.): ICOMECYME 2019, LNNS 85, pp. 51–63, 2020.
https://doi.org/10.1007/978-3-030-26991-3_6

to program the initial route of the robot; the system must automatically detect any deviations from the set route and apply the necessary corrections, as well as to avoid unexpected obstacles such as stationary cars, pedestrians, cyclists, etc. Trajectory programming must be easy enough for a user with no advanced IT knowledge; the transporter must also be able to be remote controlled by radio from a control center by an operator, which can monitor, interrupt or modify the route.

At present, manual sorting of conveyor belts is the most widely used method of separation, is very costly and expose human staff to various environmental factors. For this reason, researchers try to find solutions on various segments of waste collection and sorting. Robotic systems for waste collection of different structure [1, 2] were designed (see Fig. 1, respectively Fig. 2); some agricultural robots have a strong similarity from kinematics point of view with the waste collection robots (Fig. 3), [3].

Fig. 1. Robot for taking bags of household waste

Fig. 2. Remote control robot for waste collection

Fig. 3. The ecoRobotix weed combat robot

2 SIRAMAND Model

2.1 Construction

SIRAMAND project proposes a different concept. A mobile autonomous robot goes through a preset path, e.g. on a park alley, detects and collect the waste. It is composed mainly of: (1) chassis, (2) housing, (3) drive wheels, (4) free wheels, (5) XYZ handling system, (6) gripper and (7) waste container (see Fig. 4). A vision system, integral part of the XYZ handling system (see Fig. 5), recognizes specific objects and commands stopping the robot. This process will be detailed in the next section. A fixed control center for trajectory programming and emergency situations is also foreseen. A second camera mounted on the transporter will allow the remote control of the robot for some failure cases.

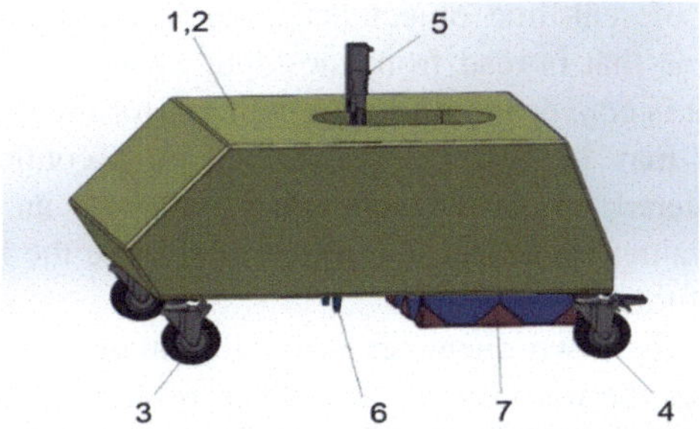

Fig. 4. The mobile autonomous robot model

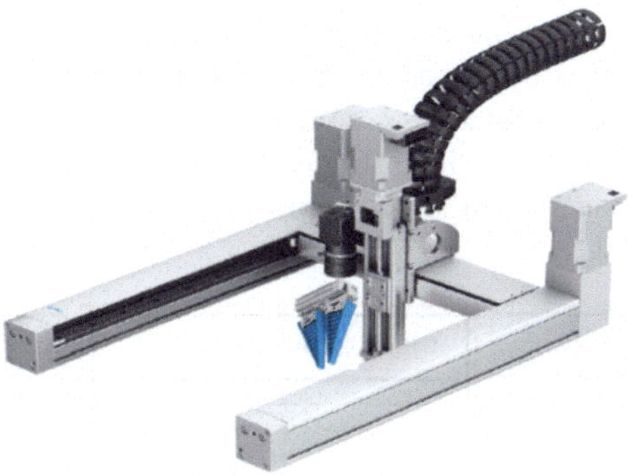

Fig. 5. XYZ handling system provided by FESTO

2.2 Object Detection

Real-time detection of predefined objects in images taken from camcorders is a current research topic in information technology field. The traditional image processing approaches use pipeline processing for the detection of compact areas in the image named blobs. The color image is converted to a gray image which is binarized according to local or global methods. The blobs must be classified and sorted according to their area, form factors and others numerical values that software can determine. This type of approach assumes that the background of the image is clearly differentiated from the useful area, so the thresholds will partition the image according to the requirements. A major impediment to these techniques for outdoor detection of objects is the uncontrollable ambient brightness of the day. Pattern recognition uses the automatic discovery of regularities in data through the use of computer algorithms for classifying the data into different categories [4]. It may use image processing as a preliminary step. The modern approach in object detection uses pattern recognition by neural networks. A deep neural network (DNN) is an artificial neural network with at least one hidden layer between the input and output layers. OpenCV is probably the most used software package for real-time computer vision. Deep neural network (dnn) module is part of actual OpenCV distribution. The dnn module allows load pre-trained models from many deep learning frameworks, including Tensorflow, Caffe,

Torch. The principle of real-time object detection using pre-trained network is as follows: (1) Each frame that is read from the input image or video stream is passed through a function that is converting it to an input blob for the neural network; (2) The classified objects that may be detected are filtered on a confidence score, a list of bounding boxes is offered; (3) If the confidence value for an object is less then a predefined threshold value, the object is rejected; otherwise the bounding box for this object is retained for further processing.

The software comprises two client-server configurations. Server Auto is a part of the software that moves the transporter. As a result of an explicit request issued by a client, server AUTO must start/stop the mechanical transporter and inform about the completion of the task. Likewise, server GC is a part of the software included in a dedicated PLC that supervises the waste removal. Server GC has to read the values of the framing rectangle of an object and to inform the client about the correctness of the object removal. Software connectivity between clients and both servers is made through sockets using two dedicated programmed thread modules, thread Auto and Thread GC. Object detection thread is continuously running for object detection. Every time an object is identified in a video frame, detected event is raised. The conceptual model of object detection system is depicted in Fig. 6.

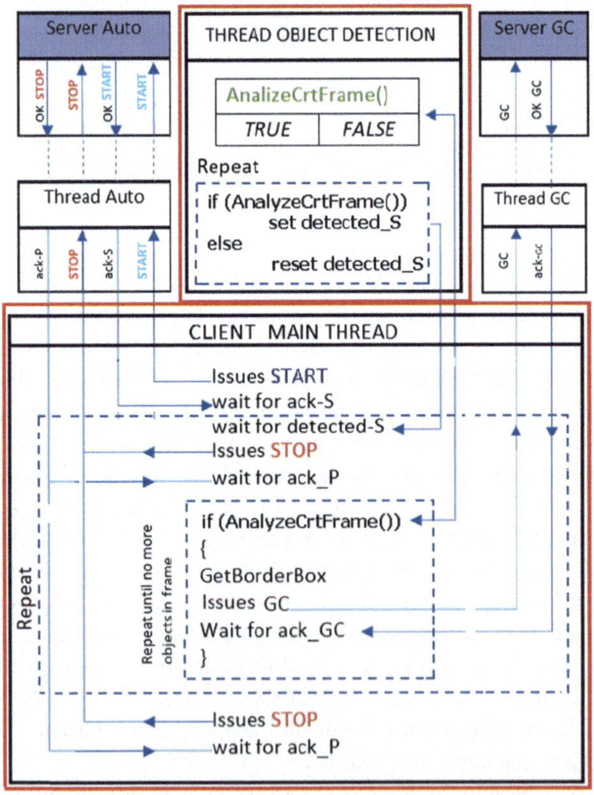

Fig. 6. Conceptual model of software system

A simplified processing flow implemented by client software is as follows:

(1) Client initializes the communication socket with automotive server – Server AUTO - using a threaded object named Thread AUTO.
(2) Client initializes the communication socket with Waste Collector server – GC Server - using a threaded object named Thread GC.
(3) Client issues a START command to Server AUTO via Thread AUTO. The mechanical transporter is started.
(4) Client waits in a non-blocking manner on START acknowledge.
(5) Client waits for an event associated with object detection.
(6) When this happens, client issues a STOP command to Server AUTO.
(7) Client waits in a non-blocking manner on STOP acknowledge.
(8) Based to the fact that the transporter does not instantly stops, a single frame then is captured and analyzed.
(9) If an object is still detected in frame, the border box of the object is extracted; client issues a Remove GC to Server GC via Thread GC and waits on GC acknowledge.
(10) Step 7 is repeated until no more objects are detected in the same frame.
(11) Processing flow is repeated beginning with Step 4.
(12) End of processing is signaled by end of video file or other specific event.

Implementation and Results. Software is implemented using Python and OpenCV 4.0.0 running on a PC with Windows 10 installed. For object detection, it is used a MobileNet pre-trained in Caffe-SSD framework that can detect 20 classes of objects. Actual implementation uses some more Event and Boolean variables that were described in the conceptual model. First phase of bottle detection is depicted in Fig. 7.

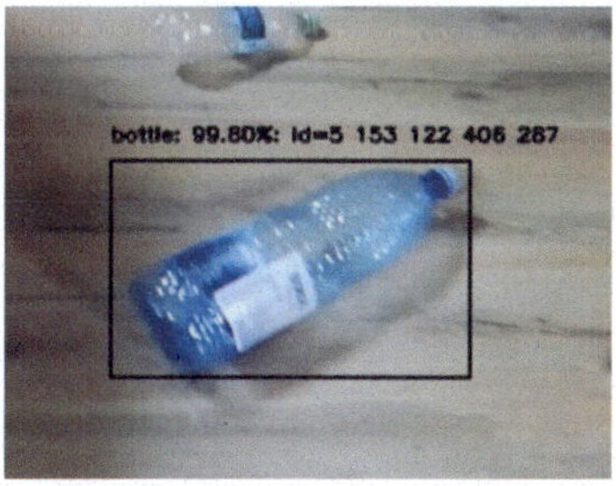

Fig. 7. First phase of bottle detection

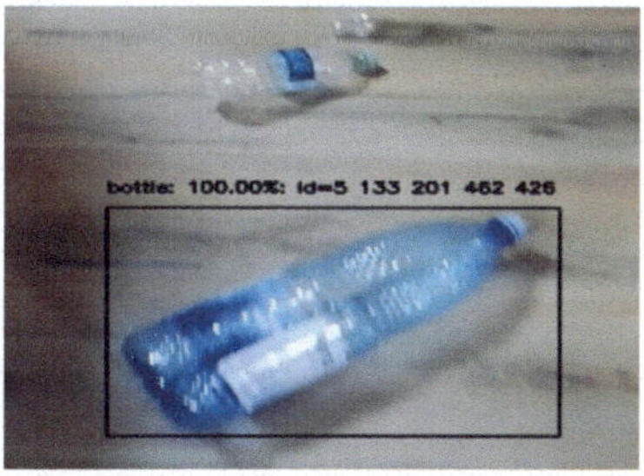

Fig. 8. Second phase of bottle detection

The software issues STOP command to the server Auto. When server Auto reports that the transporter has stopped, the current frame is analyzed. If the object is still detected (see Fig. 8), the software issues a Waste Collect command to the GC Server and waits for its completeness. The software issues then START command for transporter continues its movement.

3 Trajectory Tracking

3.1 Tracking Algorithm on a Default Route

The transporter contains two non-orientable drive wheels - the front wheels, and two free-wheel/freely adjustable wheels - the rear wheels. By correlating the speeds of the two drive wheels, the transporter travels on the desired trajectory, given by coordinates (x_i, y_i), $i = 0, ..., n$. These can be obtained in two ways:

(1) The desired route is gone at a certain speed, preferably close to the cruiser speed of the conveyor (3–5 km/h), and record the GPS coordinates with the desired sampling rate, also close to that used during the transporter displacement; the coordinates of the recorded points define the desired path;
(2) Using digital maps, with the following steps:

- Interest area is digitally downloaded from specialized applications, such as Google Maps and uploaded to vehicle control software
- Route nodes are defined by the user's preference by clicking on the cursor. Thus, the Cartesian coordinates of the defined point are recorded in the software.

In order to establish the relationship between the internal coordinates of the conveyor, i.e. the angles of rotation to left L (Left) or R (Right), q_{Li}, q_{Ri} and external coordinates x_i, y_i, the angle of rotation θ_i of the axle of the drive wheels relative to the y-axis is computed (or the longitudinal axis of the transporter with respect to the x-axis (see Fig. 9). In this figure, d is the distance between the wheels centers of radius r. The external coordinates x_i, y_i, should coincide with the x_{Ci}, y_{Ci} of the point C_i located midway between the centers of the drive wheels. The line joining the centers of the drive wheels is considered to be perpendicular to the trajectory any time.

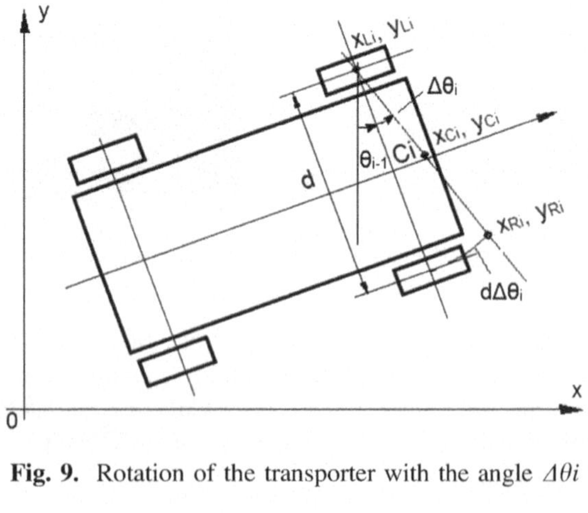

Fig. 9. Rotation of the transporter with the angle $\Delta\theta i$

It is noticed that we have the approximate relationship

$$\Delta\theta_i \cong (\omega_{Ri} - \omega_{Li}) \cdot \frac{r}{d} \cdot \Delta t \tag{1}$$

Thus, the angle of inclination of the wheel axle with respect to the y axis in point (x_i, y_i) results:

$$\theta_i \cong \sum\nolimits_i (\omega_{Ri} - \omega_{Li}) \cdot \frac{r}{d} \cdot \Delta t = \theta_{i-1} + \frac{r}{d} \cdot \Delta t \cdot (\omega_{Ri} - \omega_{Li}) \tag{2}$$

Direct Kinematics of the Transporter. It is intended to express the external coordinates as functions of the internal coordinates of the transporter:

$$\begin{cases} x_{Ci} = x_{Ci}(q_{Li}, q_{Ri}) \\ y_{Ci} = y_{Ci}(q_{Li}, q_{Ri}) \end{cases} \tag{3}$$

The initial conditions for both angles and angular velocities are considered null. For this, the values $\omega_{Li} = \dot{q}_{Li}$, respectively $\omega_{Ri} = \dot{q}_{Ri}$, $i = 0, \ldots\ n$ are considered known, with the sampling rate Δt between two successive points $i - 1, i$. By integration, the corresponding values of the angles of rotation are also obtained. The coordinates of the wheels are given by:

$$\begin{cases} x_{Li} = \sum_i \omega_{Li} \cdot r \cdot \cos\theta_i \cdot \Delta t = x_{Li-1} + \omega_{Li} \cdot r \cdot \cos\theta_i \cdot \Delta t \\ y_{Li} = \sum_i \omega_{Li} \cdot r \cdot \sin\theta_i \cdot \Delta t = y_{Li-1} + \omega_{Li} \cdot r \cdot \sin\theta_i \cdot \Delta t \end{cases} \tag{4}$$

and similar for the right wheel. Further, are obtained the coordinates of the point C_i, i.e. the points defining the trajectory of the transporter:

$$\begin{cases} x_{Ci} = x_{Ci-1} + r \cdot cos\left(\theta_{i-1} + \frac{r}{d} \cdot \Delta t \cdot (\omega_{Ri} - \omega_{Li})\right) \cdot \Delta t \frac{\omega_{Li} + \omega_{Ri}}{2} \\ y_{Ci} = y_{Ci-1} + r \cdot sin\left(\theta_{i-1} + \frac{r}{d} \cdot \Delta t \cdot (\omega_{Ri} - \omega_{Li})\right) \cdot \Delta t \frac{\omega_{Li} + \omega_{Ri}}{2} \end{cases} \tag{5}$$

Example with Simulated Data: The transporter is considered to have the distance between the centers of the wheels d = 0,7 m, wheel radius r = 0,078 m and Δt = 1 s. Simulated angular velocities (101 samples) are given in a table not shown, but having the graphs shown in Fig. 10:

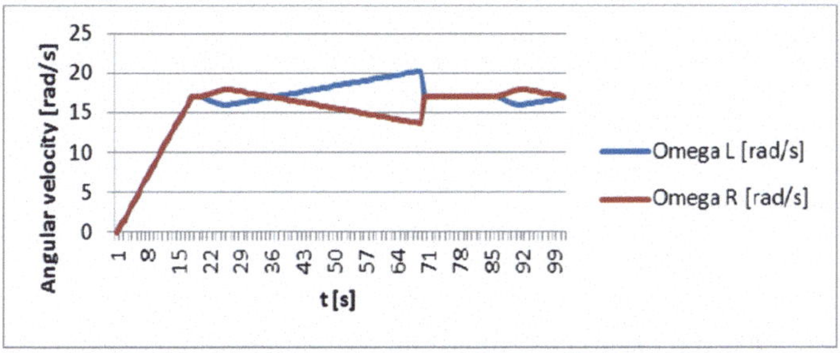

Fig. 10. Angular velocities graphs of drive wheels

Further, using the relationship (5), the coordinates of the C_i points defining the trajectory of the transporter are established, the trajectory being shown in Fig. 11:

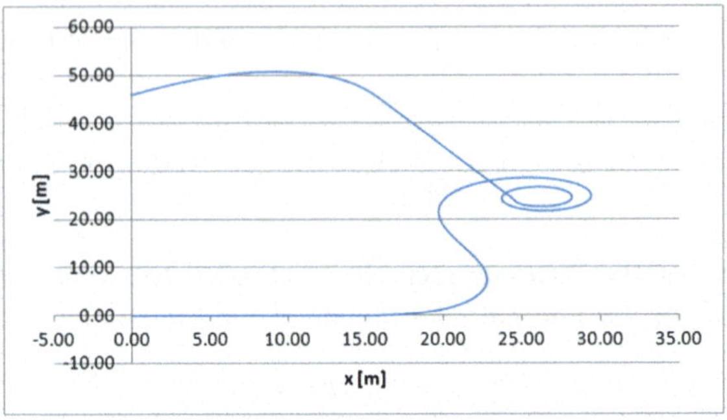

Fig. 11. The graph of the transporter trajectory

Inverse Kinematics of the Transporter. The inverse kinematic problem seeks to express internal coordinates as functions of the external coordinates, relations that are used to command the transporter:

$$\begin{cases} q_{Li} = q_{Li}(x_{Ci}, y_{Ci}) \\ q_{Ri} = q_{Ri}(x_{Ci}, y_{Ci}) \end{cases} \tag{6}$$

As in the case of direct kinematics, we are interested in setting the angular speeds of the two drive wheels $\omega_{Li} = \dot{q}_{Li}$, respectively $\omega_{Ri} = \dot{q}_{Ri}$, $i = 0, \ldots, n,$ that is, the system's command values, from which can be obtained by integrating the internal coordinates q_{Li} și q_{Ri}. However, it must be taken into account that the desired trajectory can be traveled in infinity of modes in terms of speed. For this reason, a certain cruise speed is required, which corresponds to an angular speed ω_{cr}.

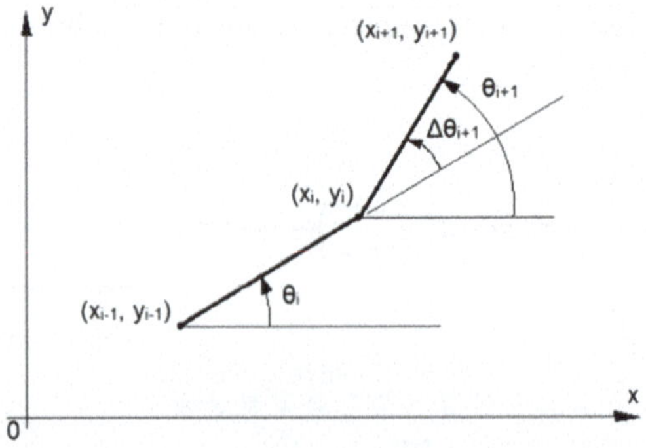

Fig. 12. Trajectory defined by two successive segments

Due to the indeterminations introduced by the arctangent function, it is preferred an algorithm based only on the variation of the angle θ_i, i.e. the angle $\Delta\theta_i$, which for physical reasons can be a sharp angle only. For calculating the θ_{i+1} angle (see Fig. 12), there are two cases:

Case 1. The points of coordinates (x_{i-1}, y_{i-1}), (x_i, y_i) and (x_{i+1}, y_{i+1}) are collinear. This condition is met if and only if the following determinant is null:

$$\begin{vmatrix} x_{i-1} & y_{i-1} & 1 \\ x_i & y_i & 1 \\ x_{i+1} & y_{i+1} & 1 \end{vmatrix} = 0 \tag{7}$$

In this case, $\Delta\theta_{i+1} = 0$ and $\omega_{Ri} = \omega_{Li} = \omega_{cr}$.

Case 2. The points of coordinates $z(x_{i-1}, y_{i-1})$, (x_i, y_i) şi (x_{i+1}, y_{i+1}) are not collinear. In this case, using the generalized Pythagoras theorem, applied in the triangle formed by the coordinate points (x_{i-1}, y_{i-1}), (x_i, y_i) şi (x_{i+1}, y_{i+1}), it is obtained:

$$\Delta\theta_{i+1} = \pm arccos\frac{\left[(x_{i+1}-x_{i-1})^2+(y_{i+1}-y_{i-1})^2\right]-\left[(x_i-x_{i-1})^2+(y_i-y_{i-1})^2\right]-\left[(x_{i+1}-x_i)^2+(y_{i+1}-y_i)^2\right]}{2\cdot\sqrt{\left[(x_i-x_{i-1})^2+(y_i-y_{i-1})^2\right]\cdot\left[(x_{i+1}-x_i)^2+(y_{i+1}-y_i)^2\right]}} \tag{8}$$

The "+" sign is chosen if the θ_{i+1} angle is in the trigonometric sense (curve to the left) and the "minus" sign is chosen if the θ_{i+1} angle is in the inverse trigonometric (curve to the right). To make the choice of the sign based on an algorithm, it is proceeding as follows: The line determined by the coordinate points (x_{i-1}, y_{i-1}) şi (x_i, y_i) divides the plane into two semi planes and has the equation:

$$y - \frac{y_i - y_{i-1}}{x_i - x_{i-1}} \cdot (x - x_{i-1}) - y_{i-1} = 0 \tag{8}$$

Combining the sign of the expression in the left side of the relations (9) with the direction of the x-axis scrolling, results in the sign of the expression (8), i.e. the algorithm automatically determines whether the curve is to left or to right. The establishment of the sign is summarized in Table 1:

Table 1. Establishment of the sign in the expression (8)

	$x_i - x_{i-1} \geq 0$	$x_i - x_{i-1} < 0$
$y_{i+1} - \frac{y_i - y_{i-1}}{x_i - x_{i-1}} \cdot (x_{i+1} - x_{i-1}) - y_{i-1} \geq 0$	+	−
$y_{i+1} - \frac{y_i - y_{i-1}}{x_i - x_{i-1}} \cdot (x_{i+1} - x_{i-1}) - y_{i-1} \leq 0$	−	+

Having calculated $\Delta\theta_{i+1}$ with relation (8) and with the sign established by the procedure described later, θ_{i+1} is calculated. This also allows us to calculate the coordinates of wheel centers:

$$\begin{cases} x_{Li} = x_i - \frac{d}{2} \cdot sin\theta_i, y_{Li} = x_i + \frac{d}{2} \cdot cos\theta_i \\ x_{Ri} = x_i + \frac{d}{2} \cdot sin\theta_i, y_{Ri} = x_i - \frac{d}{2} \cdot cos\theta_i \end{cases} \tag{9}$$

Having imposed ω_{cr}, we can choose a symmetrical distribution of the difference between angular velocities:

$$\begin{cases} \omega_{Li} = \omega_{cr} - \frac{(\omega_{Ri} - \omega_{Li})}{2} = \omega_{cr} - \frac{d}{2 \cdot r \cdot \Delta t} \Delta\theta_{i+1} \\ \omega_{Ri} = \omega_{cr} + \frac{(\omega_{Ri} - \omega_{Li})}{2} = \omega_{cr} + \frac{d}{2 \cdot r \cdot \Delta t} \Delta\theta_{i+1} \end{cases} \tag{10}$$

or, the final form of the relations of the inverse kinematic problem:

$$\begin{cases} \omega_{Li} = \omega_{cr} - \frac{d}{2 \cdot r \cdot \Delta t} (\pm) \; * \; \arccos \frac{\left[(x_{i+1} - x_{i-1})^2 + (y_{i+1} - y_{i-1})^2\right] - \left[(x_i - x_{i-1})^2 + (y_i - y_{i-1})^2\right] - \left[(x_{i+1} - x_i)^2 + (y_{i+1} - y_i)^2\right]}{2 \cdot \sqrt{\left[(x_i - x_{i-1})^2 + (y_i - y_{i-1})^2\right] \cdot \left[(x_{i+1} - x_i)^2 + (y_{i+1} - y_i)^2\right]}} \\ \omega_{Ri} = \omega_{cr} + \frac{d}{2 \cdot r \cdot \Delta t} (\pm) \; * \; \arccos \frac{\left[(x_{i+1} - x_{i-1})^2 + (y_{i+1} - y_{i-1})^2\right] - \left[(x_i - x_{i-1})^2 + (y_i - y_{i-1})^2\right] - \left[(x_{i+1} - x_i)^2 + (y_{i+1} - y_i)^2\right]}{2 \cdot \sqrt{\left[(x_i - x_{i-1})^2 + (y_i - y_{i-1})^2\right] \cdot \left[(x_{i+1} - x_i)^2 + (y_{i+1} - y_i)^2\right]}} \end{cases} \tag{11}$$

where $(\pm)*$ refers to the sign established in the manner described above. The advantage of this algorithm is that it is proper for programming.

These values are used in the transporter control. The angular velocities are controlled by the aid of Hall sensors integrated in the drive wheels.

3.2 Kalman Filtering for Reducing the Deviations from the Imposed Trajectory

The process of trajectory tracking is affected by various errors. The GPS system is disturbed by noises, which would cause the transporter's sinuous movement on the trajectory. To limit deviations from the imposed trajectory, various methods are used based in particular on sensor fusion, i.e. on combining information from multiple sensors and obtaining a better result than on each sensor. Choi et al. (2011) [5] used GPS, vision based on image matching and inertial sensors together with an algorithm that automatically switches the operation modes according to GPS status for a mobile vehicle. Nemec et al. (2019) [6] proposed for precise localization of a mobile wheeled robot the use of the next sensors: odometer, gyro, accelerometer, magnetometer and landmarks. They studied the behavior of the robot with the first sensor only, then with the first two and so on - finally fusing the five sensors. They concluded the method is an alternative for the extended Kalman filter (EKF) method. Kalman filter, together with its variants EKF and unscented Kalman filter (UKF) is one of the most applied data fusion algorithm, being used in missiles navigation, including the Apollo mission to moon, satellite navigation devices, smart phones and many others. The Kalman filter model is described by the next equations [7–10]:

$$\mathbf{x}_k = \mathbf{F}_k\mathbf{x}_{k-1} + \mathbf{B}_k\mathbf{u}_k + \mathbf{w}_k \tag{12}$$

where: $\mathbf{x}_k$ - the state vector at time k, evolving from k − 1; $\mathbf{F}_k$ - the state-transition model; $\mathbf{B}_k$ - the control-input model; $\mathbf{u}_k$ - the control vector; $\mathbf{w}_k$ - the process noise and

$$\mathbf{z}_k = \mathbf{H}_k\mathbf{x}_k + \mathbf{v}_k \tag{13}$$

where: $\mathbf{z}_k$ - the observation/measurement at time k; $\mathbf{H}_k$ - the observation model; $\mathbf{v}_k$ - the observation noise.

The covariances of the process and observation noise, assumed to have normal distribution with zero mean, are usually noted: $\mathbf{Q}_k$ – the covariance of the process noise; $\mathbf{R}_k$ – the covariance of the observation noise.

The essence of method consists in estimation of the current state based on the prediction that uses the previous state and on the update, based on the measurement at that time. Using the Kalman filter function in Matlab R2018b and the simulated data from the previous section, more specific the same path (see Fig. 11), it were obtained the state outputs in the first subplot (true state in blue and the filter response in red) and the errors in the second subplot. Varying the covariance of the process and of the observation, i.e. were simulated several cases, from which are presented two: for reduced noise (see Fig. 13) and for greater noise (see Fig. 14). It must be mentioned that for errors there are two pairs of graphs, corresponding to x and y directions.

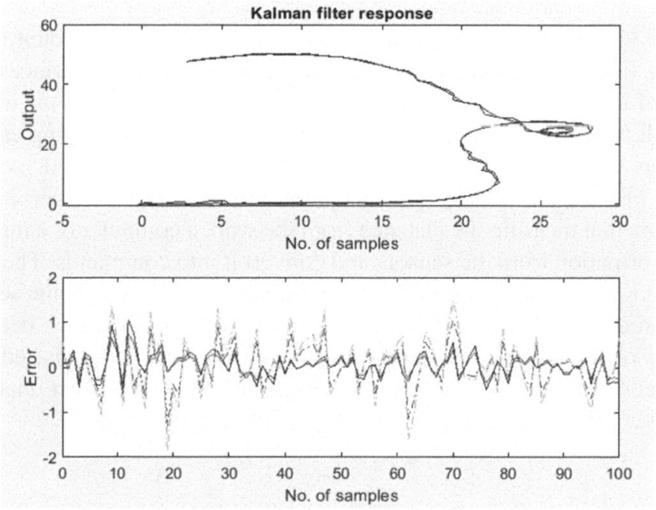

Fig. 13. Kalman filter response for Q = 0.1 and R = 0.2

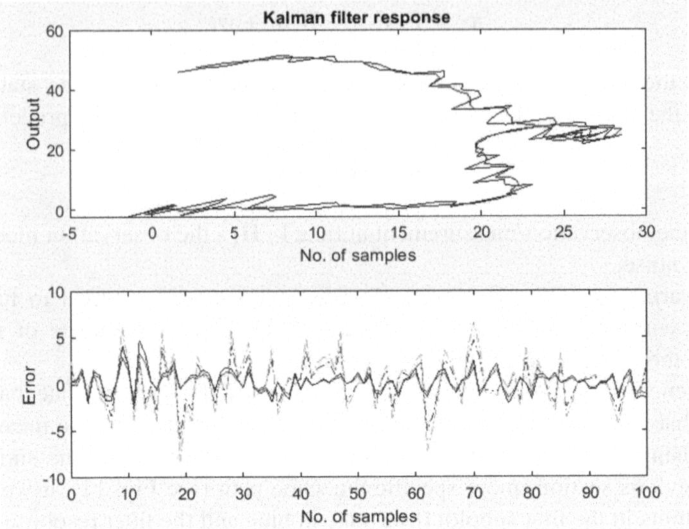

Fig. 14. Kalman filter response for Q = 2 and R = 4

It is noticed the good behavior of the filter; this instrument could be necessary in case of a noisy GPS signal.

3.3 Line or Wall Following Method

As alternative to the GPS guidance of the robot, it is taken into account the trajectory tracking using lines/strips traced on the asphalt or the side/vertical surfaces of the curbs of the alleys, if any. Of course, this requires a certain arrangement of the route. In order to test the wall follower algorithm, corresponding to the guidance using curbs, a model of autonomous vehicle, carried out at a reduced scale was developed.

The most of the wall following robots are built in a similar manner: a sensor or an array of sensors that measure the distance from the wall, a computer or a microcontroller to take the information from the sensors and convert it into commands. The autonomous model uses a closed-loop PID control algorithm based on two ultrasonic sensors for the distance measurement. It calculates the error based on the difference between the set point and the readings from the sensors, applying the corrections based on the proportional, integral and derivative terms, according to the well-known relation:

$$u_i = K_p \cdot e_i + K_i \cdot \sum_i e_i + K_d \cdot \frac{e_i - e_{i-1}}{\Delta t} \tag{14}$$

where u_i is the control variable, e_i is the error of the process variable, K_p the proportional gain, K_i the integral gain and K_d the derivative gain.

By the aid of wireless communication, data was transmitted and visualized during the process. The tests were made in three variants: (1) P - purely Proportional control, (2) PD - Proportional - Derivative control and (3) PID - Proportional - Integral-Derivative control. The differences can be observed in the next set of graphs (Fig. 15).

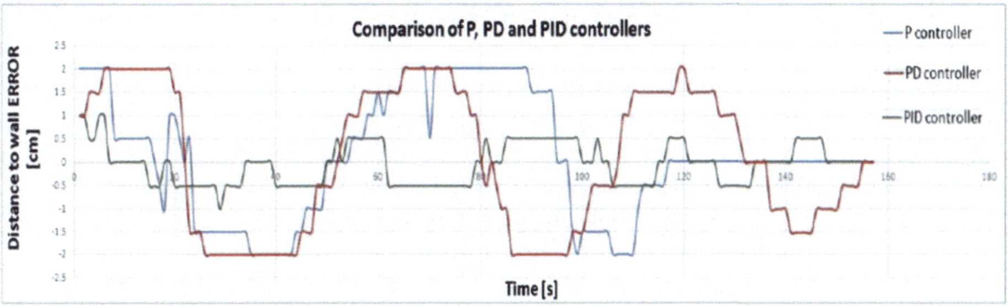

Fig. 15. Distance to wall error versus time for P, PD, PID cases

As expected, the PID controller has the best behavior and can be implemented as alternative or complementary solution for the SIRAMAND model.

4 Conclusions

This article presents the main directions of control of the SIRAMAND autonomous transporter. Other sensors, such as inertial and magnetometer sensors or the second camera placed in the front the transporter could be subsequently added to improve the trajectory tracking. The future work mainly consists in the integration of all component software.

Acknowledgements. This work was supported by a grant of the Romanian Ministry of Research and Innovation, CCCDI – UEFISCDI, project number PN-III-P1-1.2-PCCDI-2017-0086/contract no. 22 PCCDI/2018, within PNCDI III.

References

1. http://www.theoldrobots.com/dustcar.html
2. https://www.youtube.com/watch?v=860LJxp8wfk
3. https://www.youtube.com/watch?v=IKZfjtPquzs
4. Bishop, C.M.: Pattern Recognition and Machine Learning. Springer, Boston (2006)
5. Choi, J.-H., Park, Y.-W., Song, J.-B., Kweon, I.-S.: Localization using GPS and VISION aided INS with an image database and a network of a ground-based reference station in outdoor environments. Int. J. Control Autom. Syst. **9**(4), 716–725 (2011)
6. Nemec, D., Šimák, V., Janota, A., Hruboš, M., Bubeníková, E.: Precise localization of the mobile wheeled robot using sensor fusion of odometry, visual artificial landmarks and inertial sensors. Robot. Auton. Syst. **112**, 168–177 (2019)
7. https://en.wikipedia.org/wiki/Kalman_filter
8. https://www.mathworks.com/discovery/kalman-filter.html
9. Faragher, R.: https://courses.engr.illinois.edu/ece420/sp2017/UnderstandingKalmanFilter
10. Becker, A.: https://www.kalmanfilter.net/default.aspx

Smart Glasses

David Kovanda(✉) and Jan Soukal

Faculty of Mechanical Engineering, Czech Technical University in Prague, Technická 4, Prague 6, Czech Republic
{david.kovanda, Jan.Soukal}@fs.cvut.cz

Abstract. This article deals with the creation of glasses for people with a hearing impairment. Using the microphones placed in the temples, the glasses will be able to take the surrounding sounds and convert them, with the use of extended reality, into a virtual picture. We describe our display principle, which is based on the magnifying glass, and the construction of the glasses. Our display system must fulfil several points, such as good visibility for the users in the real surroundings and in different light conditions. The next descripted issue is the sound detection and its proper focus in the real surroundings. The software part, which is written in programming language C, is responsible for the visualization and processing of the received measured data. Besides these two previously metioned functions, the software is also responsible for the communication with other devices, which are able to connect to our glasses.

Keywords: Augmented reality · Smart glasses · Construction · 3D printing

1 Introduction

The concept of augmented reality is an increasingly used term in recent years. It is a technology that is not yet widespread, but yields a great potential. In theory it is complementing of the real world experience with additional digital information and virtual graphic elements. One of the techniques used are smart glasses, pioneered by the Google company, which launched its Google Glass solution to the market several years ago. However, one of the reasons why they have not caught up is their steep price. In our design of this type of glasses we try to minimize this factor and make it accessible to wider public. Availability is our primary requirement for our smart glasses as a medical device for people with hearing impairment or complete loss of hearing. There is almost half a million of people with this handicap people in the Czech Republic, 7 600 of them being completely deaf. Since only approximately 0.08% of the population can use the sign language, and other methods of communication such as lip-reading methods are not very reliable and easy, various projects are developed to identify and convert speech to text. Most often the outputs are displayed on the screens of computers and mobile phones. However, this output is not very comfortable for a daily life of a person with hearing handicap. We decided to use augmented reality and its implementation in glasses because that allows the user to keep an eye contact with the world, to have free hands and be able to do normal work. Using the built-in microphones, the ambient sound is analyzed and the display informs the user about the

G. I. Gheorghe (Ed.): ICOMECYME 2019, LNNS 85, pp. 64–73, 2020.
https://doi.org/10.1007/978-3-030-26991-3_7

direction of that sound or it displays the converted speech into text. In addition to improving the communication with others, it can alert said person to a moving car, vehicle's honks or other warning signals.

2 Basic Concept

Our goal was to design a simple device for the deaf which uses the principle of augmented reality. For a correct function of smart glasses, the construction must be with prescribed manufacturing tolerances and assembled within mount tolerances, to which the imaging part is the most sensitive. Several other factors are directly related to this problem, such as the choice of the manufacturing technology, the price of the product, or the material used. In our basic discretion, we consider the price of the device and the number of required smart glasses. We set the required number to 100 pieces.

The structure must withstand all loads resulting from normal use. We also focused on extreme conditions, such as a fall resistance and other common forms of material stress. Input audio signals are detected by a trio of microphones placed in the sides and at the front of the glasses. Identification of the direction of sound is done by measuring the time delay between audio signals received by the three microphones. We try to make the glasses a platform that can meet not only current requirements but also new demands and further SW development. Above all, it involves connecting Bluetooth to a mobile phone. Possible applications include navigation to a destination using GPS data, displaying mobile notifications, or even converting speech to text (Fig. 1).

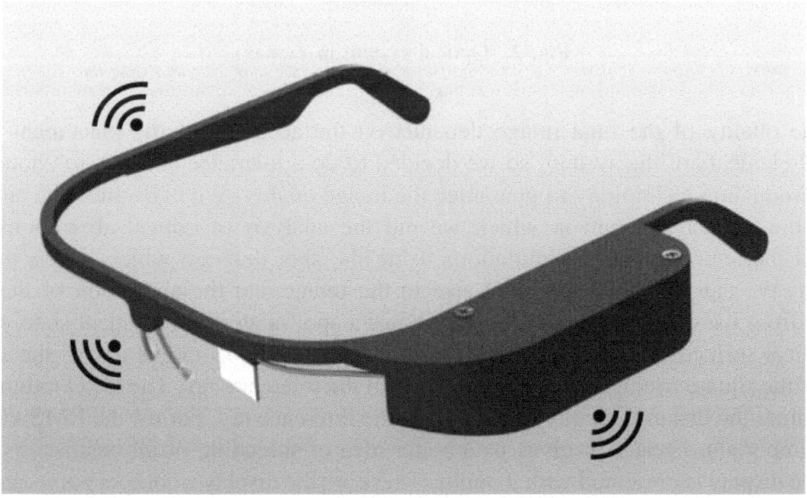

Fig. 1. Conceptual design of glasses

3 Optical Analysis

The construction is mainly influenced by the tolerances of placement of imaging system, which mediates the perception of augmented reality. The imaging system is shown in Fig. 2. It consists of the KOPIN display of 0.2″ diagonal which has a resolution of 428 × 240 pixels and an aspherized dublet with a focal length of 25 mm which forms a magnified image. The display-lens distance was calculated for the magnification 20× and the image distance of 500 mm. The last part in this system is a beam-splitter that projects the image on the retina of the eye.

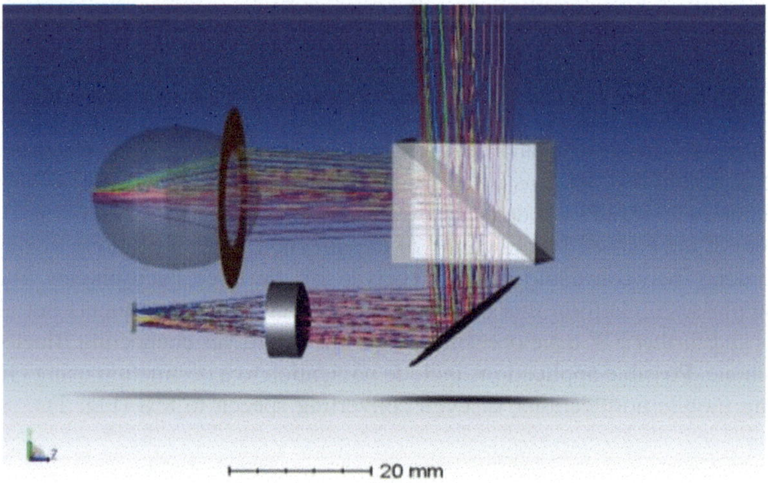

Fig. 2. Optical system in Zemax

The quality of the final image depends on the accuracy of the placement of all optical elements of the system, so we decided to do a tolerance analysis to choose the right production technology to guarantee the image quality. We performed the analysis using the Zemax program in which we did the analysis of optical aberrations. We present the results of these simulations using the spot diagram which shows the ray tracing. We can determine the spot size in the image and the aberration of the final image from the diagram. Figure 3 - left shows a spot diagram of an ideal state, where all optical surfaces are precisely estagned. In figure the RMS radius means the square root of the square averages of the distance from the reference ray. The GEO radius only determines the distance the ray farthest from the reference ray. For us, the RMS value is more important, because it gives us a better idea of spreading of all beams.

The display is presented with 9 points. These are the display´s corners points and one center point. It shows that RMS is on average 1 mm. We have chosen ±0,1 mm as a tolerance which corresponds to the accuracy class f for the length dimensions for the cable length of 6–30 mm, which is the length of the optical system. Based on this consideration, we have moved the object length by ±0,1 mm and the lens is tilted by 1°.

The results are presented using the spot diagrams. When the doublet is shifted by +0,1 mm forward there are no such large differences from the initial state. Only the edges of the display have a larger RMS as shown in Fig. 4 – left. If we move the object length by −0,1 mm, a large hole defect manifest and the RMS increases in average to 2 mm as shown in Fig. 4 – right. The last state is the lens inclination, which is shown in Fig. 3 – right, and the average RMS is 2 mm. Based on this analysis, it can be seen that the placement of optical elements of the system has a great influence on the quality of the image, because aberrations increase due to inaccuracies and the image is blurred.

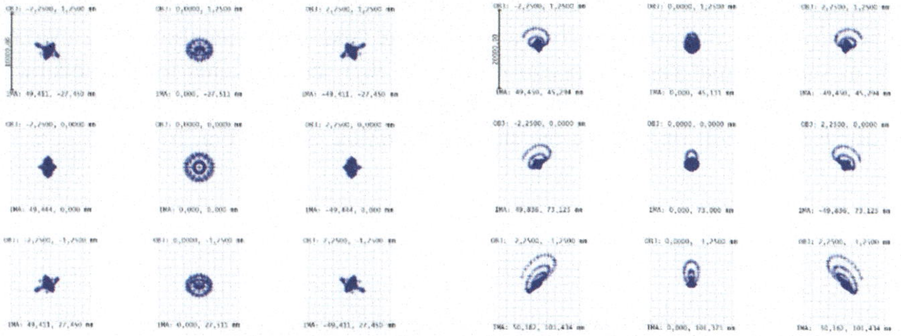

Fig. 3. Left – a spot diagram of ideal state, Right – a spot diagram of lens inclination by 1°

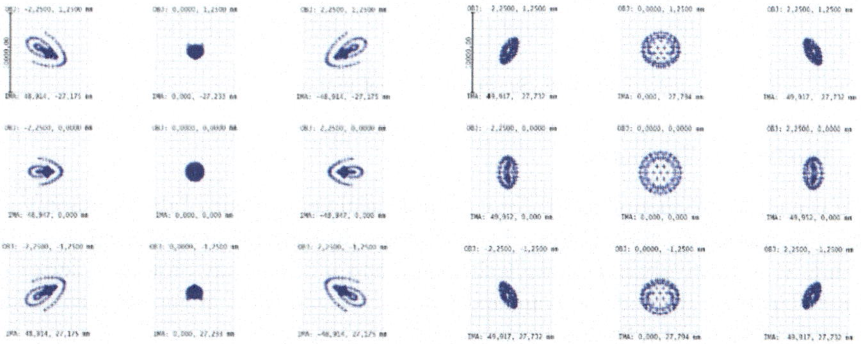

Fig. 4. Left – a spot diagram of moved lens by +0,4 mm, Right – a spot diagram of moved lens by −0,4 mm

4 Construction

Most attention was focused to the correct design of the construction of the device, that would meet the requirements mentioned above. One of the main influences on the construction itself is the technology of making the device and it goes hand in hand with the choice of material. First, we will give a balance over the determination of the material.

4.1 Material

Smart glasses for augmented reality will be in use every day, therefore must have sufficient material strength. Glasses are a relatively small device, which can easily fall off the table or may be damaged by various careless manipulation.

While grabbing one side of the glasses the bend caused by its own weight of the opposite side must be sustained. Next limitation is weight.

While in use, too big of a weight can cause discomfort and might force the user to take off the glasses. Another condition is that material must be non-conductive, because on the inside of the glasses electronics is located. From the set requirements, we had to reject the materials from the metal group because they are very heavy and conductive, even if they meet the strength, but also ceramic and glass materials. We evaluated plastic as the best material, but because of the higher requirements for the accuracy of optical elements and the possibility of designing the focusing mechanism, we chose a metal case that will only be in the field of focusing optics (Fig. 5).

Fig. 5. Front view of glasses

4.2 Production Technology

There is a lot of production technology for the plastic parts. One of the commonly used method is plastic-injection. During this method molten granulate is injected under high pressure. It is relatively expensive technology and it surely pays for large-series production. The price depends on complexity of a form and starts at hundreds and ends at thousands of Euros. So for our device is not suitable.

Another method is an increasingly developing branch – 3D printing. There are 4 basic technologies: FDM – fusion deposition modeling, SLA – stereolithography, SLS – selective laser sintering, DMLS – direct metal laser sintering.

We have 3D printer in our faculty based on FDM technology. The accuracy of this technology is about one printed layer, that means 0,1 mm. This method is also called rapid prototyping and these technological group associates common characteristics, namely that the material is not removed from the semi-finished product - trying to reach the desired shape nor adding material at the same time (casting).

3D printing is a very cheap technology in comparison with plastic-injection, therefore it is ideal for our application, however we must keep in mind limits and difficulties of this method. rom offered materials we chose the best one which is PLA.

4.3 Constructional Solution

We were inspired in the construction by design of the existing smart glasses model. The main choice is placing the electronics into the side. It is a simple solution without having to add another box attached to the glasses. It would be a very uncomfortable solution for daily use. The next decision was put to all used equipment into one side of the glasses. In case of using both sides for HW placement will increase glasses dimensions, frontal section would be bigger and design would not be as suitable (Fig. 6).

The basic dimension we designed with regard to different head sizes and follow on from the average size of sunglasses and eyeglasses. A distance between sidesis 140 mm and sides sizes are 165 mm. The smart glasses are modular and consist of 8 main parts, which can be used to modify glasses dimensions. During device designing we've put emphasis on design, ergonomics and functionality.

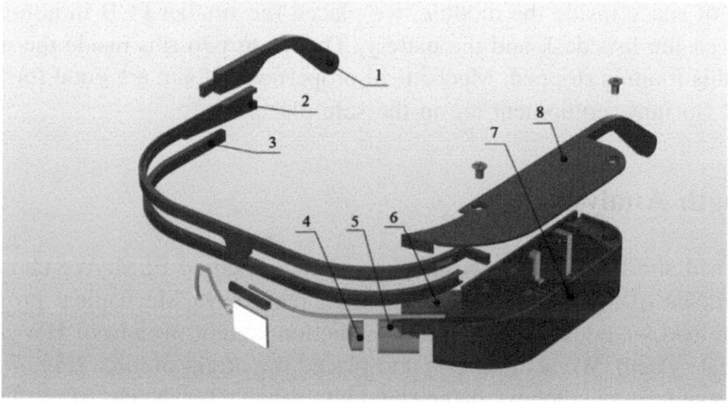

Fig. 6. A decomposed assembly of glasses

The right side is shaped insert into the frontal section, which copies the shape of the forehead. The frontal section is U-shaped. Inside this part we lead wires from the right and middle microphones to the module for the electronics. The back part is U-shaped as well with the bigger dimension. To ensure good insertion of both U profiles in one another we allowed for a backlash of 0,2 mm that delimited to heat shrinkage during the manufacturing process. Part no. 2 in the picture 8 has in the middle prominence with a thin groove where warped wire is placed with a diameter of 1,4 mm. There are nose pads at the end of the wire. The wire is easy to rebuild on personal nose width.

On the left size there is a module for electronics and optical system. This organization make the service simple and easy. Joined U-shape parts are connected with the upper cover of the module. Connection between module and upper cover is materialized as a two screws M3 and established by shape lock of the parts. Because 3D printed parts are not suitable for a cutting fine thread are in the module hexagonal hole for the nuts. It is a common practice for 3D printed parts.

Surfaces of transitions are rounded to avoid any unnecessary sharp edges. In the module there is a tube-shaped surface to input a case with doublet. In the optical axis there is small table for placing display for visualization surrounding sound. Further there is a slot for PCB and in the exact location there is a hole through the bottom which is used for USB connector. This port was designed for communication with the CPU and for loading updates. Inside the model there are two columns for placing battery with capacity 500 mAh. The value of capacity was calculated from the desired staying power.

4.4 Electronic Equipment Placing

Given preservation compact dimension, we solved a problem with inner organization and attachment of HW. The placing of printed circuit board is where control unit interfere in the margin. So we decided to place this PCB between two shallow grooves and fixed by special 3D print component. This component is stressed by the top cover which is connected with the main glasses body by nuts and screws. This PCB is connected with another PCB that process projected images by flexible clutch. Because of the lack of space inside the module, we placed the smaller PCB in non-conductive foam between the first desk and the battery. Thanks to two ribs inside the module the motion of this foam is stopped. Mechanical properties of foam are good for free fall of the glasses. So inner equipment are in the safe place.

5 Strength Analysis

We performed strength analysis with the numerical method FEM. We chose 4 of the worse examples of straining and simulated in the Ansys Mechanical program. The force of the load we counted from the construction weight, weight of HW and weight of the optical system. We set it at 1 N and placed the origin of force (Fig. 7 – point B) in the center of gravity witch we determined into module for the electronical equipment in the left side of the glasses. The point A is a place of fix.

The body of glasses we meshed fine grained tetrahedral net which was consisted of 250 000 elements.

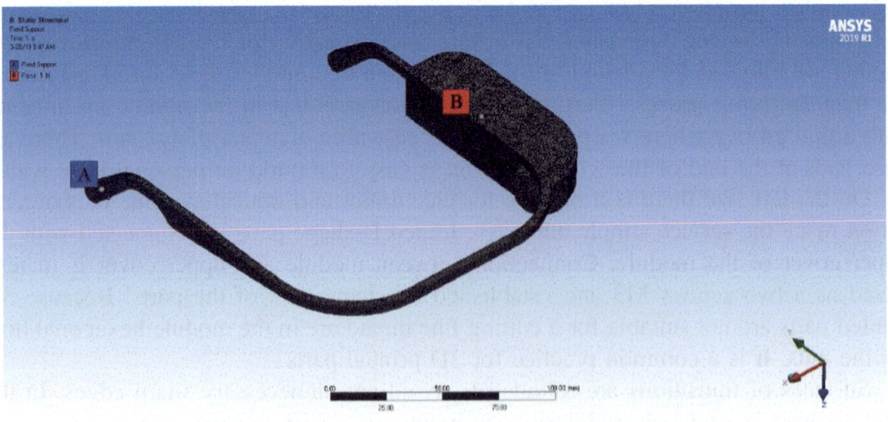

Fig. 7. A mesh grid in the galsses

The first condition which is depicted on the picture 10 introduces fixed glasses under right side in the vertical position. Construction is stressed by bending moment and the highest value 24,5 MPa. It is value which the material tolerated (Fig. 8).

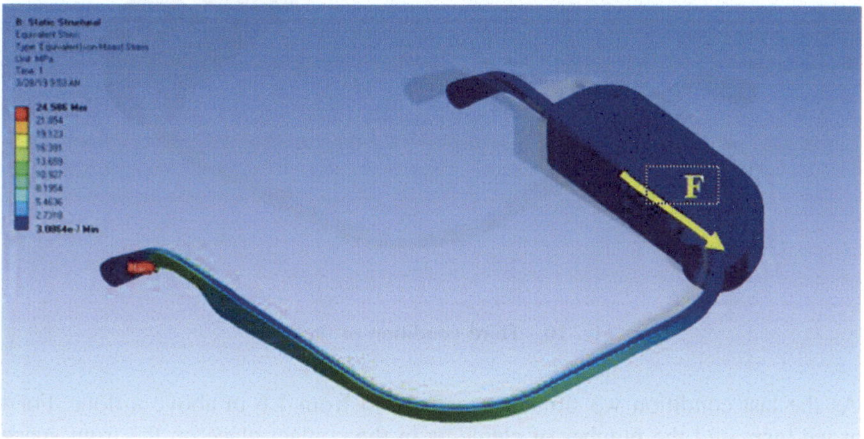

Fig. 8. First condition of the load

The second condition (Fig. 9) is simulated rotation of glasses into horizontal position. The glasses are stressed by a torque. The load peak equals to 24,1 MPa. This value is in the tolerance.

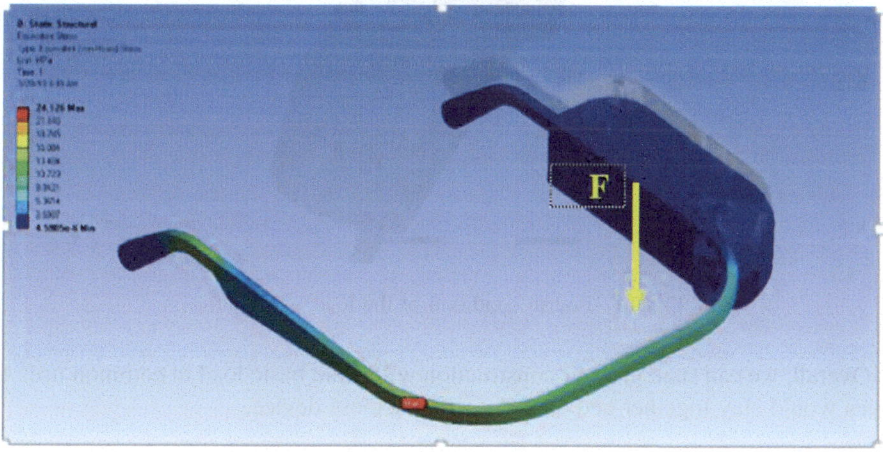

Fig. 9. Second condition of the load

In the next-to-last condition the are glasses turned with the module for electronics downwards and fixed by the same way – under the right side. The dominant constituent of load is bending moment that equals to 10,7 MPa in the critical point (Fig. 10).

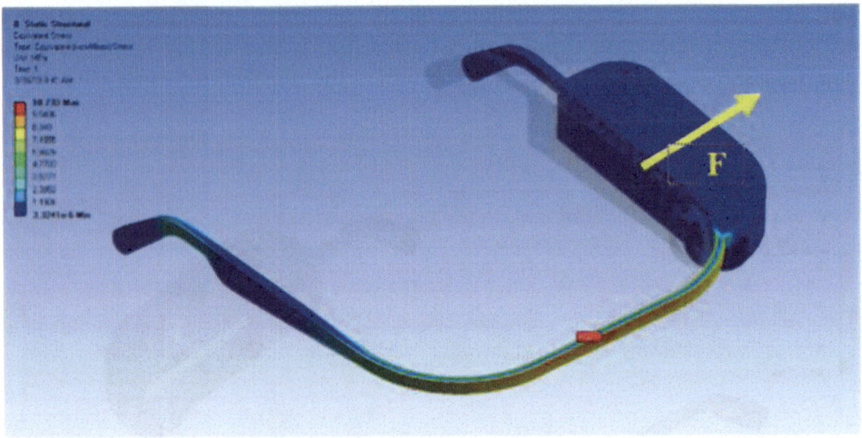

Fig. 10. Third condition of the load

As the last condition we simulated a free fall from 1,6 m above a floor. For this event we increased the number of elements in the contact place on the front space of module for electronics. The load peak equals to 60 MPa. Its high value of tension but also the construction of glasses will stand (Fig. 11).

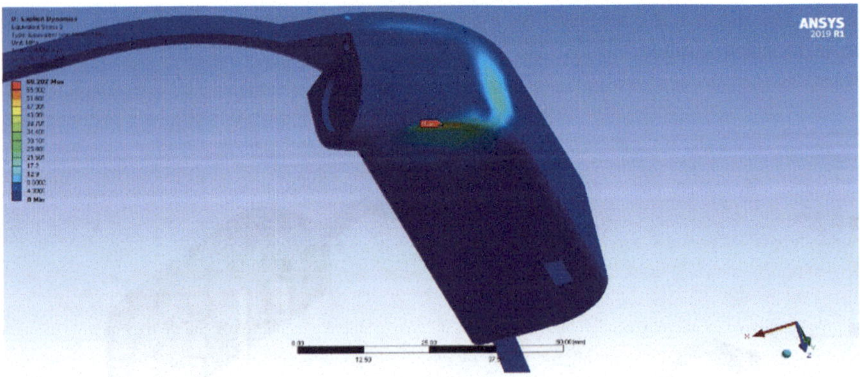

Fig. 11. Fourth condition of the load – free fall

Overall, we can state that our construction will stand basic load in common use. All pieces would stay together and as a one compactness device.

6 Conclusion

The aim of this text was to introduce the design solution of smart glasses for augmented reality. We created construction which complies with strict tolerance for optical systems, ergonomy and also as low weight as possible.

According to the optical simulations it follows that placing accuracy of members must be higher than the accuracy achieved in 3D printing. Therefore, we used a metal tube in which is doublet.

Whole body is printed from PLA that has a required properties. Finally, we did numerical simulation of load caused by daily use.

References

1. Kovanda, D.: Smart glasses. Thesis, CTU in Prague (2018)
2. Statistiky počtu osob se sluchovým postižením. Česká unie neslyšících. https://www.cun.cz/blog/2017/05/17/statistiky-poctu-osob-se-sluchovym-postizenim/. Accessed 29 Mar 2019
3. PLA filamenty. Materiálpro3D. https://www.materialpro3d.cz/materialovy-slovnik/pla/. Accessed 28 Mar 2019
4. O 3D tisku. Josef Průša. https://josefprusa.cz/o-3d-tisku/. Accessed 30 Mar 2019
5. A Guide to UnderstandinOg the Tolerances of Your 3D Printer. MatterHackers. https://www.matterhackers.com/articles/a-guide-to-understanding-the-tolerances-of-your-3d-printer. Accessed 28 Mar 2019

Noise Analyses for Rolling Bearings

Ciprian Ion Rizescu(✉), Victor Constantin, Dana Rizescu,
and Daniel Besnea

University POLITEHNICA of Bucharest,
Splaiul Independentei No. 313, 060042 Bucharest, Romania
ciprian.rizescu@upb.ro

Abstract. The paper refers to a noise analyses developed both in Matlab and Python environments in order to determine the failure of rolling bearings. The various types of vibration and sound in rolling bearings are grouped into the four categories: structural, manufacturing, handling and others (like lubricant noise, seal noise, etc.). There is known that any frequencies of vibration and sound generated in rolling bearings are related to the natural frequency of the raceway rings. Of the two raceway rings, the natural frequency of the outer ring becomes a problem more often than the inner ring due to a loose fit between it and the housing. The modes of vibration of the outer ring are roughly divided into two categories: those that consider the outer ring as a rigid body and those that consider it as an elastic body. There are several formulas for calculating the natural frequency of the bending mode in the outer ring of a ball bearing. The authors developed an experimental setup for study the noise behaviour of rolling bearings: normal or damaged bearings. Both bearings have operated on the experimental setup successively and the noises were recorded. Comparing the frequencies behaviour of the two bearings a conclusion could be set.

Keywords: Noise analyses · Rolling bearings · Frequency analyses

1 Considerations About Ball Bearings and Noise Analyses

1.1 Ball Bearings

When friction inside the bearing is of rolling nature, the bearings are called rolling bearings. The bearing is the main element of the roller bearing. Alongside the bearing, the roller bearing assembly includes shaft spindle, casing, axial fasteners, lubrication and sealing systems.

The bearings (see Fig. 1) are independent assemblies formed, in the general case, by: an outer ring (1) with an inward running track; inner ring (2) with external rolling path; rolls (3) and cage (4), which prevent contact between the rolls by their equidistant arrangement [1]. In some bearings, to reduce the radial gauge, bearings are used without the inner ring or without both rings, in which case runners are executed on the shaft spindle and possibly on the bearing housing. The various types of vibration and sound in rolling bearings are grouped into the four categories: structural, manufacturing, handling and others (like lubricant noise, seal noise, etc.).

G. I. Gheorghe (Ed.): ICOMECYME 2019, LNNS 85, pp. 74–81, 2020.
https://doi.org/10.1007/978-3-030-26991-3_8

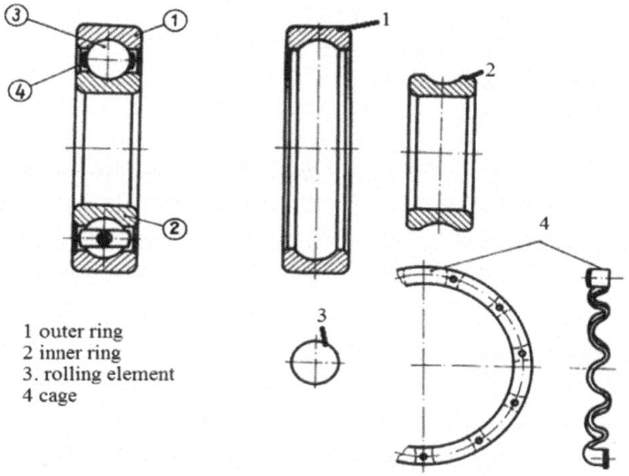

Fig. 1. Components of a rolling bearing

1.2 Noise Analyses

There is known that any frequencies of vibration and sound generated in rolling bearings are related to the natural frequency of the raceway rings. Of the two raceway rings, the natural frequency of the outer ring becomes a problem more often than the inner ring due to a loose fit between it and the housing. The modes of vibration of the outer ring are roughly divided into two categories: those that consider the outer ring as a rigid body and those that consider it as an elastic body.

There is known that any frequencies of vibration and sound generated in rolling bearings are related to the natural frequency of the raceway rings. Of the two raceway rings, the natural frequency of the outer ring becomes a problem more often than the inner ring due to a loose fit between it and the housing. The modes of vibration of the outer ring are roughly divided into two categories: those that consider the outer ring as a rigid body and those that consider it as an elastic body. There are several formulas for calculating the natural frequency of the bending mode in the outer ring of a ball bearing [2].

In order to calculate both the frequency of vibration caused by waviness or the envelope frequency of flaw noise, the orbital revolution frequency of the rolling elements (fc) and the rolling element frequency (fb) are required. These values are determined by the internal specifications of a bearing and the running speed. They can be calculated with the following formulas [3]:

$$f_c = \frac{1}{60} \cdot \frac{n_i}{2} \cdot \left(\frac{d_m - D_a \cdot cos\alpha}{d_m} \right), \; f_b = \frac{1}{60} \cdot \frac{-n_i}{2} \cdot \left(\frac{d_m}{D_a} - \frac{D_a \cdot cos^2\alpha}{d_m} \right) \tag{1}$$

- f_c: Orbital revolution frequency of rolling elements (Hz)
- f_b: Rotation frequency of rolling elements (Hz)

- n_i: Running speed of inner ring (rpm)
- D_a: Diameter of rolling elements (mm)
- d_m: PCD of rolling elements (mm)
- α: Contact angle (°).

If roller bearings have defects on the inner race, outer race or balls, it generates a series of periodic vibrations as a running roller passes over the surfaces of the defects. These vibrations occur at certain characteristic frequencies, which are determined by the rotational speed, locations of defects, and geometric parameters.

2 Experimental Setup

The mobile system consist of the shaft 1 and the disc 2 which is supported on the miniature bearings 6 (see Fig. 2). The shaft drive is frictionally driven by the wheel 3. The drive of the shaft 1 is stopped by tilting the frame 5. The possibility of rotation of the oscillating frame allows the contact between the shaft 1 and the friction wheel 3 to be established or interrupted to drive or not the moving equipment of the rotating device. The sound/noise of the rolling bearings is recorded only when the driving shaft 1 is rotating without the contact with the friction wheel 3. The pressure-field microphone 4 is a type 4144 - 1″ pressure-field microphone (from Brüel & Kjær [4]), that is, a microphone where the damping of the diaphragm is such that the pressure frequency response is flat over a wide frequency range (see Figs. 2 and 3).

Fig. 2. The experimental setup.

Fig. 3. 1″ pressure-field microphone recording bearing noise

Shaft 1 rotates in the rolling bearings 3 and the noise is recorded. During the recording period the drive motor is off. All determinations were made in an insonorized room, in our laboratory [5, 6]. There were recorded several rolling bearings: good and defective. For each type and sizes of rolling bearings was developed a data base in order to calculate f_c, f_b according to Eq. 1.

3 Evaluating Rolling Bearings

3.1 Bearings Evaluation Using Python Interface

Python is a language that offers many features and has a fast learning curve for both programers who know languages such as C and Java and for beginners. Although it is an interpreted language, this is not a member in the path of its popularity and use in numerous projects. The application is developed using the language Python 3.6 to which shall be added the libraries:

- NumPy - matrix calculation;
- PyQt4 - graphical user interface;
- MatPlotLib - graphics.

The general aspects of the two programming environments reflect the fact that although they look similar at a first consideration, they differ very much by syntaxes that are used to create the programs and the most representative aspect is by the fact that many functions of the MATLAB have no correspondence in Python, even highlighted in the two programs used to the creation of a graphic user interface, the environment of MATLAB programming being more complex in performing the functions, more powerful, but with certain disadvantage of the cost.

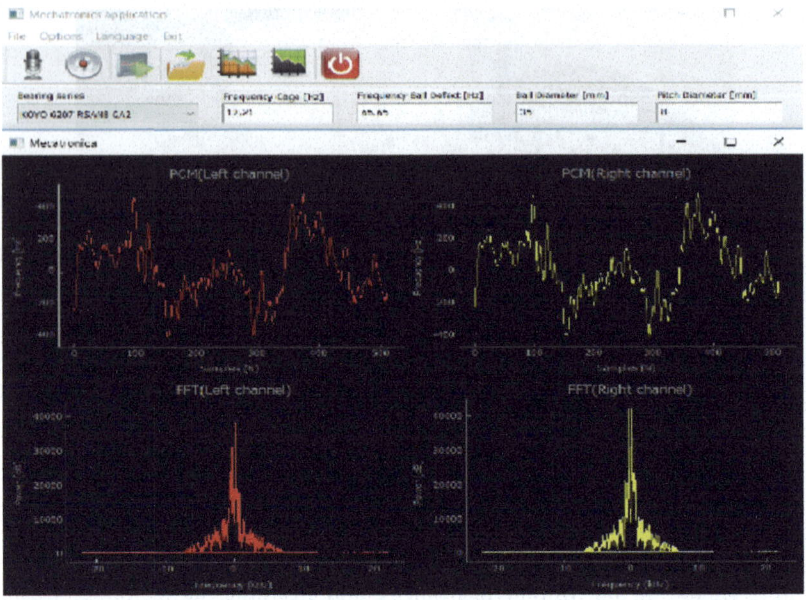

Fig. 4. Python interface for evaluating rolling bearings operation

Python is an open source with certain advantages and disadvantages with respect to Matlab solution. The study presents a comparison between a graphic user interface for rolling bearings evaluation, developed both in Matlab and Python (see Fig. 4) environments. Also some remarks, concerning the software's efficiency and an economic analysis efficiency-cost of the two software, are presented in this study which have an identical use/purpose. For both environments there were developed data bases for each type/size of rolling bearings. Then there were recorded the noise/sound of rolling bearings operating. The specific frequencies are compared with theoretical ones, as it were computed using Eq. (1).

Signal reception is performed by analog transducers in the form of a finite length time sequence. The spectral analysis of this sequence aims at identifying the harmonic components present in the signal in the form of the amplitude of that component and its frequency. This requires a new form of representation of the information gathered, obtained by eliminating the *time* variable and replacing it with the *frequency* variable. The analytical tool that fulfills this goal is Fourier transform. It is the means of passing a signal defined in the temporal domain into a signal defined in the frequency domain. In order to compute the discreet spectrum of a signal, the Python/MATLAB programming environment uses the FFT, which is a category of algorithms that implement the discrete Fourier transform, making transformation calculations at an increased speed. Although the number of points in which this is evaluated is not important, the most efficient calculation is made for a number of points equal to a natural power of 2, such as 256, 512, 1024 or 2048. The quadratic mean of the signal x(t) as a frequency function is called the signal power spectrum. It is obtained by summing the squares of the values of the average square amplitudes calculated within the program.

3.2 Evaluation Using MATLAB Interface

To generate the graphical user interfaces are used either of the six items of control for interactive graphical interface enclosed in MATLAB environment when programming with objects for the control of interactive graphics. The six predefined elements of interactive control are: push button; radio button; sliders; popup menu; editable text; check box.

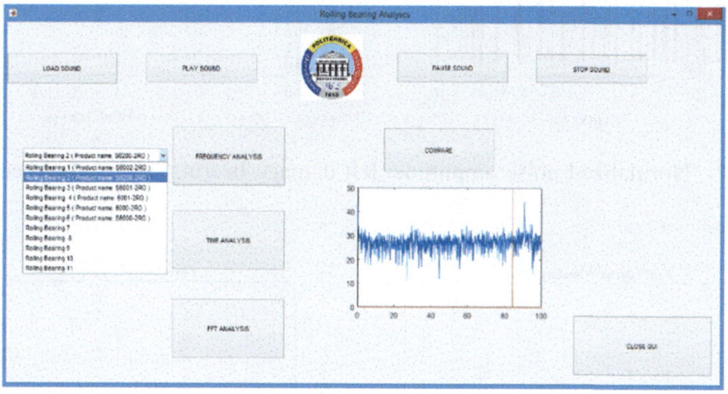

Fig. 5. MATLAB interface for evaluating rolling bearings operation

The selection of rolling bearing type is also present in Matlab interface (see Fig. 5). Here one can set/load the dimensions of the rolling bearings: shaft diameter, cage diameter, width, ball diameter. The interface was provided with a selection button, where the language can be set, English, Romanian, French and German. Another option of these interfaces is that the recorded noise of the rolling bearing operating can be played as a audio file in .wav format.

Evaluation of rolling bearings consists of: comparison of the theoretical own frequencies calculated on the basis of the dimensions of the bearings with their own frequencies determined on the basis of the audio file.

Visualization of the theoretical own frequencies, of the calculated frequencies, as well as of the "Noise Amplitude", "Normalized noise amplitude", and "Fast Fourier Transformation" (see Figs. 6, 7 and 8).

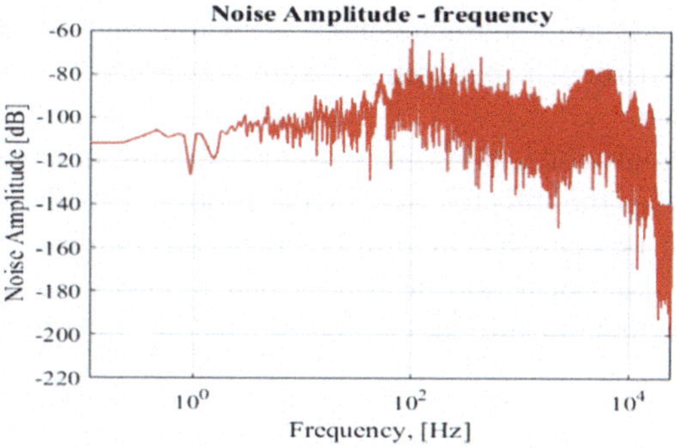

Fig. 6. Noise amplitude – frequency

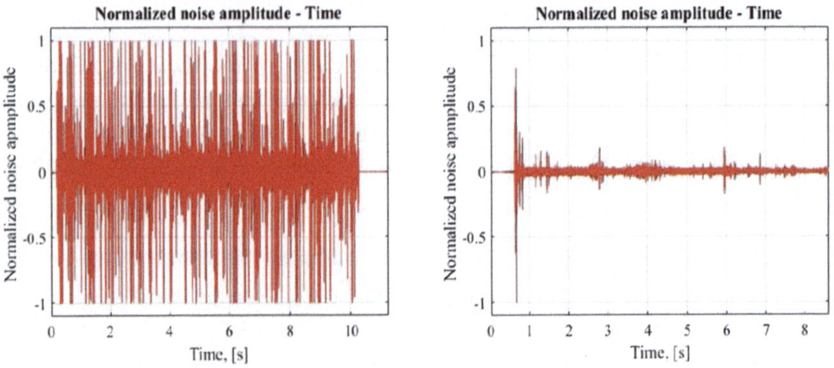

Fig. 7. Normalized noise amplitude: left damage bearing, right good bearing

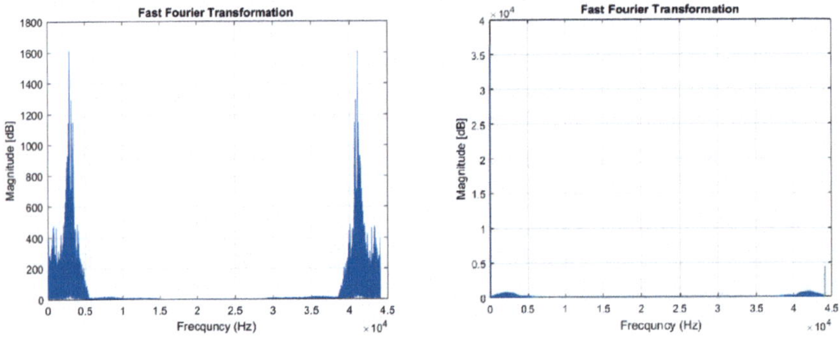

Fig. 8. Fast fourier transformation: left damage bearing, right good bearing

4 Conclusions

Experimental results show that the proposed method could effectively identify defects from real vibration signals collected from rolling bearings. There were developed two interfaces in Python and Matlab. Using Python software in achieving performance and make an graphic user interface, it was noticed that the only differences between this program and MATLAB are certain functions with different syntaxes. In the present study it stands out in particular that both software are similar in terms of programming language and interface.

Real vibration signals collected from bearings operating in an insonorized room are measured for validating the effectiveness of the proposed method. The evaluation of rolling bearing could be achieved in situ, but there are required some special conditions: it is necessary to isolate the bearing which has to be evaluate. Experimental results show that the method provides a higher accuracy for the defective frequency detection.

References

1. Cioboata, D., Dontu, O., Besnea, D., Ciobanu, R., Soare, A.: Mecatronic equipment for bearing ring surface inspection. Rom. Rev. Precis. Mech., Opt. Mecatronics **48**(48), 262–266 (2015)
2. Graney, B., Starry, K.: Rolling element bearing analyses. Mater. Eval. **70**(1), 78–85 (2012)
3. Ueno, et al.: Rolling bearing engineering, Yokendo Ltd. 139, 101 (1975)
4. Brüel & Kjær: https://www.bksv.com/-/media/literature/Product-Data/bp2031.ashx
5. SKF Homepage: www.skf.com/cmc-steyr
6. Brüel & Kjær: https://www.bksv.com/en/products/transducers/acoustic/microphones/microphone-cartridges/4144

Performance Evaluation of Different Mechanisms of Production Activity Control in the Context of Industry 4.0

Daniela Costa[1], Mariana Martins[1], Susana Martins[1], Eduarda Teixeira[1], Andreia Bastos[1], Ana Rita Cunha[1], Leonilde Varela[1], and José Machado[2(✉)]

[1] Department of Production and Systems, Algoritmi Research Centre, University of Minho, Guimarães, Portugal
leonilde@dps.uminho.pt

[2] Mechanical Engineering Department, MEtRICs Research Centre, University of Minho, Guimarães, Portugal
jmachado@dem.uminho.pt

Abstract. Production Activity Control (PAC) is fundamental to Production Management, since it allows for meeting deadlines, ensuring product quality and reducing production costs. For these reasons, it is essential for the improvement of enterprise performance to understand the production system and its integrated parts. Another production concept linked to the efficiency of enterprise performance is Industry 4.0. This is the most recent revolution of industry and one of its main goals are related with the integration of production activity control by using information technologies. The objective of this project is to implement three different mechanisms of Production Activity Control in a Flexible Flow Shop (FFS), composed of three stages with three parallel machines each. The mechanisms implemented are Workload Control (WLC), Generic Kanban System (GKS) and Drum-Buffer-Rope (DBR), and all are associated with a make-to-order (MTO) production. Additionally, three independent machine selection criteria are evaluated: Random, Load Hours and Load Units. Simio software is used for the simulation of the production system and results are given by diverse Key Performance Indicators (KPIs). After completing simulations, it can be concluded that DBR is the mechanism of PAC with the best performance for the studied scenarios. However, the scenario with the smallest value of load norm is compromising the performance of WLC. Otherwise, this mechanism would be the one with the best performance. Regarding the machine selection criteria, Load Hours is the criterion with the best performance for almost all the KPIs.

Keywords: Production activity control · Industry 4.0 · Workload Control · Generic Kanban System · Drum-Buffer-Rope · Simulation · Key Performance Indicators

G. I. Gheorghe (Ed.): ICOMECYME 2019, LNNS 85, pp. 82–103, 2020.
https://doi.org/10.1007/978-3-030-26991-3_9

1 Introduction

1.1 Problem Description

This project consists of the implementation and evaluation of three production activity control mechanisms (Workload Control, Generic Kanban System and Drum-Buffer-Rope) associated to a make-to-order production strategy. The production system is composed by a flexible flow shop where three different products (Product A, Product B and Product C) are processed at three production stages (i = 1, 2, and 3). On each stage, there are three different parallel machines (Mi1, Mi2 and Mi3) that execute the same operation for each type of product. The production system configuration is shown in Fig. 1.

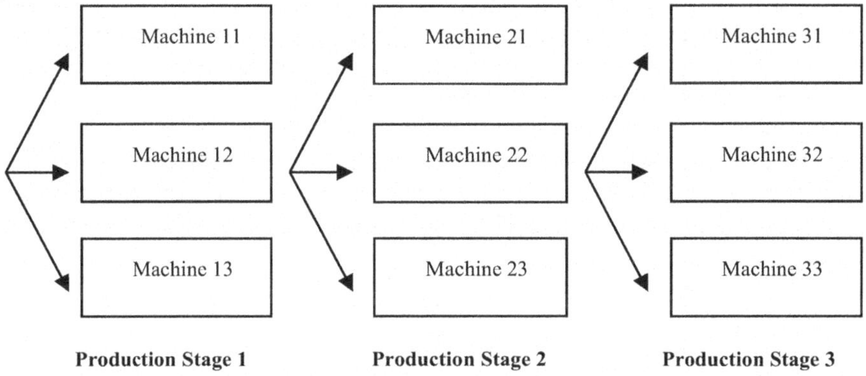

Fig. 1. Production system configuration.

Regarding the products, their proportion in the system is the same (1/3 of each one) and their processing time (hours) is represented in Table 1.

Table 1. Processing time of different products in the system

Product type	Machines M_{11}, M_{12}, M_{13}	Machines M_{11}, M_{12}, M_{13}	Machines M_{11}, M_{12}, M_{13}
Product A	1	1	1
Product B	0,5	0,5	0,5
Product C	1,5	1,5	1,5

The machines' selection can be Random (the selection is random), as a function of Load Hours (the selection is made according to the machines with the lowest workload) or as a function of Load Units (the selection is made depending on the quantity processed by each machine). To evaluate the performance of the different mechanisms, it is necessary to implement them on each different scenario of selection criteria, using the software Simio. For the modelling problem, the characteristics shown in Table 2 are required.

Table 2. Characteristics of the simulation model

Inter-arrival time	EXPO (0,37) h
Due date	TNOW + Unif (30,60) h
Processing time	Truncated 2-Erlang; average (Table 1), max = 4 * average
Ranking rule	First Come First Served (FCFS)
Uptime between failures (MTBF)	EXPO (1000) h
Time to repair (MTTR)	EXPO (3) h
Sequencing rule at the pool	Planned Release Date (PRD)

A dispatching rule (or ranking rule) defines the order in which jobs waiting in the queue of a machine are processed, as soon as the machine becomes unoccupied. In this project, the dispatching rule used is First-In-First-Out (FIFO), which dispatches jobs according to their arrival order at the queue machine [6].

Furthermore, in the pre-shop pool, it is essential to define the order in which jobs will be released into the system. One of the most important sequencing rules is the Planned Release Date (PRD), in which jobs are ordered according to the earliest planned release date, given by the difference between the job due date and the sum of lead times [7].

1.2 Objectives

The main goal of this project is to compare the different Production Activity Control mechanisms to discover which performs the best in the system. Additionally, it is pretended to identify which mechanism provides the best results among the diverse selection criteria, as well as which selection criterion is the best performing, for different condition of cards and loads.

This evaluation and comparison will be made through KPIs also implemented in the developed simulation model.

1.3 Methodology

In an early phase, the software Simio is used to model the production system, implement the different machine selection criteria and apply the different mechanisms for each criterion.

In a posterior phase, the models created are tested to extract the intended results, to compare the three mechanisms using different selection criteria and vice-versa.

2 Key Concepts Overview

2.1 Industry 4.0

According to the Consortium II, Industry 4.0 is "the integration of complex and physical machinery and devices with networked sensors and software, used to predict, control and plan for better business and societal outcomes" [9].

Therefore, Industry 4.0 has been transforming production systems. This concept is composed by several components and the key ones are Cyber-Physical Systems (CPS), Internet of Things (IoT), Internet of Services (IoS) and Smart Factory [12]. Information Technologies allows for a better management and control of production activity, namely, anticipate failures in the system, adapt the production to new scenarios and integrate all activities of the production processes. Thus, factories become smarter, flexible, efficient, dynamic and autonomous, facilitating interconnection among supply chain.

This recent revolution enables the vertical and horizontal integration, since Industry 4.0 supports decision making in a short, medium and long term, creating cross-linking in the different hierarchical levels, as simultaneously provides tools to operational decisions and the intelligent cross-linking within the company and with its partners [15].

Through this concept, a new one emerged, Collaborative Manufacturing (CM) referring to the collaboration among workers from different departments and knowledge fields. This is useful when there is a crucial decision to be made and needs various perspectives.

This paper portrays a situation where Industry 4.0 concepts can have impact in decision making, through simulation tools. This case illustrates how simulation can reproduce a real situation where three mechanisms are evaluated without being implemented in a production system.

2.2 Production Activity Control

Production Activity Control (PAC) comprises a set of rules that plans, guides and controls the production system and has the purpose of effectively using the different production resources [2]. It also considers the fulfilment of demand, deadlines and production programs of a company.

As a short-term function, PAC includes the following phases [3]:

- Job releasing to the production system;
- Load distribution through work centres;
- Job sequencing through work centres;
- Job supervision and control.

Furthermore, it is relevant to distinguish the diverse production systems' classification: Push, Pull or Hybrid. In a Push System, production only starts when the necessary materials to satisfy a production program are available to be processed. Consequently, whenever a job is concluded, it is pushed to the following work centre until all operations are finished. Contrarily, in the Pull System, jobs in the upstream production system phases are pulled by the downstream ones. Therefore, jobs are produced to fulfil demand. Lastly, the Hybrid System is a mix of both.

In this project, the system implemented is the Pull System [5].

2.3 Production Activity Control Approaches

According to Stevenson, Hendry, and Kingsman (2005), there are five different approaches to the PAC [14]:

- Material Requirements Planning (MRP);
- Workload Control (WLC);
- Just in Time (JIT)/Lean Manufacturing;
- Theory of Constraints (TOC);
- Quick Response Manufacturing (QRM).

Out of these approaches, only three are associated with the mechanisms implemented in this project: Workload Control, just in time and theory of constrains.

The main purpose of Workload Control is to reduce the work in process, avoiding the overload of the production system. The second one, just in time, associated with Generic Kanban System, targets cost reduction. Finally, the Theory of Constrains, related to the Drum-Buffer-Rope mechanism, finds the critical work centre (bottleneck). This bottleneck constrains the performance of the whole production system.

2.4 Production Activity Control Mechanisms

In PAC there are several mechanisms that can be implemented. The most referred ones are Base Stock System, Toyota Kanban System, Constant Work in Process, Paired-Cell Overlapping Loops of Cards with Authorization. However, as previously mentioned, the implemented mechanisms in this project are Workload Control, Generic Kanban System and Drum-Buffer-Rope, associated with a make-to-order production [3?].

Workload Control. Workload control is aimed at controlling the workload in the production system. Maintaining the workload of the system stable results in shorter and stabilized course times and, consequently, better delivery performances, when compared to systems without control [10].

Therefore, in this mechanism, jobs are not immediately released into the system. Instead, they wait in a pre-shop pool for authorization to be released, according to the workload levels. This authorization is only given whenever jobs follow certain Workload Control rules associated with the work centres of its course [1].

In addition, according to Wight (1970) [VER?], WLC is based on an input/output control. Thereby, the input rate of work is balanced according to the output rate and jobs are only released when the previous ones are concluded.

Generic Kanbans System. The Generic Kanban System is a mechanism based on the use of signals (denominated as Kanbans). The name of this mechanism is related to the usage of generic signals, which can be attributed to any job and not just to a specific one [5].

In each work centre, there are a number of signals available and specific to the different centres. Therefore, a job is only released into the production system when there is at least one Kanban available in every centre. Otherwise, it waits on the pre-shop pool. The Kanbans are allocated to jobs when they are released, and follow them until job processing is finished. Each time a job finishes its process on the work centre, Kanbans are dissociated and, thus, can be allocated to new jobs [11].

Drum-Buffer-Rope. The mechanism Drum-Buffer-Rope consists on identifying the constrainer work centre (bottleneck) of the production system to synchronize the production flow.

Metaphorically, once the drum (bottleneck) is identified, it defines the rhythm of the other work centres since it is the one with the highest load [4]. Thereby, the bottleneck is responsible for the performance of the whole production system [8]. Furthermore, rope controls raw material needs according to the bottleneck's capacity.

Similarly to the previous mechanisms, when jobs are not immediately released into the production system, they remain in the pre-shop pool until the authorization is given.

The bottleneck can vary over time. However, in this project, it will be constant.

3 Experimental Plan and Key Performance Indicators

To obtain simulation results, the different mechanisms are implemented in each machine selection criteria, creating nine independent models. Each model is analysed in ten scenarios. Those vary according to the mechanism: in WLC and DBR, the variable is the load norm imposed on the system (6.2, 7.3, 8.6, 10.1, 11.9, 14, 16.5, 19.4, 22.8 and infinite - unrestrictive release of jobs); in GKS, it is the number of Kanbans (7, 8, 9, 10, 12, 14, 17, 20, 23 and infinite). In addition, it is established a warm-up period of 3 000 h, a replication length of 13 000 h and a number of replications of 100.

Key Performance Indicators allow for a comparison among the mechanisms to conclude which one has the best performance:

- Shop Throughput Time (hours) is the average time between the release time of the job to the first work centre and its conclusion.
- Pool Time (hours) is the average waiting time of each job before entering the first work centre.
- Total Throughput Time (hours) is the average total time that a job remains in the system. It includes the STT and the Pool Time.
- Lateness (hours) is the difference between the conclusion time of the job and its due date, on average.
- Standard Deviation of Lateness (Lateness St. Dev.) (hours) is a dispersion measure which represents the lateness deviation from the lateness average.
- Percentage Tardy is the proportion of late jobs in relation to all jobs.

4 Results and Discussion

Results are analysed according to two different perspectives: finding the best machine selection criteria for each mechanism (First Analysis), and finding the best mechanism for each machine selection criteria (Second Analysis).

The comparison is made through the results obtained from the KPIs on the different scenarios. Additionally, a Multivariate Analysis is performed to better understand how KPIs are related and their influence [13]. On the one hand, it is concluded, through the correlation matrix, that Lateness is strongly and positively correlated to Percentage Tardy and TTT. The same situation happens between Percentage Tardy and TTT. This indicates that three KPIs have similar behaviour in the same proportions. It can be also concluded that there are no negative correlations among the KPIs. On the other hand, through the analysis of the main components, an expected result is obtained: Lateness and Percentage Tardy are considered variables with extreme importance for the study. Furthermore, it is ascertained that the total system time (TTT) is more relevant than production system time (STT).

4.1 First Analysis

In this section, the results obtained for the three machine selection criteria (Random, Load Hours and Load Units) under the mechanisms are analysed and compared.

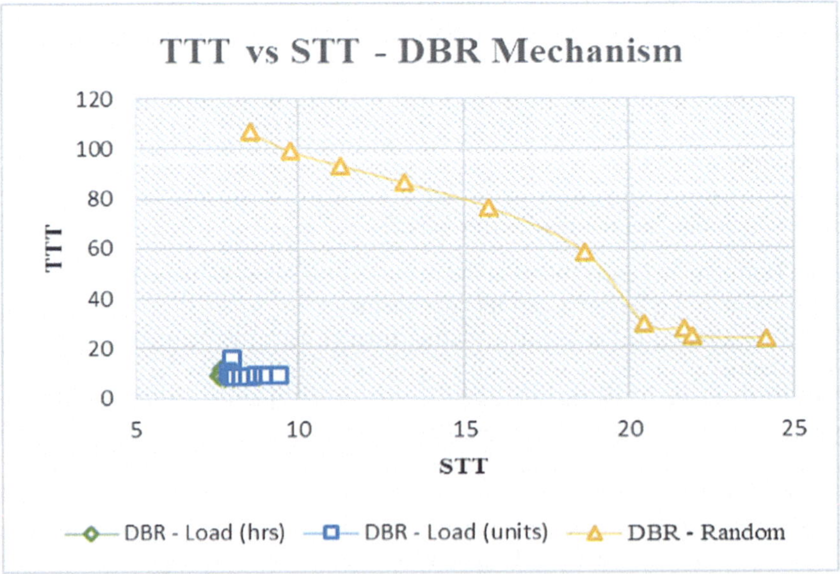

Fig. 2. Results for TTT vs. STT for the DBR mechanism.

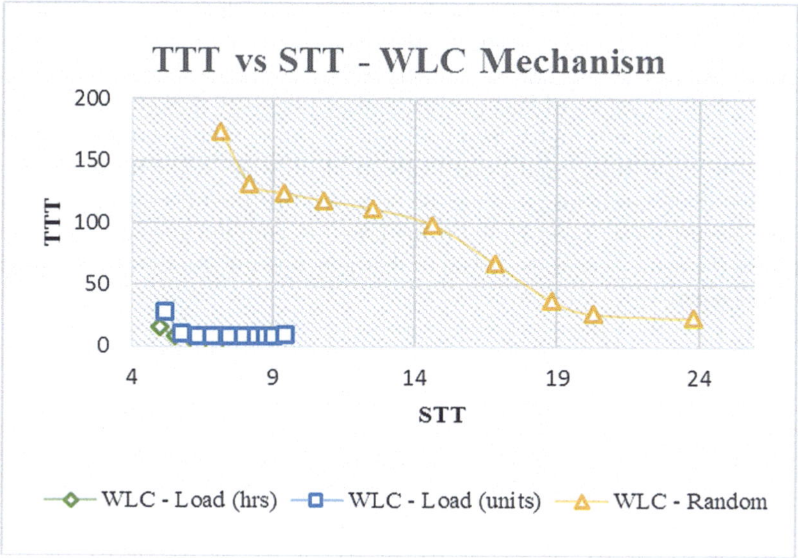

Fig. 3. Results for TTT vs. STT for the WLC mechanism.

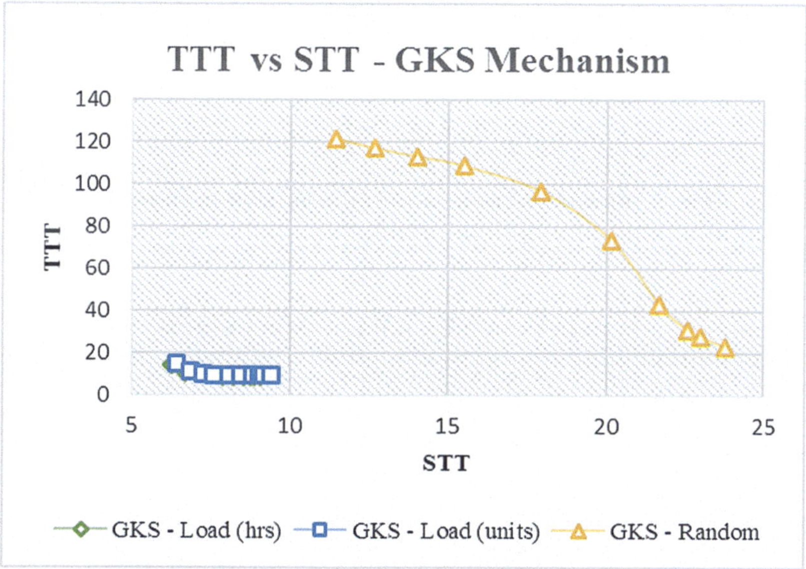

Fig. 4. Results for TTT vs. STT for the GKS mechanism.

As Fig. 2 illustrates, Random is the machine selection criterion which differs in behaviour from the other criteria. It is the criterion with the highest values of STT and TTT, and has values of TTT above 100 h for the smallest values of load norm/Kanbans, and values of STT higher than 20 h for the highest values of load norm/Kanbans. This means that jobs under this criterion remain more time in the total and the production systems than the other criteria. It is also important to refer that, under the DBR mechanism, although TTT decreases drastically with the increment of the load norm/Kanbans on the system on the Random criterion, it only reaches the highest values of the other two criteria in the highest values of load norm/Kanbans. Likewise, the smallest value of STT of the Random criterion is higher than the highest values of the Load Hours and Units.

Comparing Load Hours and Load Units to identify the one with the best performance on the three mechanisms, it can be observed that both curves present a similar behaviour. Nevertheless, Load Hours is the criterion with the smallest values of TTT and STT for the same scenarios, meaning that jobs remain less time in both total and production system.

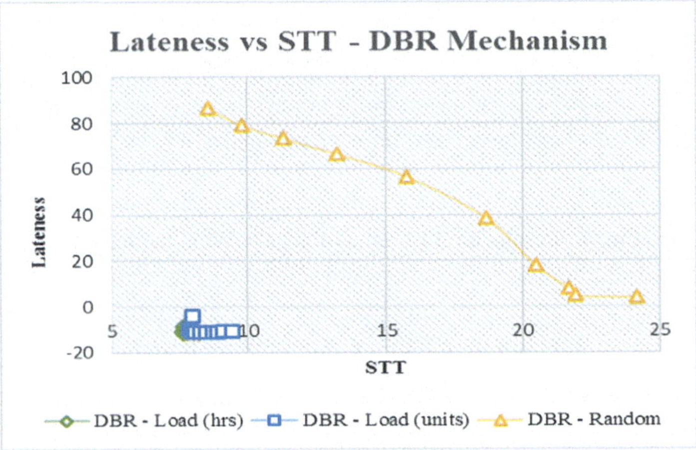

Fig. 5. Results for Lateness vs. STT in the DBR mechanism.

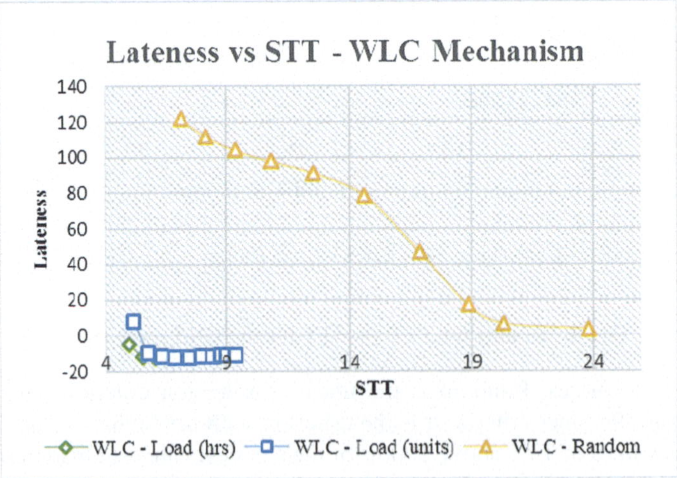

Fig. 6. Results for Lateness vs. STT in the WLC mechanism.

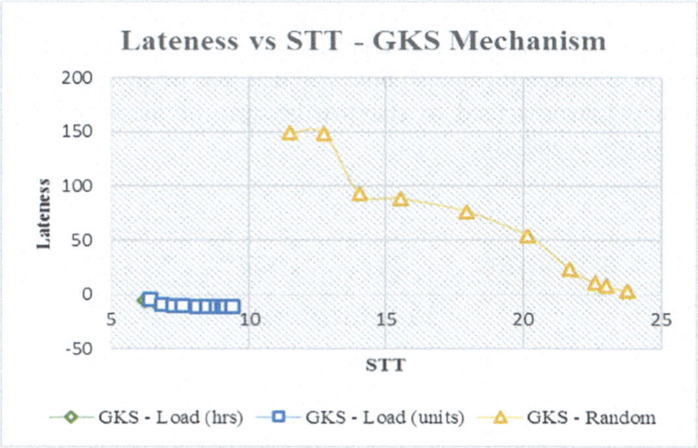

Fig. 7. Results for Lateness vs. STT in the GKS mechanism.

Regarding the average Lateness, on the one hand, Fig. 3 shows that, once again, Random is the criterion with the highest values of this KPI, with all values positive for all the mechanisms, meaning that, on average, jobs are always late. However, the curves of this criterion decreases drastically with the increase of load norm/Kanbans since there are no constraints in the production system when the load norm/number of Kanbans available is infinite. Therefore, it can be concluded that, on average, even with the highest values of load norm/Kanbans, jobs will never be on time under this criterion.

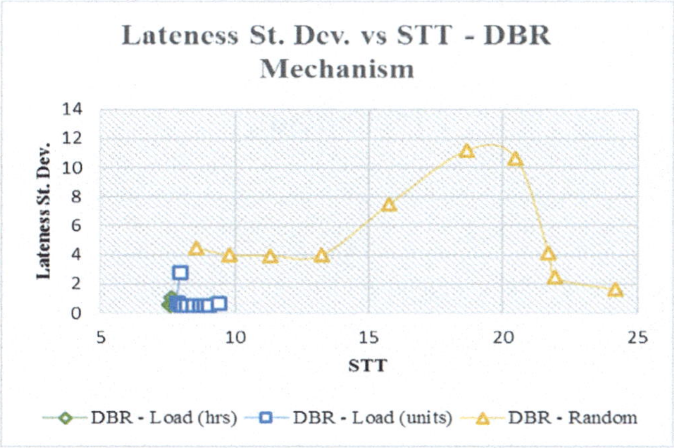

Fig. 8. Results for Lateness Standard Deviation vs. STT in the DBR mechanism.

On the other hand, Load Hours is the criterion with the smallest values of average Lateness, and is the only one with just negative values, which means that, on average, jobs are always in advance. It can also be concluded that when values tend to infinite, values of average Lateness tend to stabilize because of the increasing production system fluidity (Figs. 9 and 10).

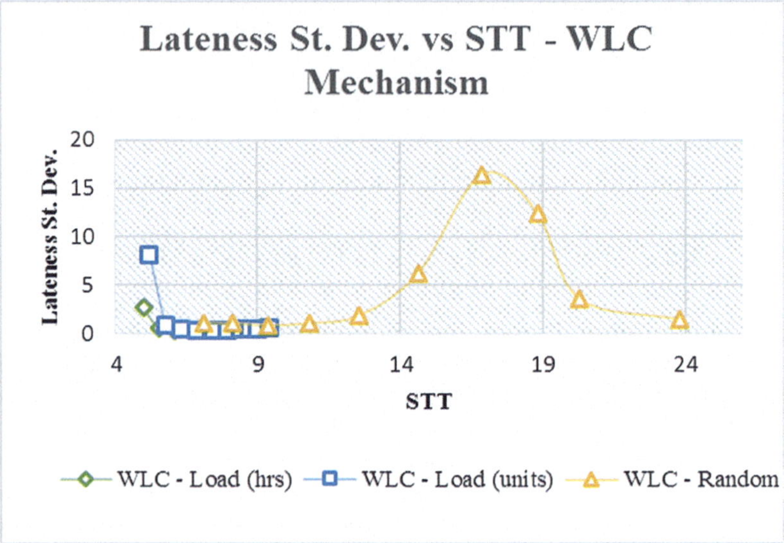

Fig. 9. Results for Lateness Standard Deviation vs. STT in the WLC mechanism.

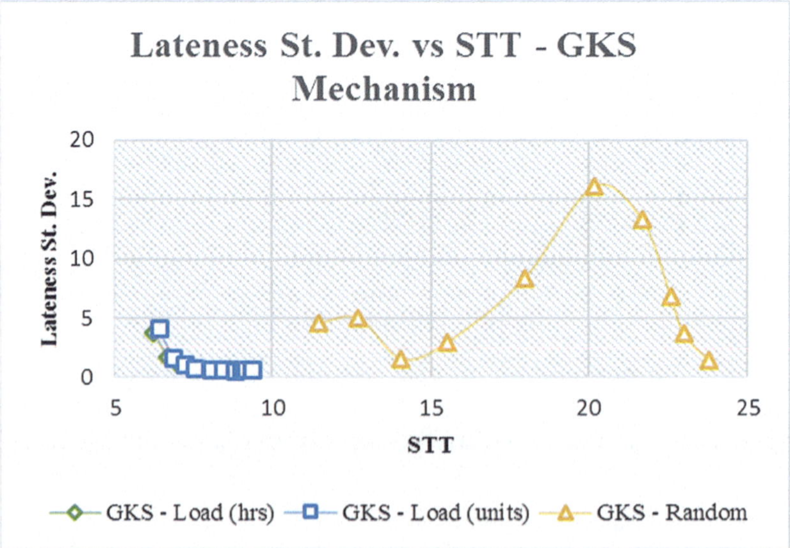

Fig. 10. Results for Lateness Standard Deviation vs. STT in the GKS mechanism.

Referring to Lateness Standard Deviation, as seen in Fig. 4, Random is also the criterion with the highest values and variability of values, meaning that there is a large dispersion of Lateness values in relation to the average. It is important to notice that these values increase until the scenario where the load norm is 16.5 for WLC and 14 for DBR and GKS, decreasing afterwards, meaning that for higher values of load norm/Kanbans, Lateness Standard Deviation is lower.

In addition, Load Units has a high value of Lateness Standard Deviation on the smallest value of load norm under the WLC and DBR mechanisms, meaning that, in this scenario, Lateness has a large dispersion of the values relatively to the average. This value is significantly different from the value observed on the same scenario for the other criterion. However, the remaining values are on the same range. It can be concluded that, for Load Hours and Load Units, values of Lateness Standard Deviation are low (Figs. 11, 12 and 13).

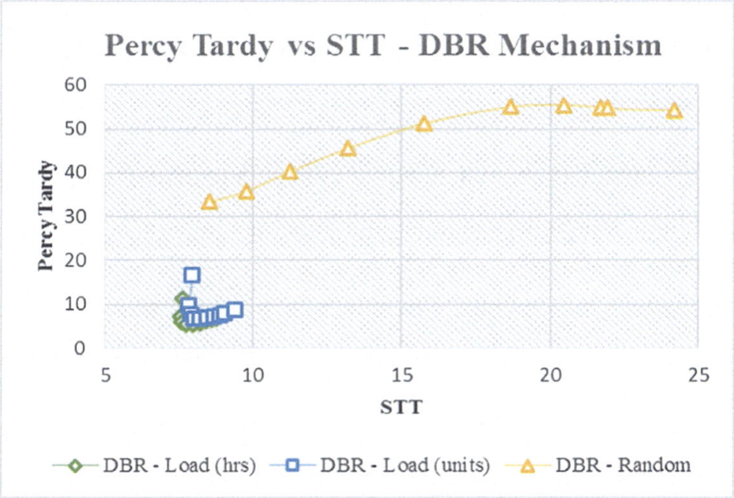

Fig. 11. Results for Percentage Tardy vs. STT in the DRB mechanism.

Lastly, Fig. 5 shows that the values of Percentage Tardy for Random are always above 50%, except for the DBR mechanism. Therefore, more than a half of the jobs are always late. Concerning the DBR mechanism, for this machine selection criterion, the registered values of Percentage Tardy increase with the increase of the load norm on the system, only decreasing in the last three scenarios, contrarily to the other mechanisms, and the highest value of this KPI founded is 55.46%. However, for GKS, the highest value is 97.95%.

Once again, Load Hours registers the smallest value of Percentage Tardy, and for DBR and WLC, Load Units has always higher values than Load Hours in all scenarios.

To conclude, while Random is the criterion with the worst performance for all the mechanisms, Load Hours is the one with the best results and performance in this case study.

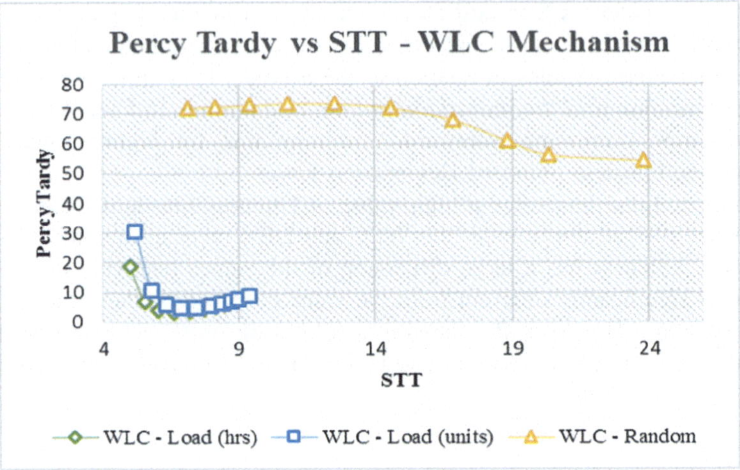

Fig. 12. Results for Percentage Tardy vs. STT in the WLC mechanism.

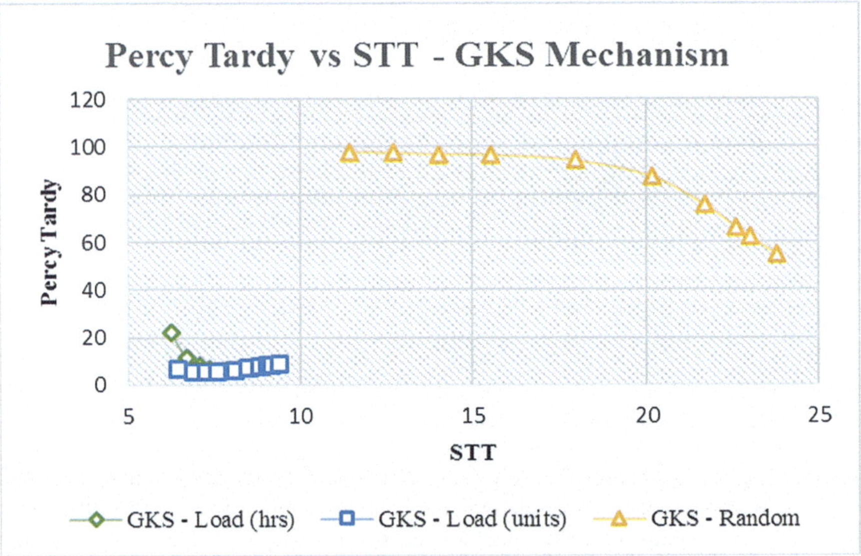

Fig. 13. Results for Percentage Tardy vs. STT in the GKS mechanism.

4.2 Second Analysis

Firstly, it is important to refer that in the following analysis and for all graphics, curves converge to the same point. This occurs due to the fact that as the production time of jobs increases (STT), the total time in the system (TTT) tends to decrease and stabilize.

Theoretically, when the load norm and signals increase, jobs remain less time in the pre-shop pool because they do not have to wait for authorization to be released.

Therefore, there is a decrease of the total time in the system and, in infinite, TTT equals STT (because of the dissolution of pool time).

Consequently, the mechanisms present all the same behaviour (convergence of curves) independently of the machine selection criteria implemented.

Random. Subsequently, the results from the Random as a machine selection criterion will be explained (Figs. 14, 15, 16 and 17).

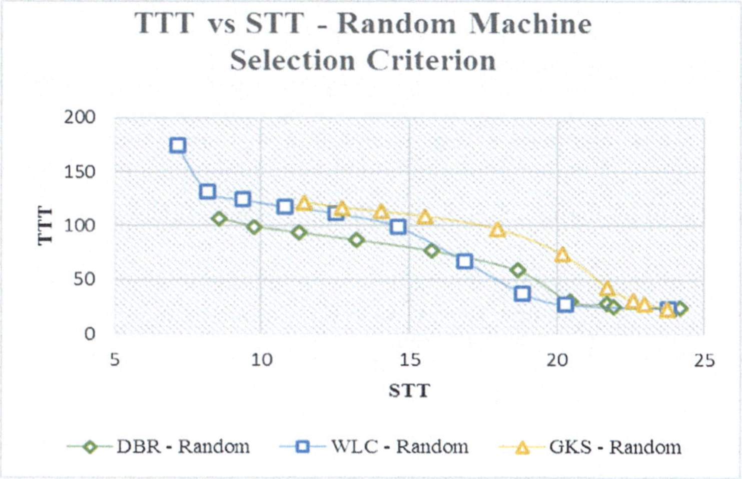

Fig. 14. Results for TTT vs STT with random machine selection criterion.

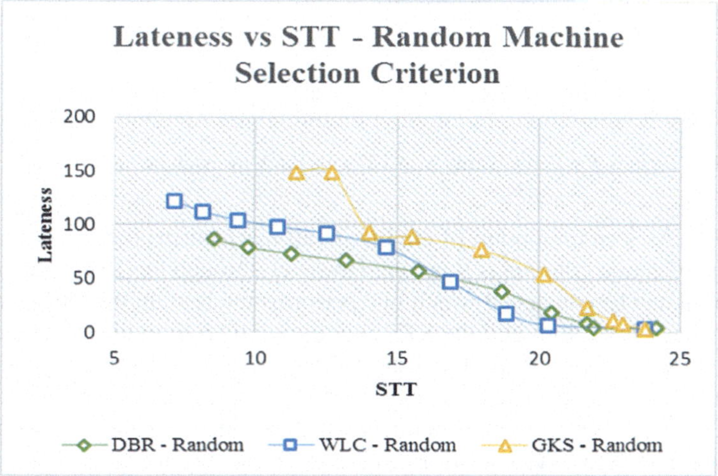

Fig. 15. Results for Lateness vs STT with random machine selection criterion.

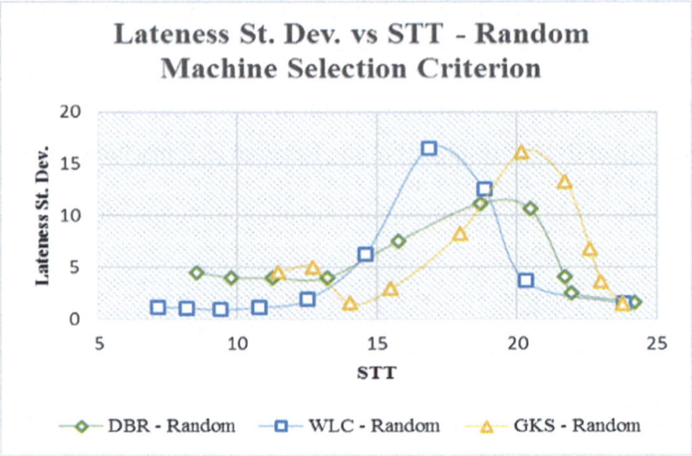

Fig. 16. Results for Lateness St. Dev. vs STT with random machine selection criterion.

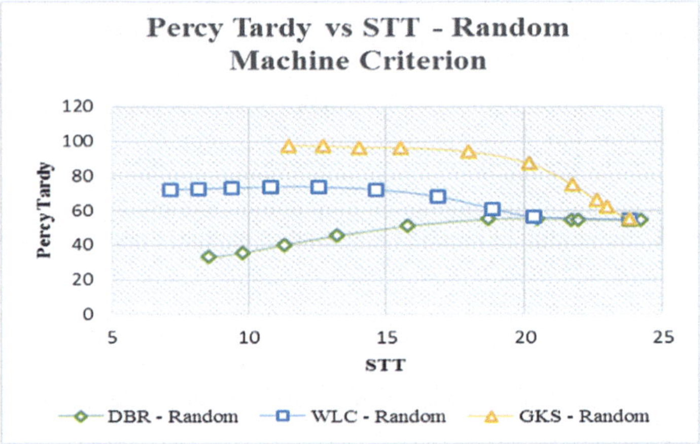

Fig. 17. Results for Percy Tardy vs STT with random machine selection criterion.

TTT vs. STT. As seen on top left of the Fig. 6, all mechanisms behave similarly when the machine selection criterion is Random. However, through the TTT average for each mechanism, it can be concluded which one performs the best. In the DBR mechanism, the TTT average is 62.90 h and the STT average is 16.56 h; in the WLC mechanism, the TTT average is about 91.46 h and the STT average is 14.25 h; lastly, in the GKS mechanism, the TTT average is about 75.92 h and the STT average is 18.29 h. Therefore, the DBR is the mechanism under which products remain the least time in the total system. The WLC mechanism is where products stay the least time in the production system, but the most time in the total system. This indicates that products are waiting more time in the pre-shop pool under this mechanism in comparison to the others.

Lateness vs. STT. Once again, the different curves that represent the different mechanisms behave analogously (top right of the Fig. 6). As the load norm imposed to system increases, Lateness values decrease significantly. When the system has a higher load norm, it is possible to have jobs finished at the right time, registering minimums of Lateness of 3.80 h (WLC), 3.78 h (GKS) and 4.19 h (DBR). On average, the mechanism that has the least number of late jobs is DBR (43.71 h).

Lateness Standard Deviation vs. STT. The graphic on down left of the Fig. 6 reveals how Lateness values vary across different load norm values. The highest values for all mechanisms are registered in the range of load norm 14–16.5 h. It can be seen that WLC and GKS present a higher dispersion of Lateness values from the average, while DBR presents constant values of Lateness Standard Deviation.

Percentage Tardy vs. STT. Regarding the Percentage Tardy results (down right of the Fig. 6), it is possible to conclude, once again, that all mechanisms perform similarly. However, on average, DBR performs the best, since it presents the lowest value of Percentage Tardy (48.09%). This result is expected, since the DBR mechanism has the lowest value of Lateness.

As the result of this brief analysis, DBR is considered the best one when the criterion is Random.

Load Hours. In this section, the results obtained for the three mechanisms on the Load Hours machine selection criterion are studied and compared (Figs. 18, 19, 20 and 21).

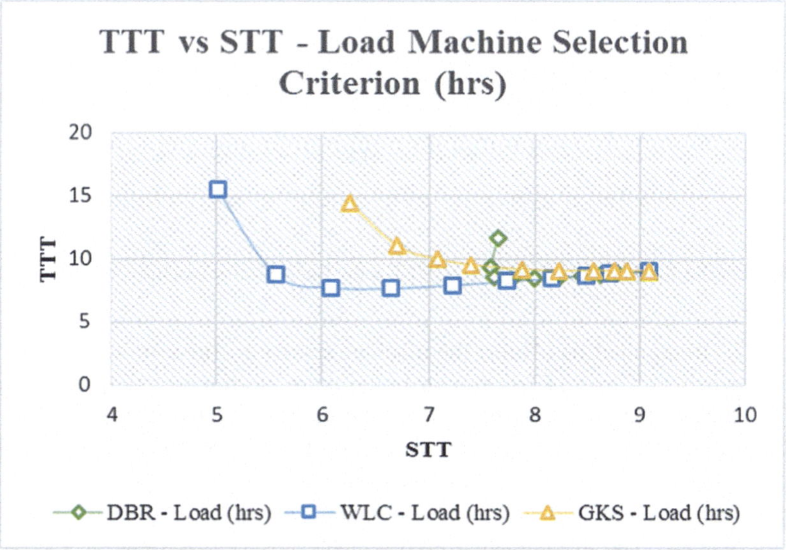

Fig. 18. Results for TTT vs STT with Load Hours machine selection criterion.

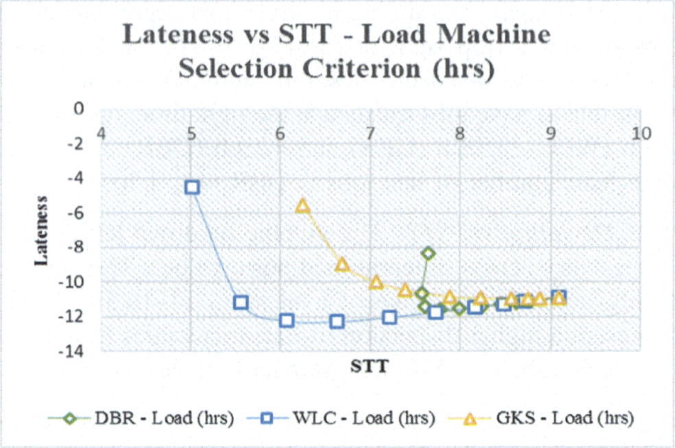

Fig. 19. Results for Lateness vs STT with Load Hours machine selection criterion.

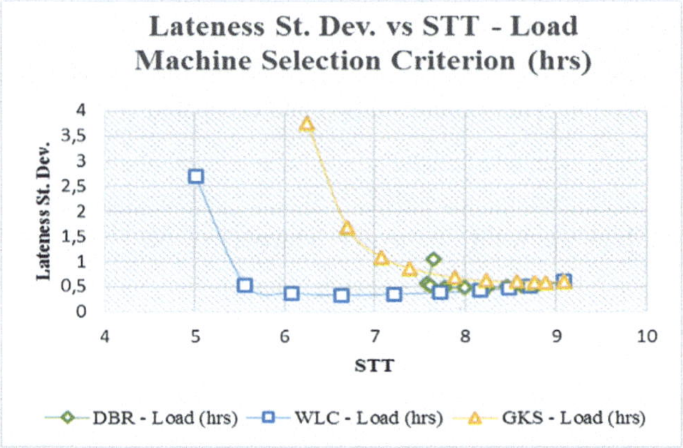

Fig. 20. Results for Lateness St. Dev. vs STT with Load Hours machine selection criterion.

TTT vs. STT. Being the Load Hours the machine selection criterion, it is shown that the relation between TTT and STT is similar in the three mechanisms despite the range of the TTT values of each mechanism, as it can be seen on top left of the Fig. 7. In all of them, the TTT average values are similar (TTT in DBR mechanism: 9.05 h, TTT in WLC mechanism: 9.11 h and TTT in GKS mechanism: 9.97 h). Therefore, it can be concluded that products remain more time in the total system under the GKS mechanism, on average. The mechanism which registers the lowest value of TTT is WLC (7.76 h). Despite that, it is under the DBR mechanisms that products remain, on average, the least time in the total system, although they remain the longest in the production system. This fact allows for the conclusion that the products remain less time in the pre-shop pool under the DBR than under the other mechanisms.

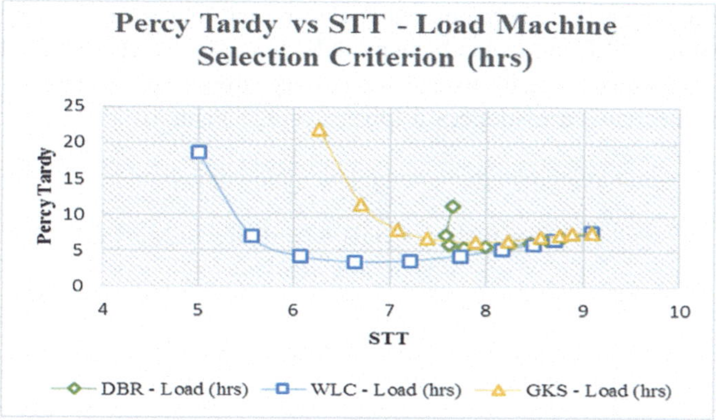

Fig. 21. Results for Percy Tardy vs STT with Load Hours machine selection criterion.

Lateness vs. STT. Regarding KPI Lateness (top right of the Fig. 7), a strong correlation with KPI TTT (ρ = 1) is observed. Therefore, the minimum value of Lateness registered originates from the WLC mechanism (−12.30 h). However, on average, the mechanism which results in a lower value of Lateness is DBR. All mechanisms present only negative values, indicating that products are in advance.

Lateness Standard Deviation vs. STT. In relation to Lateness Standard Deviation (graphic on down left of the Fig. 7), the minimum value is registered under the WLC mechanism. Nevertheless, it can be visually observed that Lateness Standard Deviation moves further away from the average in the WLC and GKS mechanisms, leading to the conclusion that stable values of Lateness are obtained through the DBR mechanism.

Percentage Tardy vs. STT. Since the lowest Lateness values arise from DBR mechanism and the products remain the least time in the total and production systems, it is expectable that this mechanism has a lower percentage of late jobs. However, the WLC mechanism has, on average, a slightly lower percentage of late jobs (6.68%) in comparison to the DBR (6.87%). Regarding the GKS mechanism, it can be concluded that the percentage of late jobs is the highest in this mechanism (down right of the Fig. 7).

Therefore, when the machine selection criterion is based on the Load Hours of the production system, the best performing mechanism is the DBR.

Load Units. Lastly, the results of Load Units machine selection criterion are presented (Figs. 22, 23, 24 and 25).

TTT vs. STT. Comparing the performances of the different mechanisms (graphic on top left of the Fig. 8), the WLC mechanism has the lowest STT value (7.55 h) and the highest value of TTT (10.88 h), on average, implying that products remain more time in the pre-shop pool. However, this can be justified by the existence of an atypical value from the first scenario, where load norm is very low, which results in the high TTT value and explains why WLC has, simultaneously, the highest value of TTT and the lowest STT. In the other mechanisms, the same situation occurs: the averages of TTT are influenced by the first scenario. To evaluate the performances without that influence, the averages were recalculated excluding the first scenario, allowing for the conclusion that the WLC is the mechanism under which products remain the least time in the entire system (TTT: 9.95 h) and also in the production system (STT: 7.78 h). Though, DBR mechanism evidences a lower average value of TTT.

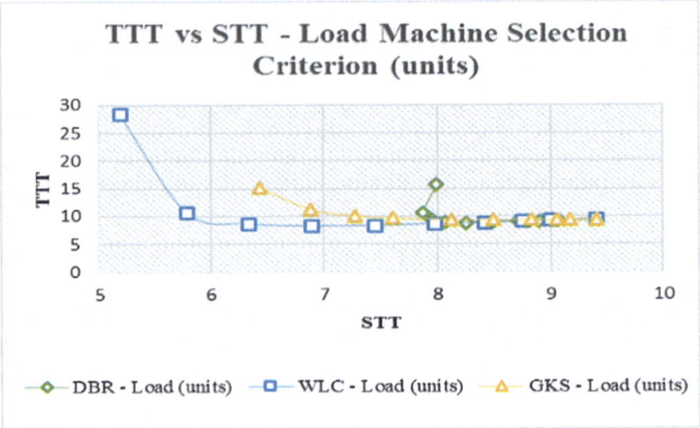

Fig. 22. Results for TTT vs STT in Load Units machine selection criterion.

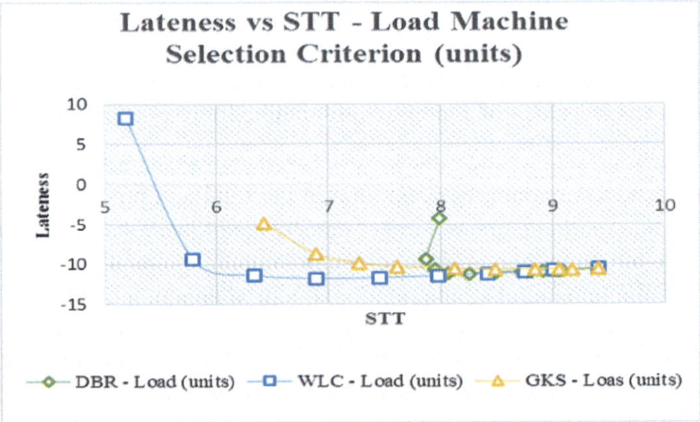

Fig. 23. Results for Lateness vs STT in Load Units machine selection criterion.

Lateness vs. STT. Through the analysis of the graphic on top right of the Fig. 8, it can be verified that almost all jobs are concluded before their due date. It is through the WLC mechanism that the minimum Lateness value is obtained (−11.81 h).

Similarly to the previous situation, if the first observation of the WLC mechanism were to be omitted, this mechanism would have the lowest average value of Lateness. However, considering the general panorama, it is under the DBR mechanism that products are the most advanced.

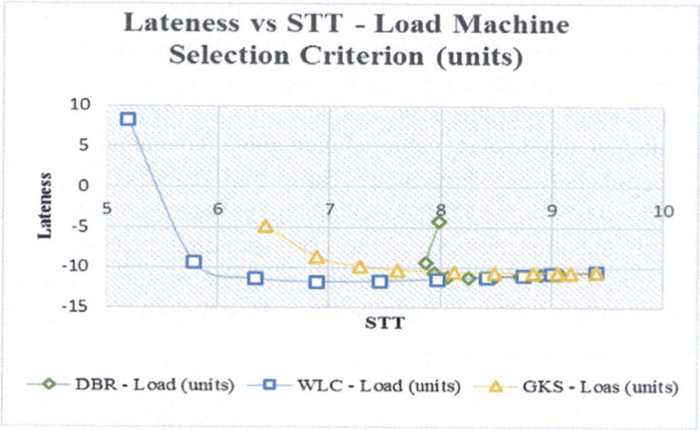

Fig. 24. Results for Lateness St. Dev. vs STT in Load Units machine selection criterion.

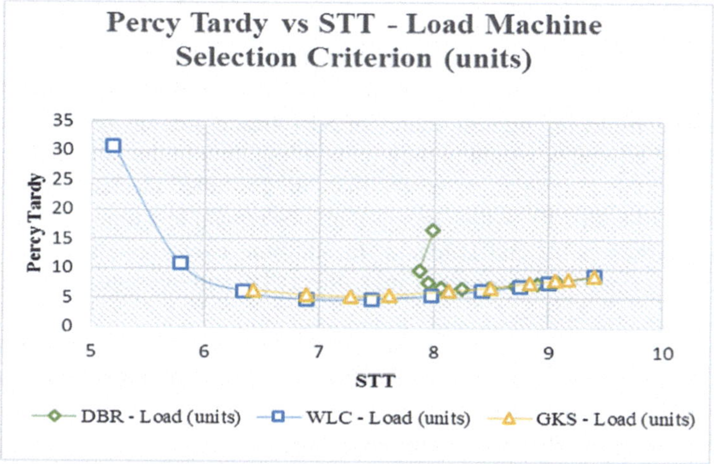

Fig. 25. Results for Percy Tardt vs STT in Load Units machine selection criterion.

Comparing the curves TTT vs. STT and Lateness vs. STT, a similarity can be observed. Using the Spearman Correlation between TTT and Lateness, it can be concluded that in the DBR mechanism $\rho = 1$, in the GKS mechanism $\rho = 1$ and in the WLC mechanism $\rho = 1$. These correlations indicate the existence of a linear and positive relation between these variables, since TTT is the total time in the entire system and Lateness represents the subtraction of the instant when a job is completed from its due date.

Lateness Standard Deviation vs. STT. In all mechanisms (down left of the Fig. 8), the first value presents the highest Lateness variation and the remaining values mostly have a Lateness Standard Deviation under one hour. That result indicates that, except for the first scenario, there is a little dispersion of Lateness values regarding the average.

Percentage Tardy vs. STT. Although the WLC mechanism presents some values lower than other mechanisms, the GKS mechanism is the one with the lowest average percentage of late jobs, as it can be seen on down right of the Fig. 8. This does not match the previous conclusions regarding Lateness, since it was previously observed that the DBR mechanism had the lowest value of late jobs. To analyse this situation, the Spearman correlation between Percentage Tardy and Lateness is used: DBR − ρ = 0.98; GKS − ρ = −0,61. In the DBR mechanism, as Lateness increases or decreases, the percentage of late jobs change in the corresponding proportion. The opposite situation occurs in the GKS mechanism: the increase of the Lateness of the jobs corresponds to a decrease in Percentage Tardy, and vice-versa. Despite having a lower Percentage Tardy, the GKS presents higher job lateness when compared to the DBR.

Considering the Load Units as a machine selection criterion, it is concluded that the DBR has the best performance. However, WLC would be better in all KPIs if the first scenario was excluded.

5 Conclusion

After all the analysis performed, it can be concluded that the DBR is the mechanism with the best performance in almost all the studied scenarios. Even though it is the mechanism which has the highest values of STT (since jobs do not have to wait for authorization of all stages, meaning they remain less time in the pool and are released into the production system more easily than in the other mechanisms), it has the smallest values of both TTT (which means that jobs remain less time in the total system and in the pre-shop pool) and Lateness. In addition, even though it is not the best mechanism in terms of Percentage Tardy, the DBR still presents quite satisfactory values.

The previous conclusion can be related to the fact that this mechanism has a fixed bottleneck. In this project, stage 1 is considered the permanent bottleneck, since it is the stage in which products queue up to enter the production system. After that, the products flow without influence of any mechanism.

However, it is also important to note that the scenario with the lowest load norm is compromising the performance of WLC, since it has better performance in the other scenarios.

Regarding the machine selection criteria, Load Hours performs the best since it has the smallest values of almost all KPIs. Contrarily, Random is the criterion with the worst performance. This result is expected because Load Hours is the only criterion that has the workload of the machines in consideration. Therefore, in similar contexts companies should consider the implementation of Load Hours instead of the other criteria.

Suggestions for further developments include the implementation of DBR with a non-constant bottleneck, verifying which stage has the highest load throughout the production time to identify the bottleneck in real time. In addition, using the Optquest study in Simio can be important to obtain the optimum number of Kanbans and load norm.

References

1. Akillioglu, H., Dias-Ferreira, J., Onori, M.: Characterization of continuous precise workload control and analysis of idleness penalty. Comput. Ind. Eng. **102**, 351–358 (2016). https://doi.org/10.1016/j.cie.2016.05.026
2. Burbidge, J.L.: Production control: a universal conceptual framework. Prod. Plan. Control **1**(1), 3–16 (1990)
3. Carmo-Silva, S.: Textos de Gestão da Produção, v01_10. Publicação Interna Departamento de Produção e Sistemas, Universidade do Minho, Braga
4. Chakravorty, S., Atwater, J.: The impact of free goods on the performance of Drum-Buffer-Rope scheduling systems. Int. J. Prod. Econ. **95**, 347–357 (2005). https://doi.org/10.1016/j.ijpe.2004.01.001
5. Chang, T., Yih, Y.: Generic Kanban Systems for dynamic environments. Int. J. Prod. Res. **32**, 889–902 (1994). https://doi.org/10.1080/00207549408956977
6. Chen, B., Matis, T.: A flexible dispatching rule for minimizing tardiness in job shop scheduling. Int. J. Prod. Econ. **141**, 360–365 (2013). https://doi.org/10.1016/j.ijpe.2012.08.019
7. Fernandes, N.O., Thürer, M., Silva, C., Carmo-Silva, S.: Improving workload control order release: Incorporating a starvation avoidance trigger into continuous release. Int. J. Prod. Econ. **194**, 181–189 (2016)
8. Georgiadis, P., Politou, A.: Dynamic Drum-Buffer-Rope approach for production planning and control in capacitated flow-shop manufacturing systems. Comput. Ind. Eng. **65**, 689–703 (2013). https://doi.org/10.1016/j.cie.2013.04.013
9. Hermann, M.P.: Design principles for Industrie 4.0 scenarios. In: 2016 49th Hawaii International Conference on System Sciences (HICSS), pp. 3928–3937. IEEE, January 2016
10. Kingsman, B., Hendry, L.: The relative contributions of input and output controls on the performance of a workload control system in make-to-order companies. Prod. Plan. Control 579–590 (2002). https://doi.org/10.1080/0953728021000026285
11. Lage Junior, M., Godinho Filho, M.: Variations of the Kanban System: literature review and classification. Int. J. Prod. Econ. **125**, 13–21 (2010). https://doi.org/10.1016/j.ijpe.2010.01.009
12. Posada, J., Toro, C., Barandiaran, I., Oyarzun, D., Stricker, D., de Amicis, R., Vallarino, I.: Visual computing as a key enabling technology for Industrie 4.0 and industrial internet. IEEE Comput. Graph. Appl. **35**, 26–40 (2015)
13. Reis, E.: Estatística Multivariada e Aplicações. Edições Sílabo, Lda, Lisboa (1997)
14. Stevenson, M., Hendry, L., Kingsman, B.: A review of production planning and control: the applicability of key concepts to the make-to-order industry. Int. J. Prod. Res. **43**, 869–898 (2005). https://doi.org/10.1080/0020754042000298520
15. Stock, T., Seliger, G.: Opportunities of sustainable manufacturing in Industry 4.0. Procedia CIRP **40**, 536–541 (2016). https://doi.org/10.1016/j.procir.2016.01.129

Development of an Equipment and Calibration Method for Bearing Rings Multi-parametric Inspection

Cioboată Daniela[1](✉), Soare Adrian[2], Stanciu Dănuț[1], Abălaru Aurel[1], and Logofătu Cristian[1]

[1] National Institute of Research and Development in Mechatronics and Measurement Technique, Pantelimon Road No. 6-8, 021631 Bucharest, Romania
cioboatadoina@yahoo.com, danutstanciu@yahoo.com, aurel.abalaru@gmail.com, cristilogofatu@yahoo.com
[2] SC COMIS SRL, Road Nicolae Iorga No. 83, 106400 Câmpina, Prahova, Romania
adyse_prod@yahoo.com

Abstract. The new manufacturing processes have to undertake a continuous product quality improvement due to major economic involvement and global technical progress. The focus of this paper is to present a dimensional and geometric inspection equipment developed for roller bearing rings with medium sizes (45…200 mm) and a method for calibration using a small number of standards for the entire measuring range. The experiments made on functional model proves the capability and functionality of the system. The obtained results will be presented and discussed.

Keywords: Bearing rings · Dimensional inspection · Calibration · Modular structure · Mechatronics

1 Introduction

Bearings are central parts of nearly all applications involving precision rotary motion. In the context of the global progress, rolling-element bearings are required to operate at tougher conditions (operating speeds and temperatures) and consequently the quality of the dynamical performance of bearings need to be improved. The dynamic performance of bearings depends on the dimensional and geometric precision of their components.

The radial roller bearings generally consist of two rings, a number of rolling elements and a cage [1] (Fig. 1).

To improve the uniformity and repeatability in the manufacturing process is very important for bearings quality. Besides the progress of manufacturing technology, a well-organized quality control system based on smart, flexible and precision measuring machines is the other key element in the production of reliable and high-performance products. Among the most important characteristics that determine the quality of the rolling bearings are: dimensional and geometrical accuracy of the bearing rings

G. I. Gheorghe (Ed.): ICOMECYME 2019, LNNS 85, pp. 104–117, 2020.
https://doi.org/10.1007/978-3-030-26991-3_10

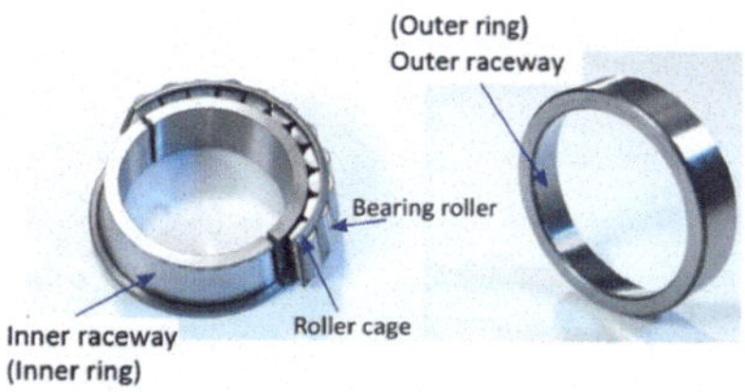

Fig. 1. Roller bearing components

raceway, the outer diameter of the outer ring, the inner diameter of inner ring, the rings width, the rotation accuracy. The values of all these deviations, for different types of bearings, determine the precision class of a bearing [3].

The manufacturing of bearings usually consists of consecutive operations when primary shape of the ring is obtained via machining followed by hardening and grinding or honing. Dimensions, shapes and surfaces quality of turned rings are important to the quality of finished products. In the absence of rigorous inspections during the turning process, the quality of the bearing components can be compromised. Predictable repeatability in the manufacturing process is crucial to ensuring the performances of bearings.

The diversity of the constructive bearings forms and dimensions requires choosing some adequate criteria for measurement instrumentation design. Modular construction and flexibility of measurement instrumentation are two of the important design criteria. By building devices for multiple diameters measurement on parts with two or more cylindrical features, into a single clamping, it is possible to eliminate the expense of duplicate work-holding devices and save time and space.

With the evolution of CNC manufacturing machines increased requirements for quality control in terms of accuracy, frequency, measurement cycle duration and statistical data processing. Development of smart, flexible and high level of automation equipment for manufactured parts quality inspection is one of the main requirements for the technological improvements of the manufacturing systems [2].

Based on analysis of manufacturing technology of bearing rings, methods and instruments for their dimensional measurement, this paper recommends a multiparametric measurement equipment, with large measurement range, which uses the measurement method based on a rotary table and that needs a small number of calibration standards.

2 General Characteristics of Roller Bearing Rings

The equipment presented in this paper is designed for multiparametric measurement of the inner and outer bearings rings, especially of cylindrical roller bearings of medium size (minimum inner diameter of 45 mm and maximum outside diameter of 200 mm, maximum width of 53 mm). The measured parts are characterized by the surfaces of revolution that allow centering and clamping on inner or outer diameter and supported on a flat end surface (Fig. 2).

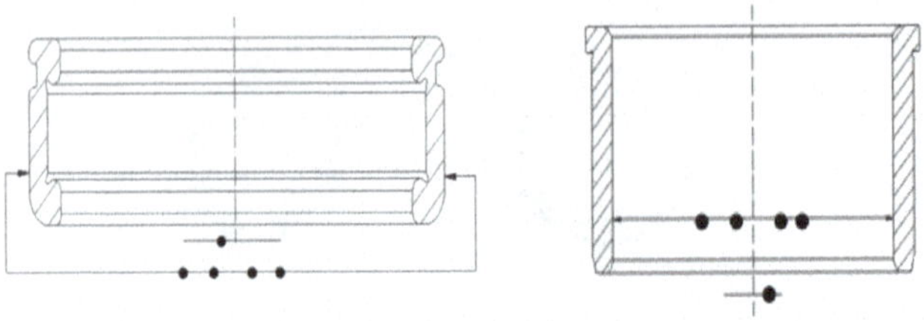

Fig. 2. Supporting and centering elements of bearing rings

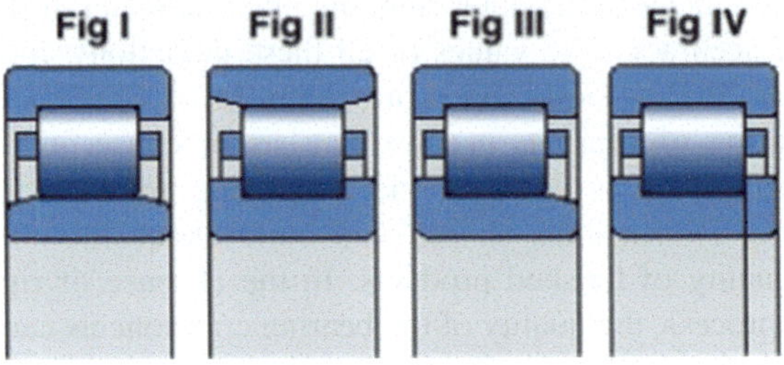

Fig. 3. Cylindrical roller bearing types

The main forms of cylindrical roller bearing inner and outer rings are presented in Fig. 3.

An important feature of the bearing rings which influence the fixturing device and measurement scheme are the maximum chamfer or fillet dimensions in axial directions. According with Din 620-6 (Rolling bearings - Rolling bearing tolerances - Part 6: Chamfer dimension limits) and Schaeffler Technologies [4] we considered that for measured bearing rings which can be measured on the presented equipment the maximum value for $r_{max\ a}/r_{1max\ a}$ is 3,5 mm (see Fig. 4).

① Bore or outside diameter
② Lateral face
a = Measurement distance a (Start of inspection area for bore or outside diameter tolerances) [mm]
r_{min}, $r_{1\ min}$ = Symbol for smallest chamfer dimension in radial and axial direction [mm]
$r_{max\ r}$, $r_{1\ max\ r}$ = Largest chamfer dimension in radial direction [mm]
$r_{max\ a}$, $r_{1\ max\ a}$ = Largest chamfer dimension in axial direction [mm]

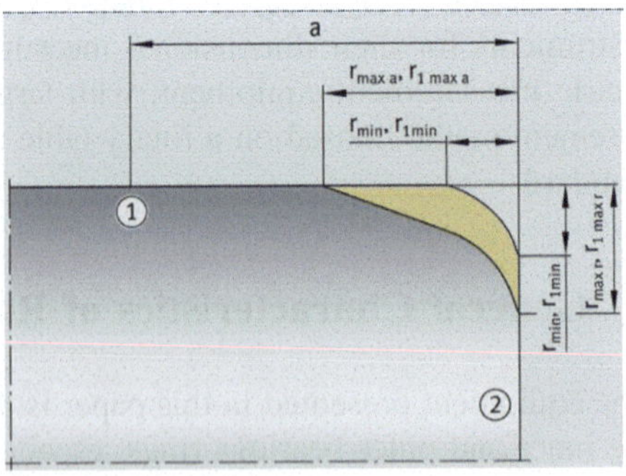

Fig. 4. Limit chamfer/fillet dimensions [4]

3 Overview of the Bearing Measurement Instruments

Due to the wide variety of shapes and sizes, the measurement of dimensional and geometric deviation of bearing rings is today generally made with universal equipment and devices and less often with specialized equipment. Bearing rings are bodies of revolution. As a result, the measurement systems are based on the classic circular profile measurement or coordinate measurement schemes. Coordinate measuring machines, roundness measuring machines, contour measurement machines or length measuring machines are commonly used to inspect the quality of bearing rings, especially in metrological laboratories (Fig. 5).

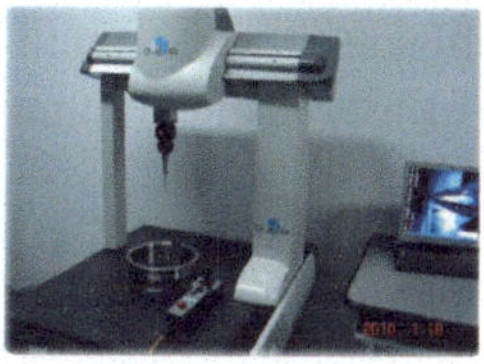

a. Coordinate measuring machine – DEA

b. Roundness measurement Surtronic R series - Taylor Hobson

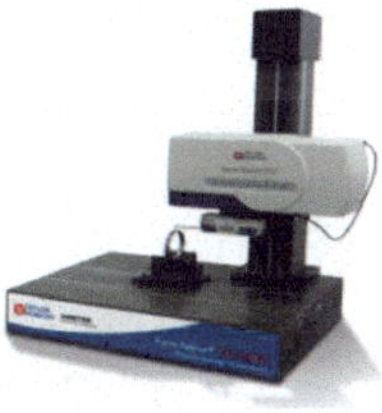

c. Form Talysurf PGI Taylor Hobson

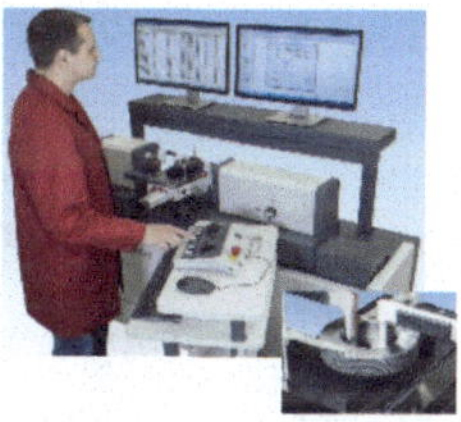

d. Mahr Universal lengths measuring machine

Fig. 5. Universal measuring equipment

Many manufacturers use for measurement on production flow bench devices with dial indicators which need to be calibrated periodically. This control technology requires a great number of standards, is time consumption and does not allow the statistical control. Figure 6 shows some measuring devices currently used by bearings manufacturers.

Due to the rapid technological development, as well as stronger competition on global market, the trend is to replace these inspection technologies with innovative, intelligent control techniques to achieve greater productivity, higher quality of bearings, lower manufacturing costs.

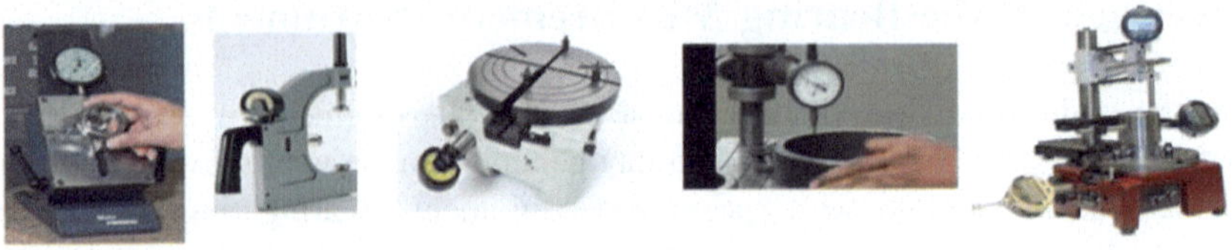

Fig. 6. Bench measurement devices

4 Architecture of the New Bearing Rings Measuring Equipment

In this paper we present a novel architecture of an intelligent, flexible, multi-functional equipment which can be used for rolling bearing rings measurement. The measurement tasks are (see Table 1):

Table 1. Measurement tasks

Inner and outer diameter measurement in multiple planes	Inner and outer diameters roundness measurement in multiple planes	Runout of inner or outer diameters in multiple planes	Axial circumferențial runout

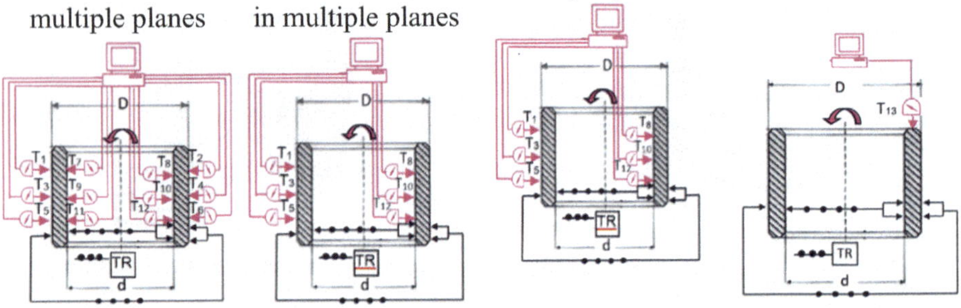

T1 …T13 – digital linear measurement transducers; TR – incremental encoder

- Measurement of the inner and outer diameters (deviations and absolute values) (simultaneously in maximum 3 sections)
- Measurement of roundness of inner and outer diameters (simultaneously in maximum 3 sections)
- Measurement of run-out of inner or outer diameters (simultaneously in maximum 3 sections)
- Measurement of the bearing rings thickness deviation (axial circumferențial runout).

The technical problem to be solved during development of this equipment was to make it flexible enough to allow measurement of rings with various sizes and shapes, in multiple sections simultaneous, without the need for a large number of calibration standards. The equipment has a modular hardware and software structure for increasing functional flexibility to be able to adapt rapidly to changes in production [6].

The developed measurement equipment consists of the following systems as shown in Fig. 7: base Table (1); rotary Table (2); adjustable centering system (3); outer diameters measurement modules (4); inner diameters measurement modules (5); axial measurement module (6); vertical column (7); data acquisition, processing and display unit (8); console with emergency stop button (9); console with buttons for manual controls (10); pneumatic system (11); electrical box; integrated software for the entire calibration and measurement process.

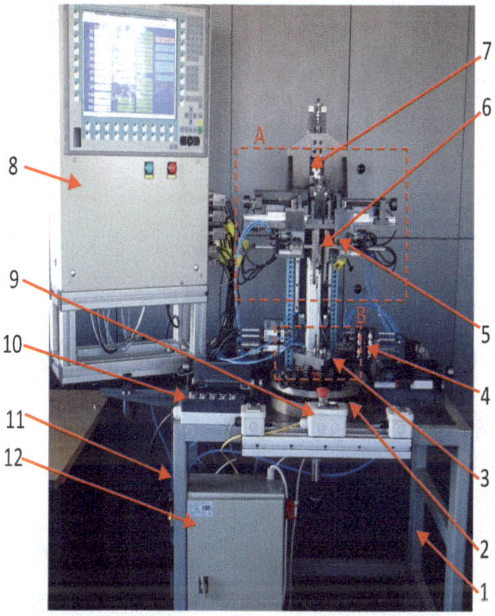

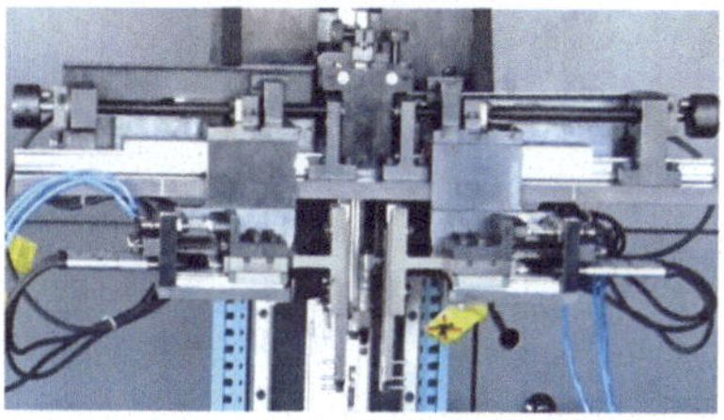

Detail A: Inner and axial measurement modules

Detail B: Centering system

Fig. 7. Equipment for bearing ring measurement

The arrangement of the probes and measurement principle is presented in Fig. 8. Centering and fastening of the bearing ring on the rotary table is made manually, with a four points flexible device which allows fixing the rings both on the inner or outer diameter (see Fig. 7 - Detail B and Fig. 8). Fixing must ensure controlled gripping forces to prevent rings deformation [3]. Gripping force control is achieved by elastic force limiting coupling element integrated in hand fastening wheel (Fig. 7 - Detail B).

For constructive reasons, the radial measurements are made with intermediate probing elements (Fig. 9), adjustable in both radial and axial directions, to allow the measurement of a wide range of bearing rings.

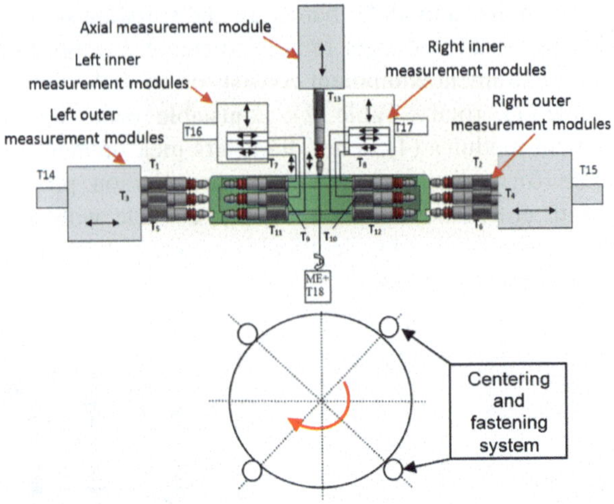

Fig. 8. Probes arrengement and part centering in four points

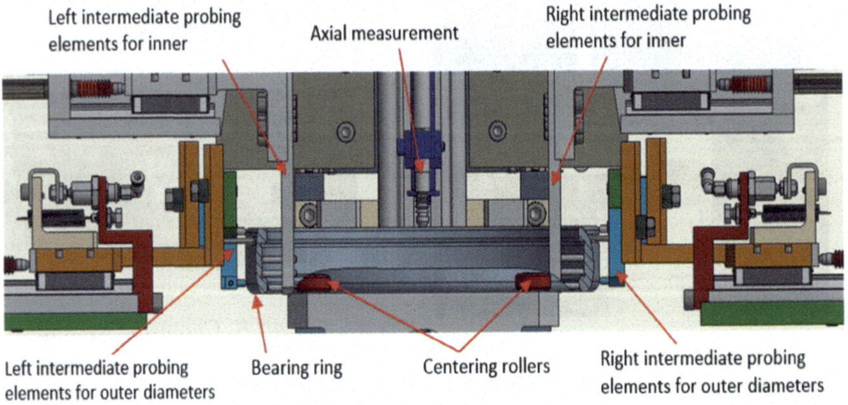

Fig. 9. Probing elements

Retraction of the contact probes from part surfaces, at the end of the measurement cycle, is done with miniature pneumatic cylinders with a stroke of 10 mm. The purpose is to protect the probes when the ring is placed into the measuring station or when it is removed. The upward or downward movement of sub-assemblies for inner diameters measurement and axial measurement is also done pneumatically. The pneumatic scheme contains elements for adjusting the speed of moving parts, position sensors and air silencers.

In order to measure a large range of bearing rings without the use of a large number of calibration standards as well as for the measurement of the absolute diameters, in parallel with the direction of inner and outer diameters measurement were placed four digital absolute scales (T_{14} … T_{17}) with a measuring range of 100 mm and a resolution of 0,001 mm (see Fig. 8).

5 Calibration and Experimental Results

The presented equipment may be used to measurement the outside and inside diameters and circular profiles in maximum three radial sections simultaneous. It can also be used to determine the radial width deviation of the bearing rings. The main concept of the measurement method is to place the bearing element on a rotary table, centered on inner or outer diameter, rotate the element and capture the data from points on the outer or inner surfaces and on the frontal face, at a specified sample rate.

Inner and outer diameters measurement is based on two point measurement method (Fig. 10a). Diameters are measured by the distance between a pair of two probes aligned in opposition to each other and working in tandem at a specified sample rate, while the part is rotated through 360°. An average (mean) value is calculated from the sum of the measured diameters at each angular position.

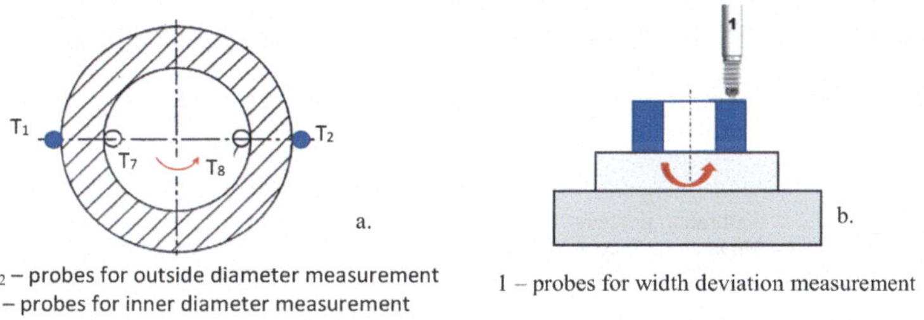

Fig. 10. Measurement method for: a. inner/outer diameters measurement; b. radial width deviation measurement

Axial measurements are made with a single probe which determine the deviation at each angular position (Fig. 10b). The variation of inner ring width is determined by diference from maximum and minimum measured deviations.

Radial and axial measurements are made with 13 digital linear sensors (T_1…T_{13}) with measurement range of 10 mm (see Table 1 and Fig. 8). These transducers measure only deviations from the nominal value. All intermediate probing elements can be adjusted for measurement of circular profiles into different planes, depending on the form and sizes of the measured part.

Calibration of the equipment is done with a standard ring. The absolute values of the measured outside diameters are calculated based on the value of the outer diameter of the standard, the deviations measured by the pairs of transducers T_1–T_2, T_3–T_4 or T_5–T_6 and the value measured by the digital linear transducers T_{14}, T_{15}.

The absolute values of the measured inner diameters are calculated based on the value of the inner diameter of the standard, the deviations measured by the pairs of transducers T_7–T_8, T_9–T_{10} or T_{11}–T_{12} and the value measured by the linear digital transducers T_{16} and T_{17}.

So, for the measurement Sect. 1:

$$D_{1i} = C_1 + T_{14} + T_{15} + \Delta_{T1}(\theta_i) + \Delta_{T2}(\theta_i) \tag{1}$$

$$D_1 = \frac{\sum_{i=1}^{n} D_{i1}}{n} = C_1 + T_{14} + T_{15} + \frac{\sum_{i=1}^{n} (\Delta_{T1}(\theta_i) + \Delta_{T2}(\theta_i))}{n} \tag{2}$$

$$d_{1i} = C_2 + T_{16} + T_{17} + \Delta_{T7}(\theta_i) + \Delta_{TS}(\theta_i) \tag{3}$$

$$d_1 = \frac{\sum_{i=1}^{n} d_{i1}}{n} = C_2 + T_{16} + T_{17} + \frac{\sum_{i=1}^{n} (\Delta_{T7}(\theta_i) + \Delta_{T8}(\theta_i))}{n} \tag{4}$$

where: D_{1i} = Outer diameter in Sect. 1, measured at angle θ_i

i = number of sample points for a complete rotation (360°)

$C_{1,2}$ = constante that take into account the position of the scales T_{14}, T_{15} respectively T_{16}, T_{17} relative to the plane containing the axis of the rotation and perpendicular to the measuring direction; they are determined experimentally.

Similarly are determined the diameters in Sects. 2 and 3.

Calibration scheme for outer diameters measurement is presented in Fig. 11.

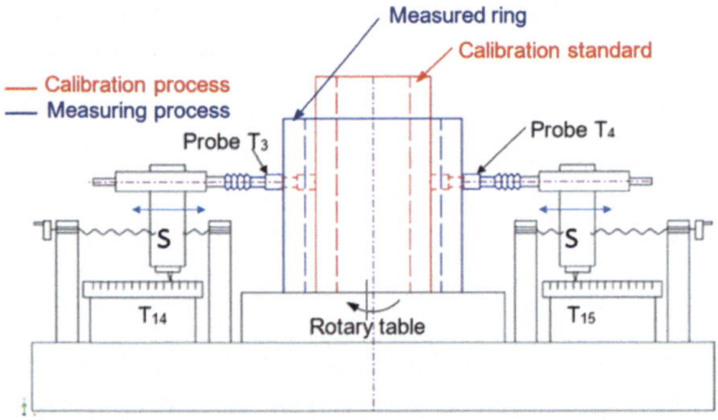

Fig. 11. Scheme of calibration and measuring processes

Initially, the equipment calibration was performed using a standard ring with inner diameter of 45 mm and outside diameter of 85 mm. This standard ring was measured in metrology laboratory with CMM measurement machine (Leitz Reference 600) and roundness measurement machine (Roncorder EC 2500) (see Table 2 and Fig. 15)

To determine the coefficients C_1 and C_2, were used the pairs of probes T_3–T_4 and T_9–T_{10}.

The equipment allows operation in 2 working modes: manual and automatic. The manual working mode allows adjustment and calibration of the equipment, ensuring the following functions (Fig. 12): actuating the pneumatic cylinders for retracting and

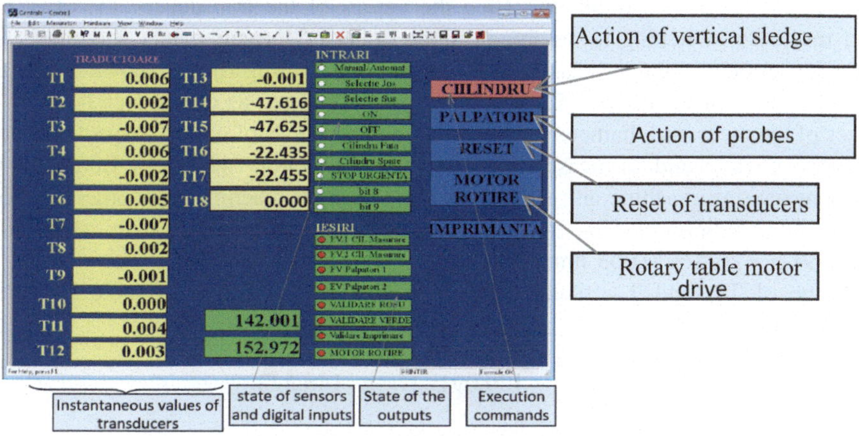

Fig. 12. Menu for manual working mode

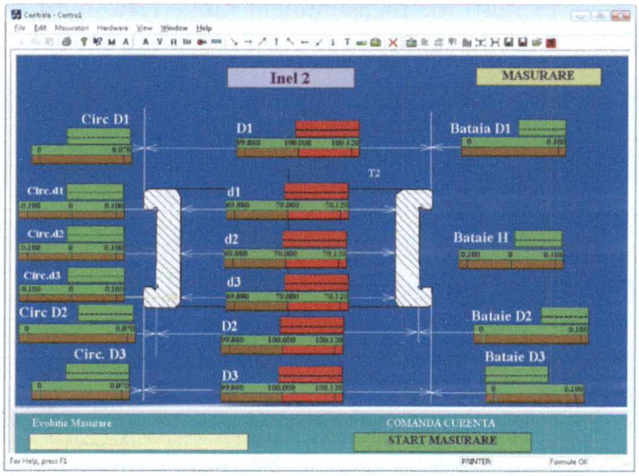

Fig. 13. Menu for automatic measurement mode, for an inner bearing ring

returning the radial probes T1…T12 on the part; pneumatic lowering and lifting of the vertical carriage wich are supporting the inner and axial measurement modules; rotary table driving; reset of T1…T13 transducers.

For calibrating the outer and inner diameters measurement modules, standard ring was placed on rotary table and centred on inner diameter. The horizontal sledges S (Fig. 11) were moved until the probes for outer diameters measurement touched the standard. The transducers T14 and T15 must indicate approximately the same value. This is the value at which the calibration of the equipment will be done whenever required, with the mentioned standard. Then, the sledges S have been locked. The transducers T1…T6 have been set up in the measurement field, and then they were locked.

Similarly, were adjusted transducers T7…T2 for measuring inner diameters, moving the slides which are carrying the readheads of the T16 and T17 scales and modules for inner measurements. Then the values of the transducers T1…T12 have been reset. By switching to automatic mode, a set of measurements was made. The menu for automatic mode measurement of an inner bearing ring is shown in Fig. 13.

The values of coefficients C_1 and C_2 were determined as differences between the values of the measured diameters D1 and d1 and the values of outers, respectively inner diameters of the standard measured in metrology laboratory with Leitz Reference 600 CMM measuring machine and roundness measurement machine Roncorder EC 2500.

Results: Value of T_{14} and T_{15} scales for calibration with standard 1 (Table 2) = −26,5 ± 0,003 mm

Value of T_{14} and T_{15} scales for calibration with standard 1 (Table 2) = −36,5 ± 0,003 mm

$$C_2 = 129{,}091;\ C_1 = 148.523$$

Experimental results have shown that a single standard is not sufficient to achieve the desired accuracy (±0,01 mm) across the entire measuring range. Based on the analysis of the measurement error distribution (Fig. 14) of 11 rings, in two sections (h_1 = 10 mm and h_2 = 17 mm) the measurement range (200 mm) was divided into 3 sections and 3 standards were made (one for each section):

Standard 1: d = 45 mm; D = 85 mm – for inner measurement range of 45…70 mm and outer measurement range of 50…110 mm
Standard 2: d = 70 mm; D = 110 mm – for inner measurement range of 70…105 mm and outer measurement range of 110…145 mm
Standard 3: d = 105 mm; D = 145 mm – for inner measurement range of 105…180 mm and outer measurement range of 145…200 mm

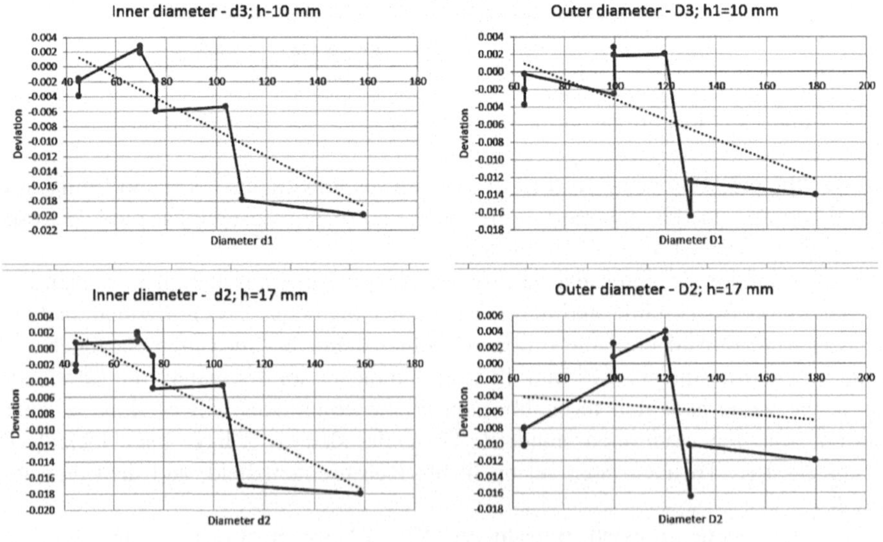

Fig. 14. Error distribution charts

Table 2. Standards dimensions

Standard	d (mm)			D (mm)			d roundness (mm)	D roundness (mm)
	h1 = 10 mm	h2 = 17 mm	h3 = 37 mm	h1 = 10 mm	h2 = 17 mm	h3 = 37 mm	h2 = 17 mm	H2 = 17 mm
1	45,0032	45,0031	45,0015	85,003	85,0026	85,003	0,0007	0,0033
2	69,9954	69,9957	69,9953	110,0079	110,0067	110,0086	0,0015	0,0016
3	104.9910	104.9914	104.9920	145.0022	155.006	145.009	0.0018	0.0032

For each of these ranges, were determined the values of the C_1 and C_2 coefficients and the values of the $T_{14}...T_{17}$ scales for calibration.

The dimensions of the standards measured in the metrological laboratory with the Leitz 600 CMM measurement machine and the Roncorder EC 2500 are shown in Fig. 15 and Table 2.

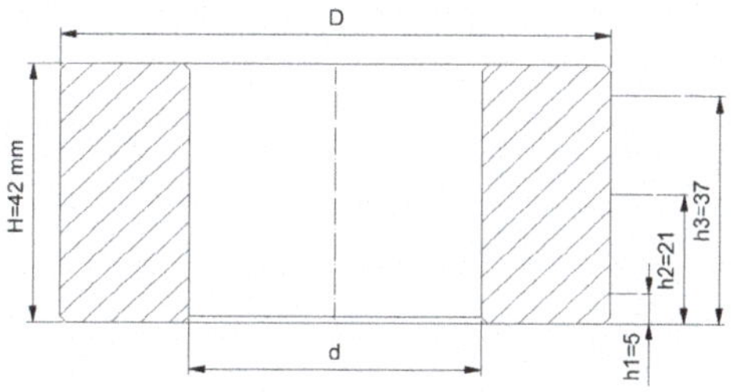

Fig. 15. Standard rings

As we mentioned, this equipment may be used also to measurement the roundness of inner and outer bearing rings circular profiles. The least square is the most commonly method used in practice for roundness measurement. Measurement software accuracy is influenced by various factors: number of measured points, workpiece centering precision, the filters used, workpieces dimensions [7].

The measured radial data include both the radial form error of the part and the radial errors of the rotary table spindle. Error separation is important for reducing the measurement errors.

The rotary table spindle has random and systematic components to its radial error motion. We use the reversal method to isolate and remove the systematic component of

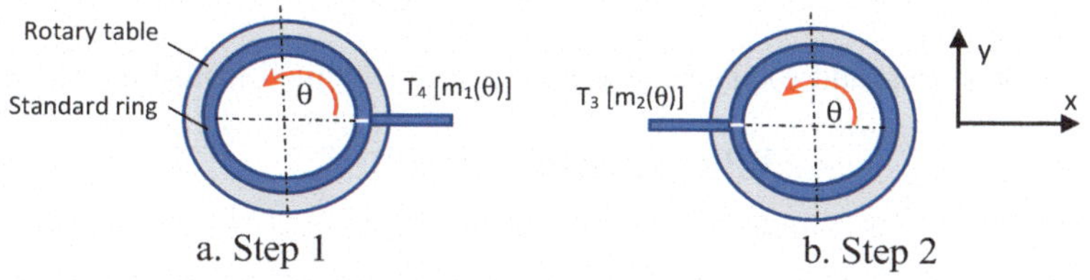

Fig. 16. The reversal method for separating part error from spindle error

the radial errors of the rotary table movement. This method for separating part error from spindle error involves measuring the part in two positions of the probe and the part (Bryan et al. 1967; Donaldson 1972) [5].

For rotary table error separation we used the pair of probes T3, T4 (Fig. 8) and the standard 1 (Table 2). First the standard ring was centered on inner diameter on the rotary table in a predetermined position, and a set of measurement data was acquired while the table was rotated through 360°. We considered only measured data of T4. Then, the standard ring was rotated by 180° and a new set of data was acquired, with the same starting position of the rotary table as in step 1. We considered only measured data of T3 (Fig. 17).

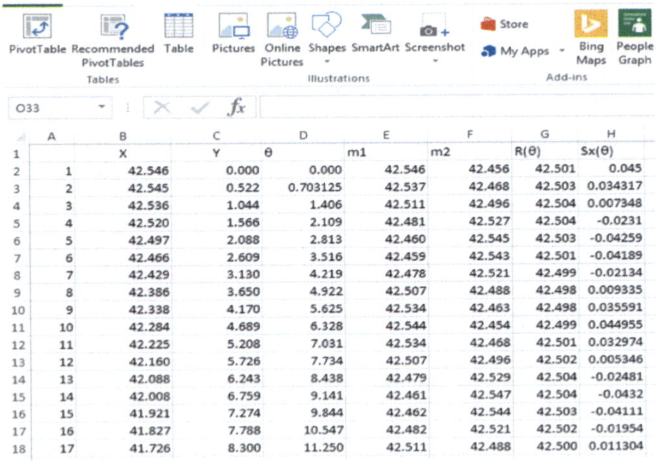

	A	B	C	D	E	F	G	H
1		X	Y	θ	m1	m2	R(θ)	Sx(θ)
2	1	42.546	0.000	0.000	42.546	42.456	42.501	0.045
3	2	42.545	0.522	0.703125	42.537	42.468	42.503	0.034317
4	3	42.536	1.044	1.406	42.511	42.496	42.504	0.007348
5	4	42.520	1.566	2.109	42.481	42.527	42.504	-0.0231
6	5	42.497	2.088	2.813	42.460	42.545	42.503	-0.04259
7	6	42.466	2.609	3.516	42.459	42.543	42.501	-0.04189
8	7	42.429	3.130	4.219	42.478	42.521	42.499	-0.02134
9	8	42.386	3.650	4.922	42.507	42.488	42.498	0.009335
10	9	42.338	4.170	5.625	42.534	42.463	42.498	0.035591
11	10	42.284	4.689	6.328	42.544	42.454	42.499	0.044955
12	11	42.225	5.208	7.031	42.534	42.468	42.501	0.032974
13	12	42.160	5.726	7.734	42.507	42.496	42.502	0.005346
14	13	42.088	6.243	8.438	42.479	42.529	42.504	-0.02481
15	14	42.008	6.759	9.141	42.461	42.547	42.504	-0.0432
16	15	41.921	7.274	9.844	42.462	42.544	42.503	-0.04111
17	16	41.827	7.788	10.547	42.482	42.521	42.502	-0.01954
18	17	41.726	8.300	11.250	42.511	42.488	42.500	0.011304

Fig. 17. Radial errors of the rotary table spindle

If Sx(θ) is the rotary table error along the probing direction and R(θ) is the standard ring error, then, in step 1 (Fig. 16a), the probe T4 reads a measured signal m1(θ) given by:

$$m_1(\theta) = R(\theta) + S_x(\theta) \tag{5}$$

In the step 2 (Fig. 16b), the probe T3 reads a measured signal m2(θ) given by:

$$m_1(\theta) = R(\theta) - S_x(\theta) \tag{6}$$

The standard ring and rotary table errors can be obtained from Eqs. (5) and (6):

$$R(\theta) = (m_1(\theta)) + (m_2(\theta))/2$$

$$Sx(\theta) = (m_1(\theta)) - (m_2(\theta))/2 \tag{7}$$

6 Conclusions

In this paper was presented a multifunctional equipment for bearing measurement. This equipment meet the following requirements: rapid and precise calibration; a small number of standards for calibration; rapid adaptation to the requirements of the manufacturing process; individual adjustment of the probes in radial and axial direction; multi-parametric control of a wide range of bearing rings; simultaneous measurement of multiple parameters in multiple horizontal planes; decisional capacity based on predetermined criteria; statistical calculation program for control and streamlining of manufacturing process.

To increase measurement accuracy the measuring range was divided into three areas, for each of them being determined the calibration parameters. To increase the accuracy of the roundness measurement was performed separation of part errors from rotary table errors.

Acknowledgements. This work was co-financed by the European Regional Development Fund, through the Competitiveness Operational Programm under the project POC-A1-A1.1.4-E-2015 "Knowledge Transfer Partnerships to Enhance Business Competitiveness in the Field Automotive Industry and Components and Improve Road Traffic Safety – KTAutoComp", subsidiary contract 2791/2018.

References

1. Lancaster, A., Dury, M.: Good Practice Guide No. 148. Measurement of the Surface Texture of Large Roller Bearings. The National Physical Laboratory (NPL), 2017 Version 1.0 (2017)
2. Luciana, C.: The improvement of performances in automatic dimensional inspection for bearing production, an important way to quality assurance in mechanical engineering. In: Proceedings of the 8th WSEAS International Conference on Instrumentation, Measurement, Circuits and Systems, pp. 79–82 (2009)
3. Bucur, G., Moise, A., Pană, I., Orhei, D.: Integrating PLCs in automatic systems for dimensional inspection. CNC bearings rings inspection. In: Electronics, Computers and Artificial Intelligence, 8th edn. (2016). https://doi.org/10.1109/ecai.2016.7861186
4. Schaeffler Group Industrial: Super Precision Bearings. Spindle bearings. Super precision cylindrical roller bearings. Axial angular contact ball bearings (2011)
5. Muralikrishnan, B., Raja, J.: Computational Surface and Roundness Metrology. Springer, London (2009). https://doi.org/10.1007/978-1-84800-297-5. ISBN 9781848002968
6. Cioboata, D., Dontu, O., Besnea, D., Ciobanu, R., Soare, A.: Mecatronic equipment for bearing ring surface inspection. Rom. Rev. Precis. Mech. Opt. Mechatron. **2015**(48), 262–266 (2015)
7. Cioboată, D., Palade, D.-D., Stanciu, D., Abălaru, A.: Considerations regarding roundness measurement of closed profiles and open profiles. Maaa. **75**(2), 129–140 (2013)

Constructive Solution for Multi-filament 3D Printing

Daniel Besnea[1(✉)], Octavian Dontu[1], Edgar Moraru[1], Ciprian Rizescu[1], Gheorghe I. Gheorghe[2], and Elena Dinu[1]

[1] University Politehnica of Bucharest, Splaiul Independentei no. 313, District 6, Bucharest, Romania
d_bes@yahoo.com, octavdontu@yahoo.com, eddy_milan91@yahoo.com, elena_dinu@yahoo.com, ciprian.rizescu@upb.ro

[2] National Institute of Research and Development in Mechatronics and Measurement Technique, Pantelimon Road, no. 6-8, District 2, Bucharest, Romania
incdmtm@incdmtm.ro

Abstract. The article presents a new 3D printer constructive model using an extruder that can work with three types of thermoplastic filaments as raw material. The designed and realized 3D printer has the ability to produce multilayer multi-colour parts of materials with different properties and characteristics due to the extrusion head with a special construction - diamond type, which allows the control and combination of the three types of filaments.

Keywords: Additive technologies · Multi-filament 3D printing · Thermoplastic extrusion

1 Introduction

The latest years have brought new developments to these new technologies, which makes it possible today to find a wide variety of products manufactured by these new technologies and many in development, making them more and more competitive in wider areas. Potential users of additive technologies are asking themselves questions before choosing one or another technology: which process is best suited to a particular application, what kind of raw material the piece will be constructed, which will be the physical and mechanical properties of the product, how will it be the precision and quality, how much it will cost and how long the 3D model can be realized.

As any technology, additive technologies undergo optimization processes, the more they are at the beginning of the road. Optimization is primarily aimed to increase the efficiency of using technologies in manufacturing a product by reducing manufacturing times, and secondly by improving the quality of at least some of the surface of the piece. Also, a lot of emphasis is put on the aesthetic aspect and personalization, and the additive technologies are very good at this point of view [1].

In the paper, the authors designed and realized a 3D printer with possibility of multicolour printing due to a new model of the extruder – diamond type, represented in

G. I. Gheorghe (Ed.): ICOMECYME 2019, LNNS 85, pp. 118–123, 2020.
https://doi.org/10.1007/978-3-030-26991-3_11

Fig. 1. The system has three independently controlled filament lines that are combined into a single nozzle output of 0.4 mm. Among the advantages of this printer can be called easy calibration, higher precision than printing with multiple extruders, printing with different colours and minimal waste material. The key feature of this printer is the diamond-shaped extruder. It is designed to have the smallest possible mixing chamber to make colour changes as fast as possible and to avoid unnecessary use of material. Also, to ensure a quick and efficient heating of the nozzle, it has been built as compact as possible. The paper will present the printing possibilities of this new type of 3D printer with the features of the parts obtained and the software used [2].

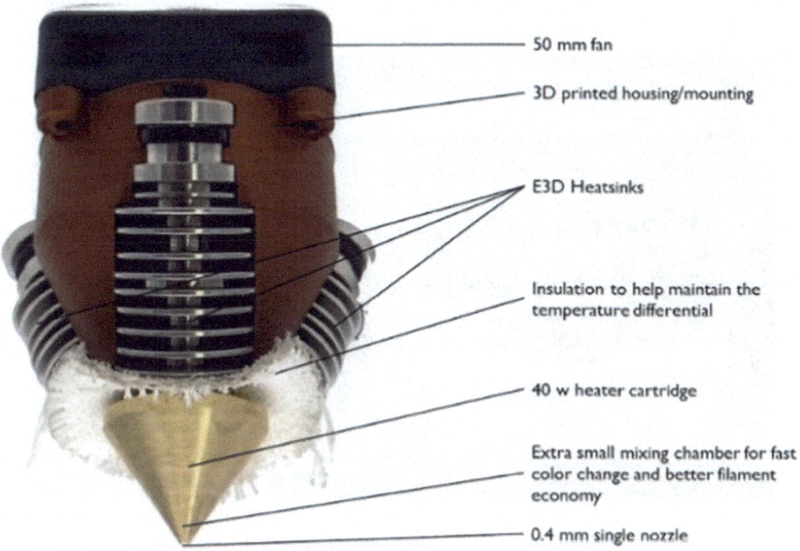

Fig. 1. Diamond type extruder [2]

2 Materials and Methods

The 3 in 1 extruder, as its name suggests, has three independent feed channels, with a single supply chamber at the end of the three feed channels. On the basis of the principles of the three primary colours, the three filaments of different colours will be fed according to a certain feed rate, given by step by step motors of the leading of filaments, then they will be melted and mixed into the nozzle chamber to generate another colour or mixed colour. Among the main features of the printer are: the 3 in 1 diamond-shaped extruder supports 3 filaments of different colours and different materials, creating pieces with a wide range of colours and materials, the 3 in 1 diamond print head is made of a high temperature resistant metallic material required in the printing process (max. 240 °C), so it is possible to print multiple materials such as ABS, PLA, PETG, Nylon and more others. The designed and constructed 3D printer is shown in Fig. 2. As the command module, the Arduino MEGA 2560 was chosen for projects that require multiple I/O lines, more memory and more RAM. With 54 digital I/O pins, 16 analogue inputs and a larger program space is the recommended board for 3D printers and robotics projects [3].

3 Experimental

The 3D printer used in the article together with the computing unit is shown in Fig. 2. The additive technology [4, 5, 7, 8] by thermoplastic extrusion (FDM) was used in the study, and three filaments of the same material - PLA (polylactic acid) of different colours (Fig. 3).

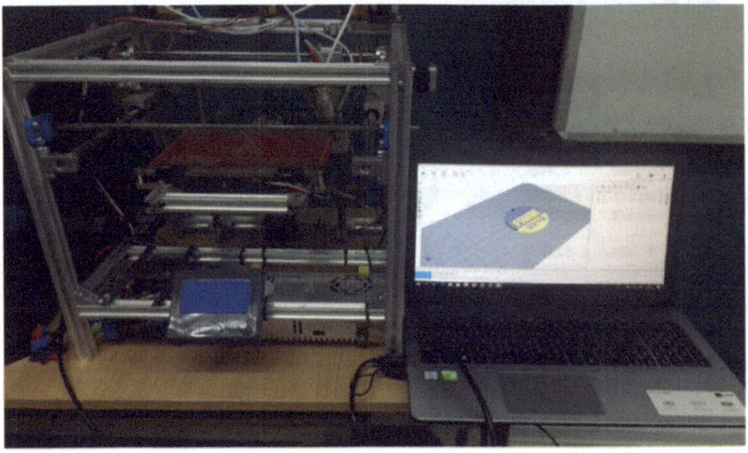

Fig. 2. Multi-filament 3D printer with PC

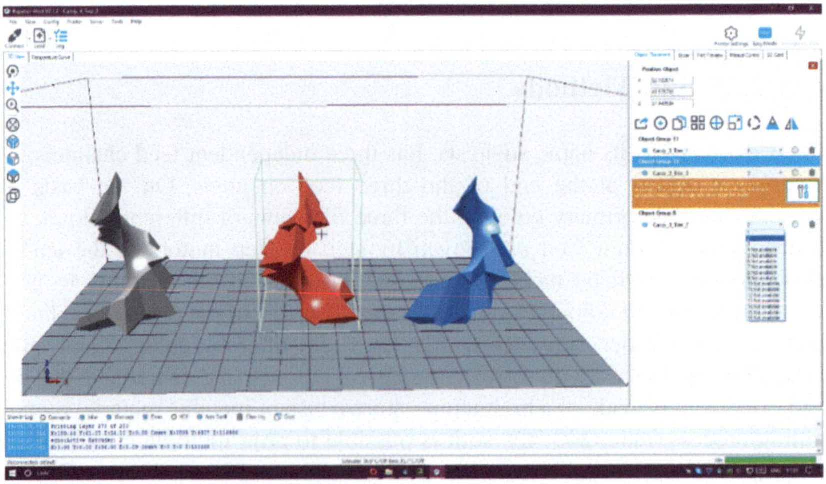

Fig. 3. Repetier Host program interface

Repetier-Host has been used as a software that has a simple interface that makes USB connection to the 3D printer possible, displaying pieces in .stl format and transforming into a GCode file using Slic3r. Repetier-Host has a friendly graphic interface and offers more options for controlling and displaying .stl and GCode formats, the program also having the option to directly save the code generated on the SD memory card. The software also offers the possibility to print multiple models simultaneously, as well as their positioning on the work platform (Fig. 4).

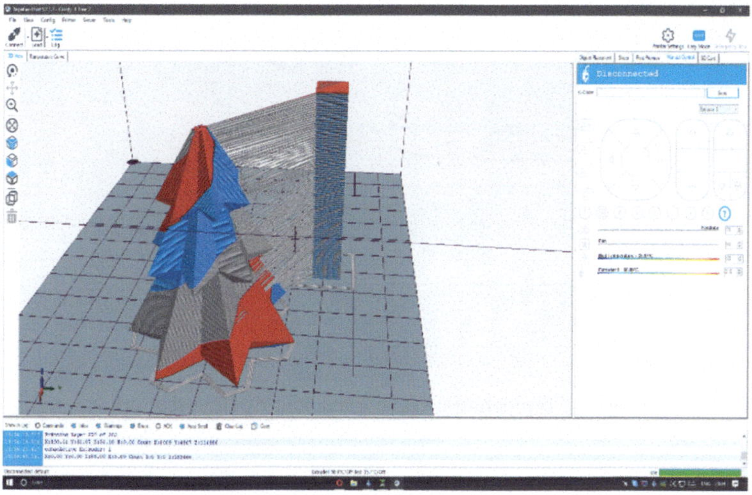

Fig. 4. Simulation of print process from the program

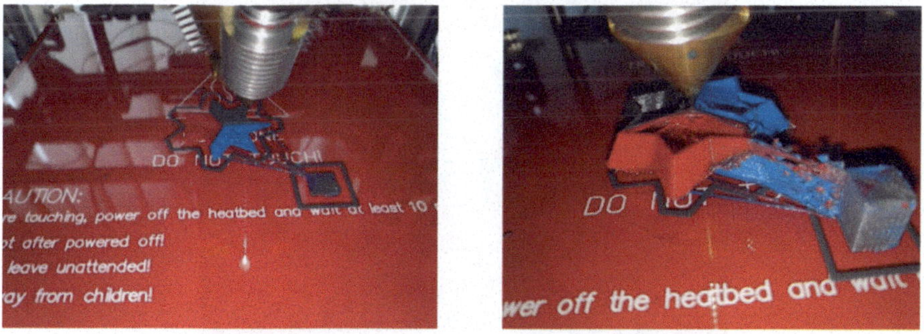

Fig. 5. Steps of realizing a piece on the 3D multi-filament printer

It can be seen from Fig. 5 that a sacrificial area is constructed where the extruder changes its material or colour when needed. From the program, it is also possible to control and visualize bed and extruder temperature according to time (Fig. 6). In Fig. 7 there are various pieces made on this three-filament printer, the resulting diamond type extruder having several advantages over multiple extrusion systems. The unique nozzle eliminates the calibration problems associated with separate nozzles for each colour, eliminating the non-productive time with replacing the filaments. The extruder also has the ability to combine colours or materials together to create new combinations, allowing more than three colours to print.

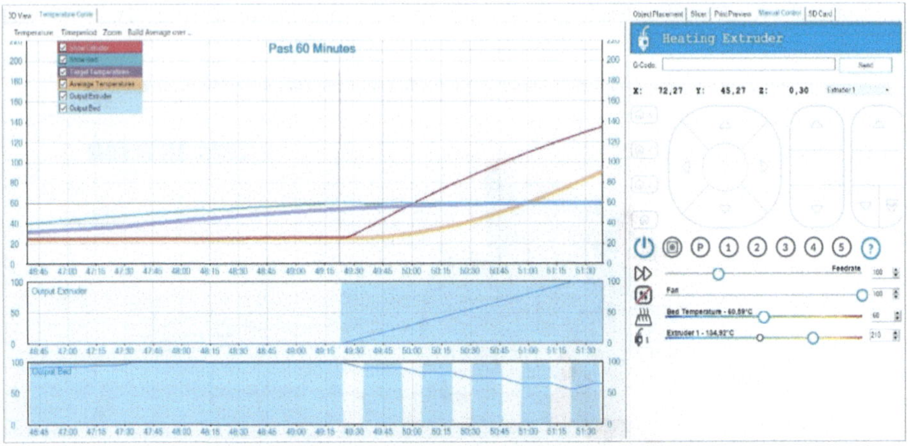

Fig. 6. Temperature curve generated by the program

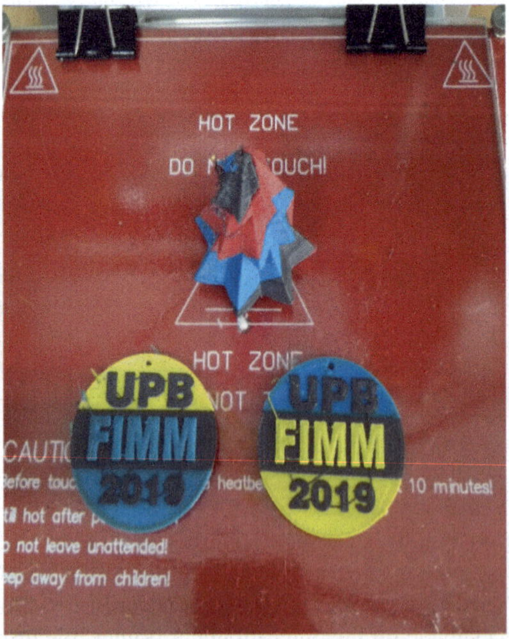

Fig. 7. Parts realized on the 3D printer with 3 filaments

4 Conclusions

In recent years, the issue of printing multicolour pieces has come to the forefront. One of the major problems is that traditional extrusion methods are based on printing using monochrome thermoplastic filaments. However, creators and innovators have designed a range of solutions to allow manufacturers to print in different colours through this technology. In the paper was presented such a constructive solution designed and realized to have the possibility of multicolour printing through FDM technology. So far, the main use of 3D printing has been the creation of prototypes for components and products. These physical models are important because they provide information that a simulation cannot replace. However, most physical objects contain many colours and materials. This makes the most common 3D printing technologies inadequate to produce them. It also means that the prototyping process requires extra time to extend time and increase spending. Multi-colour printing is a step towards creating pieces and products with the desired features and aspect. This multicolour printing approach makes the printing process more efficient and allows for more detailed models. Model colouring during production also contributes to reducing the number of post-processing steps, increasing cost efficiency. Also, with FDM thermoplastic extrusion technology, cost-effective prototypes can be achieved at a much lower cost than other additive technologies with possibility of multicolour printing like binder jetting technology [6].

References

1. Berce, P., Balc, N., Pacurar, R., Bratean, S., Caizar, C., Radu, S.A., Fodorean, I.: Tehnologii de fabricatie prin adaugare de material si aplicatiile lor. Editura Academiei Romane, Bucuresti (2014)
2. https://reprap.org/
3. https://www.arduino.cc/
4. Besnea, D., Rizescu, D., Rizescu, C., Dinu, E., Constantin, V., Moraru, E.: Additive technologies and materials for realization of elastic elements. Int. J. Mechatron. Appl. Mech. **3**, 249–254 (2018)
5. Dontu, O., Ioana, M.C., Tanase, B., Baran, N., Gheorghe, I.G., Moraru, E.: Researches regarding the use of additive technologies in the construction of water aeration elements. Int. J. Mechatron. Appl. Mech. **3**, 7–12 (2018)
6. https://www.lboro.ac.uk/research/amrg/about/the7categoriesofadditivemanufacturing/binderjetting/
7. Besnea, D., Gheorghe, G.I., Dontu, O., Moraru, E., Constantin, V., Moga, I.C.: Experimental researches regarding realization of wastewater treatment elements by means of modern technologies. Int. J. Mechatron. Appl. Mech. **4**, 61–65 (2018)
8. Rizescu, C.I., Besnea, D., Rizescu, D., Moraru, E., Constantin, V.: Mechanical analysis of leaf springs realized by additive technologies. Lecture Notes in Mechanical Engineering, pp. 307–318 (2019)

Thermal Analysis of Some Prosthetic Dental Biomaterials Processed by Selective Laser Melting

Lyubov Shpakova[1], Gheorghe Ion Gheorghe[2], Constantin Nitu[3], Octavian Dontu[3], Edgar Moraru[3(✉)], Daniel Besnea[3], and David Dragomir[3]

[1] Karaganda State Technical University, Nursultan Nazarbayev Avenue, No. 56, Karaganda, Kazakhstan
lu.shpakova@ya.ru

[2] National Insitute of Research and Development in Mechatronics and Measurement Technique, Pantelimon Road, No. 6-8, District 2, Bucharest, Romania
incdmtm@incdmtm.ro

[3] University Politehnica of Bucharest, Splaiul Independentei No. 313, District 6, Bucharest, Romania
constantin_nitu@yahoo.com, octavdontu@yahoo.com, eddy_milan91@yahoo.com, d_bes@yahoo.com, david_dragomir@yahoo.com

Abstract. The paper presents a new technique for obtaining dental prostheses with special properties – selective laser deposition. A mathematical model will be analyzed regarding the thermal characteristics of the additive technology by selective laser melting, which will deduce the temperature, the thermal velocity and the thermal acceleration during the process. Three different biocompatible materials - titanium alloy, Co-Cr alloy and stainless steel will be considered. It will also study the influence of important parameters on thermal characteristics in order to obtain prostheses with conforming properties.

Keywords: Additive technologies · Dental prosthesis · Selective laser melting

1 Introduction

The use of powdered materials as a raw material in the manufacturing processes by adding material has been a huge advance in the ever more efficient and diversified use of these new technologies. These technologies are similar to other additive technologies from the point of view of the working principle, with the exception that the raw material is in the form of powder and the energy source used to materialize a section by melting the particles is a high power laser beam. Selective laser melting is a superior stage of laser selective sintering. The selective laser melting process also has a three-dimensional model as a starting point, but uses as a power source a more powerful laser beam. The raw material is the metallic powder of great diversity (alloy steel, stainless steel, titanium, etc.) and a fine granulation. The process begins with sectioning the

G. I. Gheorghe (Ed.): ICOMECYME 2019, LNNS 85, pp. 124–132, 2020.
https://doi.org/10.1007/978-3-030-26991-3_12

piece between 20–100 μm, creating 2D images for each section, and transferring them to a fabrication preparation file. The virtual 3D model materializes with the first section and the first layer of powder deposited on the table. The process is repeated until the object is completed in volume. Selective laser melting systems are equipped with 200 W high power lasers and even more. This makes the process and manufacturing steps different from those known from selective laser sintering, meaning that the metallic powder is not only partially sintered, but melted, and the bonding of the particles between them is made in the melted state due to the larger laser power and due to the smaller thickness of a section, compared to selective laser sintering. Consequently, the resulting piece is obtained in a single operation (no further sintering operation is required in the furnace), with a density of almost 100% and physico-mechanical properties practically identical to the properties of the same parts manufactured by conventional technologies. The applications of this new technology are very broad and result from the fact that by its use can be manufactured functional parts with a density of almost 100%. It is possible to manufacture parts with particularly complex and detailed geometric shapes, which are non-technological or even impossible to achieve through classical technologies [1]. A field of extremely promising and important applications is the medical one, namely dental prosthetics implantology, where this technology creates the possibility of obtaining implants or prostheses from biocompatible materials (titanium alloys, cobalt, etc.). Such a prosthesis obtained by laser selective melting is shown in the Fig. 1. After realization of the prosthesis, it has to go through a post-processing stage that consists of surface finishing [2].

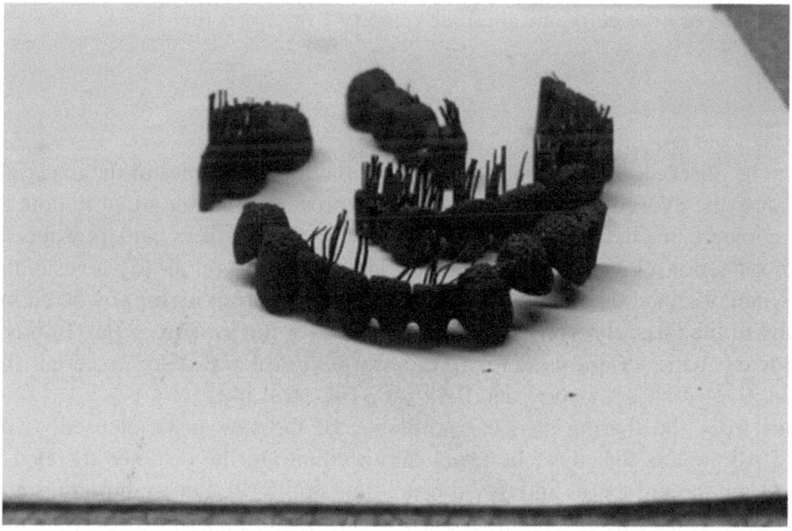

Fig. 1. Dental prostheses made by selective laser melting in rough condition [2].

In order to estimate temperature and other thermal characteristics during the laser selective melting process, the mathematical model of heat transfer should be analysed. This analysis should be performed to investigate the influence of important process parameters on the temperature distribution in the area of interest in order to obtain optimal conditions for achieving biocompatible prostheses with appropriate properties.

2 Mathematical Model

The most of the theoretical approaches generally used to estimate temperatures during this additive manufacturing method are based on Rosenthal's primary work [3]. Rosenthal proposed a method to achieve the temperature at a mobile energy source assuming several hypotheses: the characteristics of the material do not depend on temperature, the energy and laser velocity is considered to be constant [3, 4]. Then the heat transfer solution can be expressed by next equation:

$$T = T_0 + \frac{nQ}{2\pi kr} e^{-\left(\frac{v(x+r)}{2\alpha}\right)} \tag{1}$$

$$r = \sqrt{x^2 + y^2 + z^2} \tag{2}$$

Where n represents the level of laser absorption by the material, Q - power, v –velocity of source, k, α - thermal features of the material, x, y, z - coordinates in a moving coordinate system, r - distance from the heat source, T_0 – initial temperature.

This approach can also be used to calculate the thermal velocity, differentiating according to time [3, 4].

It is also possible to introduce in this field the thermal acceleration, which determines the rate of change of the thermal velocity over time, that is:

$$T'' = \frac{\partial^2 T}{\partial t^2} \tag{3}$$

It can be observed from above relations that thermal properties of the material affect significantly the evolution of temperature of the process. It is important to note that we cannot compare the level of laser absorption for flat surfaces and powders due to multiple reflection phenomena in the latter [4]. Rubenchik et al. [5] investigated this problem and obtained the relationship between laser absorptivity for powdered material according to the laser absorptivity of a flat surface for the same material. In this study, the following values were used for laser absorptivity of powdered material: 0.71 for TiAl6V4, 0.64 for Co-Cr alloy and 0.68 for 316L steel [4, 5].

Apart from the thermo-physical properties of the raw material used, two other essential parameters that must be taken into account are the distance traveled by the laser between the pulses (p) and its exposure time (t_e) [5–9]. For a continuous emission system, the two parameters mentioned above are substituted with laser velocity [4].

The method developed by Rosenthal [3, 4] expressed through the relationship (1) can be applied to describe the heat transfer phenomenon during the additive process through selective deposition. However, this method can be put into practice just for a moving laser. Bajaj [4] assumed that if the interval traveled by the laser between pulses is insignificant compared to the interval over which the heat diffuses during exposure, it can be assumed that the laser moves continuously. By making this assumption, it is possible to replace Eq. (1) with the Eq. (4) for non-continuos laser type starting from the following considerations:

- the apparent or real velocity (v_a) is introduced which is equal to the ratio between the length of the crossed laser when it is inactive and the sum of the exposure time and the inactive time
- the power is substituted with the apparent power (Q_a) equal to the ratio of the nominal power product to the exposure time and the sum of the active and inactive times.
- the model described above will need to be transformed in terms of time and penetration depth (z). Therefore, coordinate on x can be expressed as the product of apparent speed and time, i.e. $x = -v_a t$. Assuming the laser moves only on the ox axis, then the value for y in relation (2) is zero. The general relation (4) results from the inclusion of the relation (2) in relation (1) and according to the described considerations.

This relationship can be applied to estimate the temperature at a certain point depending on the penetration depth and time after the laser has left this zone.

$$T = T_0 + \frac{nQ_a}{2\pi k\sqrt{(-v_a t)^2 + z^2}} e^{-\frac{v_a\left(-v_a t + \sqrt{(-v_a t)^2 + z^2}\right)}{2\alpha}} \tag{4}$$

3 Results

Equation (4) can be differentiated with respect to time to obtain the thermal velocity. This relationship and differential equations were solved in the Matlab Simulink environment, and the block diagram of the mathematical relationship is illustrated in Fig. 2. Table 1 shows values for the thermal and physical properties of the materials used in the calculation.

The following are assumed as initial data: $Q_a = 250$ W, $z = 5 \cdot 10^{-5}$ m, $v_a = 4$ m/s.

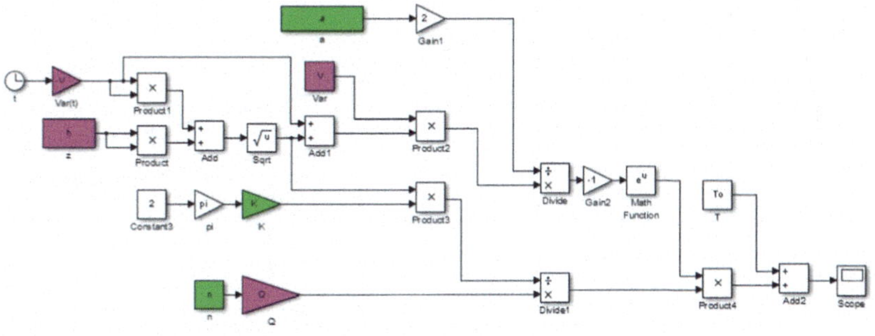

Fig. 2. Block diagram of the mathematical model for heat transfer during the selective laser melting process

After solving Eq. (4), it can be determined the temperature over time during the selective laser deposition. Figure 3 shows the temperature over time for different materials - titanium alloy, chromium and cobalt alloy, austenitic 316L stainless steel.

Table 1. Properties of materials used in the study

Material	Laser absorbtivity n [-]	Thermal diffusivity α [m^2/s]	Thermal conductivity k [W/m $\cdot$ K]	Melting point T_m [K]
TiAl6V4	0.71	$3.02 \cdot 10^{-6}$	7.2	1878
Co-Cr	0.64	$3.73 \cdot 10^{-6}$	9.4	1603
316L	0.68	$3.69 \cdot 10^{-6}$	15	1673

Also on the graph are the values for the melting temperature for each material. It can be seen that at a laser velocity of 4 m/s and a power of 250 W, the 50 micron powder layer will melt in the case of TiAl6V4 and Co-Cr, the curves exceeding the melting temperature for the respective materials.

Instead, the 316L powder will not melt under these conditions to a maximum of 1250 K, with more than 400 K under its melting temperature. This means that this material requires more energy to melt, that is, a higher laser power.

It can be seen that the laser power is directly proportional to the temperature during the selective laser melting process. Stainless steel, compared to the other two investigated materials, requires a higher power of the laser to melt, the calculations showing that only at a power of 300 W the temperature approaches the melting point of the material.

The laser velocity is very important during the process, the temperature rising as the laser velocity decreases. With the help of the laser velocity, it can also be compensated by the energy, decreasing the speed decreases and the power of the laser. However, special attention must be paid to this aspect; the speed must not be too low, and it can reach very high temperatures - even beyond the boiling point of the materials [4, 9].

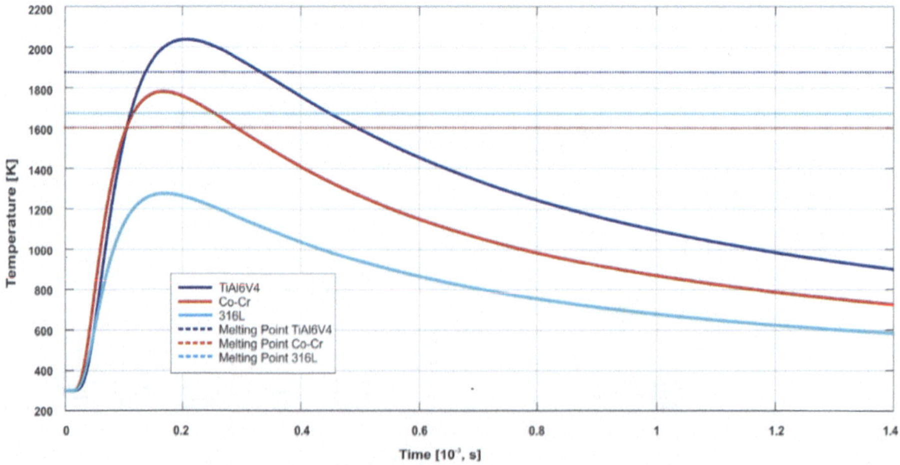

Fig. 3. Temperature vs time for different materials during the selective laser melting process

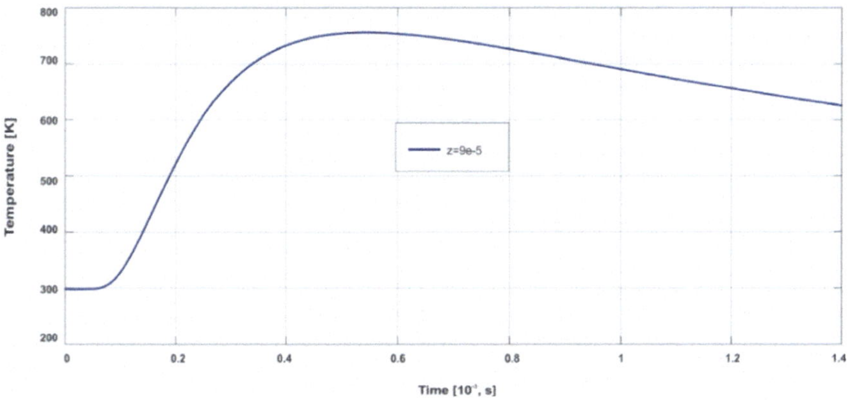

Fig. 4. Temperature vs. time for Co-Cr alloy at z = 90 μm

Regarding the penetration depth, with its increase, the temperature decreases. For example, for the Co-Cr alloy, for a layer thickness of 40 μm to about 10^{-4} s, the maximum temperature is almost 2600 K and for a layer thickness of 90 μm (Fig. 4) the maximum temperature of nearly 750 K is obtained at $5 \cdot 10^{-4}$ s. To find out the cooling rate or how the material cools, the Eq. (4) must be differentiated in relation to time. The solutions of the equation were also obtained in the Matlab environment. The results are presented in Fig. 5. From Fig. 5 it can be seen that the maximum thermal velocity is reached at about $0.5 \cdot 10^{-4}$ s for stainless steel at a value of $1.45 \cdot 10^7$ K/s and the titanium alloy at a value of $2.23 \cdot 10^7$ K/s. The Co-Cr alloy reaches a maximum of $2.2 \cdot 10^7$ K/s later at $0.6 \cdot 10^{-4}$ s.

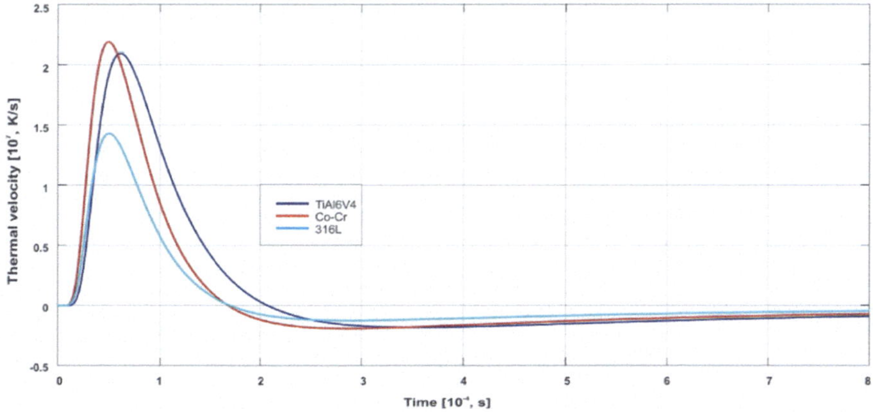

Fig. 5. Cooling rate for different metals

The graph can be explained as follows: the material suddenly heats to reach the maximum thermal velocity, after which the material is heated, but with lower thermal velocity values. The negative values in the graph already show that the material has

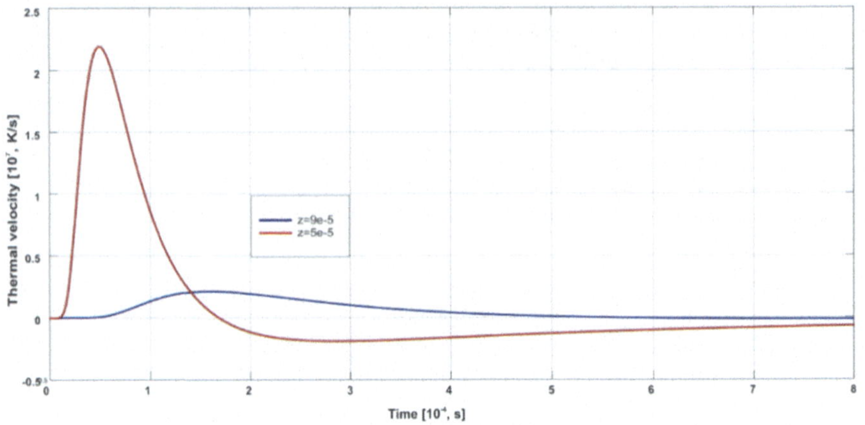

Fig. 6. Cooling rate at different thicknesses for Co-Cr alloy

started to cool. Cooling takes place slowly and gradually compared to heating until the 0 is reached for the thermal velocity, so the temperature stabilizes (Fig. 5).

As in the case of temperature, the laser power has a similar influence on the thermal velocity or the cooling rate of the materials. The laser power influences proportionally the thermal characteristics of the process, while the influence of laser velocity and thickness of the powder layer is described by more complex physical laws. By changing the laser power, reaching the maximum value of the thermal velocity almost does not change in terms of the time it has reached, and the material is cooled similar and uniformly for different powers of the applied laser. Instead, changing the laser velocity will result in an absolutely different behaviour of the cooling rate over time.

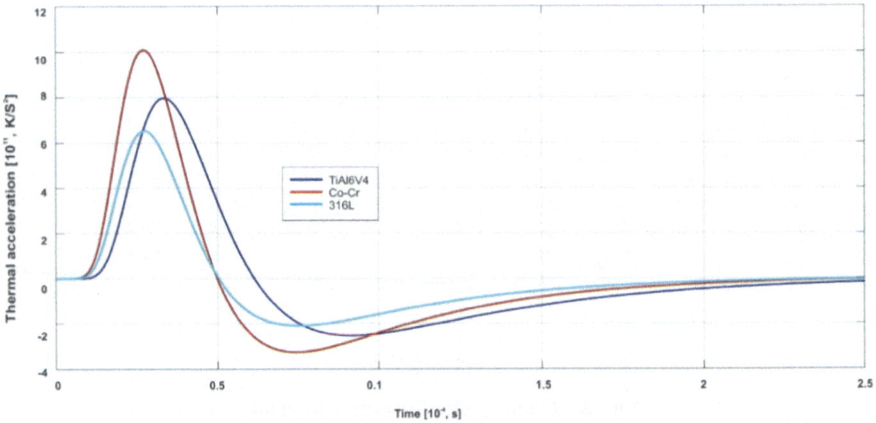

Fig. 7. Thermal acceleration vs. time for different materials

If maximum values are obtained almost at the same time, the way the material is cooled at different laser velocities is different. Regarding the layer thickness or the laser penetration depth, the maximum thermal velocity is obtained earlier as the layer is thinner. This is due to the fact that the inertial properties are proportional to the dimensional characteristics.

Figure 7 represents the solutions for the differential equation for thermal acceleration during the selective laser melting process.

Very high values of the order 10^{11} K/s^2 for thermal acceleration are obtained, the titanium alloy having the highest maximum of this characteristic at heating (over 10^{12} K/s^2 at $0.35 \cdot 10^{-4}$ s) and cooling ($-3 \cdot 10^{11}$ K/s^2 at $0.7 \cdot 10^{-4}$). But over time, the average Co-Cr cooling thermal acceleration value is higher. Stainless steel has the lowest heating and cooling accelerations. The power and laser velocity have a similar influence on thermal acceleration, as in the case of other thermal investigated characteristics Fig. 8 shows the thermal acceleration at different thicknesses for titanium alloy processed material.

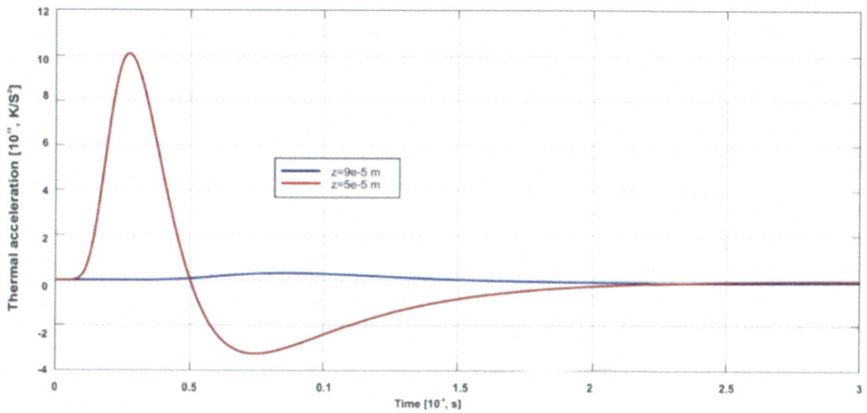

Fig. 8. Thermal acceleration at different thicknesses for TiAl6V4

4 Conclusions

The article presented a method for determining the temperature field, thermal velocity and acceleration during this metallic additive technology. According to the modelling described, it can be concluded that the properties of the powders used significantly influence the temperature fields, the study of material properties being very important. Titanium alloy and cobalt-chromium alloy have better general properties and are easier to process compared to stainless steel that needs more power to be manufactured under optimal conditions. This analytical modelling can be useful for prognosis of the temperature over time in a certain point, thermal velocity and thermal acceleration. It is also possible to study the influence of the main parameters that affect this process in order to achieve finite parts with improved features. This approach can be explored to develop certain process parameters in order to produce components with specific microstructures

[4]. The cooling rate and thermal acceleration resulting from solving differential equations with values above 10^7 K/s and over 10^{11} K/s^2 will result in the formation of structures with very good chemical homogeneity and some better mechanical properties than structures obtained by conventional fabrication methods [10].

References

1. Berce, P., Balc, N., Pacurar, R., Bratean, S., Caizar, C., Radu, S.A., Fodorean, I.: Tehnologii de fabricatie prin adaugare de material si aplicatiile lor. Editura Academiei Romane, Bucharest (2014)
2. Moraru, E., Dontu, O., Petre, A., Vaireanu, D., Constantinescu, F., Besnea, D.: Some technological particularities on the execution of dental prostheses realized by selective laser deposition. J. Optoelectron. Adv. Mater. **20**(3–4), 208–213 (2018)
3. Rosenthal, D.: The theory of moving sources of heat and its application to metal treatments. Trans. ASME **68**, 849–866 (1946)
4. Bajaj, P.: SLM manufacturing of molybdenum. Disertation, The University of Sheffield (2016)
5. Rubenchik, S., Boley, C.A., Mitchell, S.C., Wu, S.S.Q.: Metal powder absorptivity: modeling and experiment. Appl. Opt. **55**(23), 6496–6500 (2016)
6. Kruth, J.P., Wang, X., Laoui, T., Froyen, L.: Lasers and materials in selective laser sintering. Assem. Autom. **23**(4), 357–371 (2003)
7. Huang, Y., Yang, L., Du, X., Yang, Y.P.: Finite element analysis of thermal behavior of metal powder during selective laser melting. Int. J. Therm. Sci. **104**, 146–157 (2016)
8. Fu, G., Zhang, D., He, N., Mao, Z., Zhang, K.: Finite element analysis of interaction of laser beam with material in laser metal powder bed fusion process. Materials **11**(5), 765 (2018)
9. Du, Y., You, X., Guo, L., Liu, Z.: A model for predicting the temperature field during selective laser melting. Results Phys. **12**, 52–60 (2019)
10. Qiu, C., Al Kindi, M., Aladawi, A.S., Al Hatmi, I.: A comprehensive study on microstructure and tensile behaviour of a selectively laser melted stainless steel. Nat. Sci. Rep. **8**, 7785 (2018)

Fabrication Technologies of Aeration Systems for the Ecological Treatment of Wastewater

Edgar Moraru[1(✉)], Daniel Besnea[1], Octavian Dontu[1], Gheorghe I. Gheorghe[2], Ioana Corina Moga[3], and Georgiana Elena Popescu[1]

[1] University Politehnica of Bucharest, Splaiul Independentei, no. 313, District 6, Bucharest, Romania
eddy_milan91@yahoo.com, d_bes@yahoo.com, octavdontu@yahoo.com, georgiana_ela_popescu@yahoo.com

[2] National Institute of Research and Development in Mechatronics and Measurement Technique, Pantelimon Road, no. 6-8, District 2, Bucharest, Romania
incdmtm@incdmtm.ro

[3] DFR Systems SRL, Drumul Taberei, no. 46, District 6, Bucharest, Romania
corinamoga@yahoo.com

Abstract. The paper presents new technological solutions for realization of aerators used for ecological treatment of wastewater. It was presented the CAD-CAM design steps of the plates with microorifices in order to optimize their further processing on a CNC machining center. The diffussers on which the plates are mounted were obtained by additive technologies - the FDM process. Also, it was presented a new process for microorifices cleaning.

Keywords: Additive technologies · Aeration diffusers · Wastewater treatment

1 Introduction

Wastewater treatment requires the use of advanced processes in sewage treatment plants that use modern, reliable and efficient technologies and equipment. The biological or ecological treatment process is particularly complex and solving it involves a series of physical, chemical, biochemical and hydraulic phenomena. The main goal today is to improve these particularly important processes, and the optimization of technologies and exploitation of sewage treatment plants can be analyzed from at least four points of view: (a) minimizing the costs of exploitation processes and installations; (b) optimizing the energy consumption of installations; (c) minimizing the consumption of reagents and materials; (d) achieving maximum water purification performance at minimal cost. It is obvious that in the end what matters is the costs and therefore the correct evaluation of them will lead to a well-defined and driven process [1]. Previous studies [2, 3] presented a system of aeration/oxygenation of wastewater formed mainly: water tank, compressor, process parameter measurement and control system and the essential system - the fine bubble generator. In this study, the fine bubble generator is

G. I. Gheorghe (Ed.): ICOMECYME 2019, LNNS 85, pp. 133–141, 2020.
https://doi.org/10.1007/978-3-030-26991-3_13

designed as follows: the plates will have the dimensions of 360 × 90 × 6 mm with microorifices diameter under one millimetre. The plate will be mounted on a cassette realized by additive technologies [5]. This process has been chosen for making the cassettes because of the low price and execution time (Fig. 1).

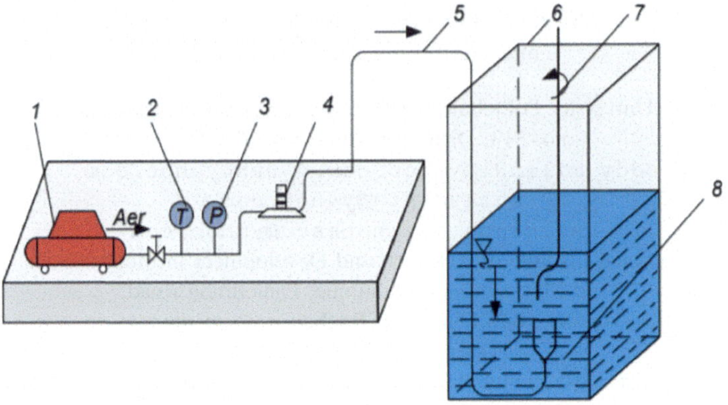

Fig. 1. Installation for wastewater oxygenation: 1 - Air compressor; 2 - Thermometer; 3 - Manometer; 4 - Rotameter; 5 - Feed pipe; 6 - Water tank; 7 - Oxygenometer; 8 - Diffuser [2]

2 Materials and Methods

As already accepted worldwide, CAD (Computer-Aided De-sign) is computer-assisted technology to help design, analyze, and optimize a product. So any computer program that includes computerized graphics and an application program that facilitates engineering functions from the design process is classified as CAD software. Computer-Aided Manufacturing (CAM) is computer-assisted technology to plan, manage and control manufacturing operations through a direct or indirect interface of the computer with the enterprise's production resources. One of the most important areas of the CAM is the numeric or NC command. This is the technique of using the programmed instructions to control the manufacturing systems.

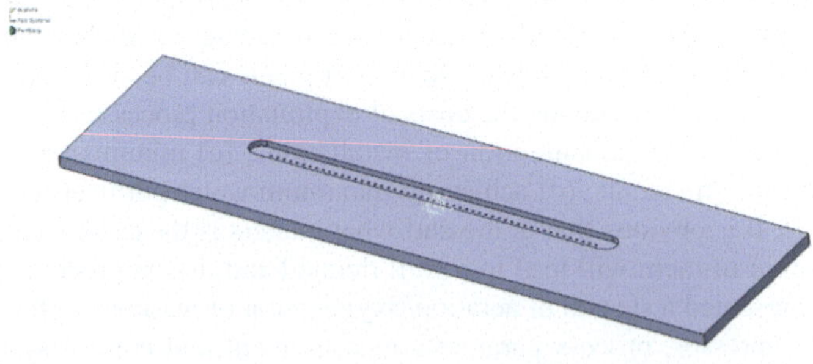

Fig. 2. Three-dimensional model of the plate with microorifices

The practical application of CAD/CAM software will be presented - in the case of 3D design of aluminium alloy plates to be processed on a CNC machining center, as well as 3D design for the implementation of cassettes by means of additive technologies on which will be mounted the plates. Figures 2 and 3 show the CAD model for the microorifces plate and the new cassette model with the possibility of interchangeability of the plates.

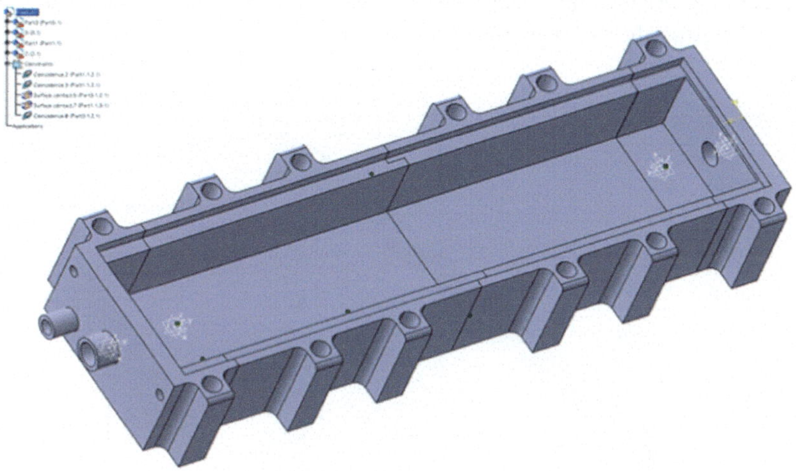

Fig. 3. CAD model of the cassette

For the execution of plate holes on CNC machining centers it is necessary to generate the program in machine language. This technological process has been chosen due to the higher execution precision compared to the electro-erosion process, the reduced execution time and the reduced costs.

Fig. 4. Check of the tool trajectory for the milling operation

Also, with the same design software, where three-dimensional CAD models [4] have been obtained, the milling technology process can be simulated in order to optimize the finished product and to correct the possible execution errors. Figure 4 shows the tool trajectory for the milling operation. The steps to be taken are quite similar for all types of operations of the advanced manufacturing module in the program. For example, for the realization of microorifices it is necessary to perform the following steps: centering the holes, choosing the centering tool, choosing the drill (Fig. 5), defining the drilling operation, simulating and visualizing the drilling process (Fig. 6), generating the G code (Fig. 7).

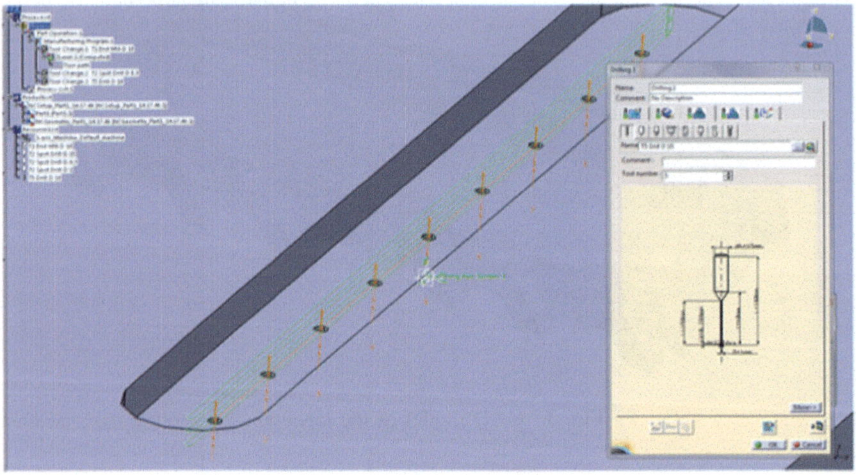

Fig. 5. Choosing of the drill

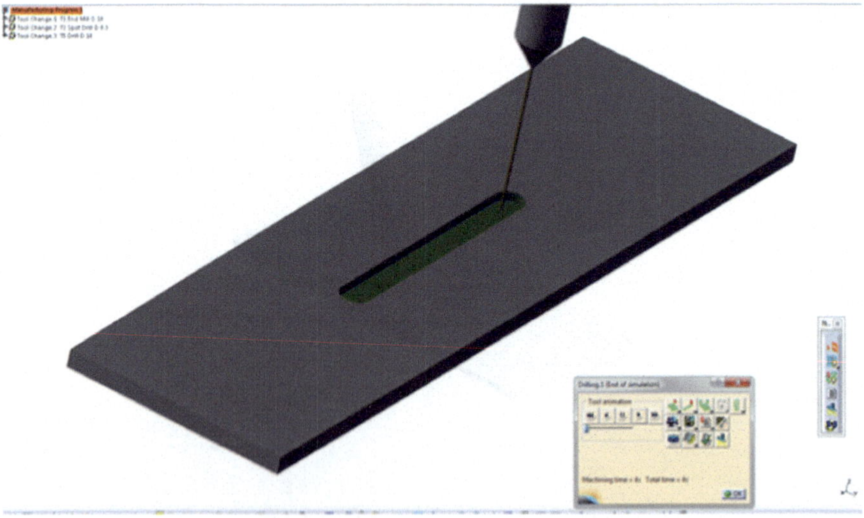

Fig. 6. The visualization of drilling process

```
:0006 (piesa 6)
N1 G49 G64 G17 G80 G0 G90 G40 G99
N2 Z100.
N3 T0001 M6
N4 X-45.2157 Y.0381 S4000 M3
N5 G43 Z15. H1
N6 G1 Z5. F300.
N7 X-45.1931 Y-.106 F1000.
N8 X-45.1347 Y-.1747
N9 X-44.8206 Y-.2487
N10 X44.7171 Y-.25
N11 X45.0631 Y-.2109
N12 X45.2073 Y-.0739
N13 X45.2145 Y-.0257
N14 X45.1708 Y.1398
N15 X45.1005 Y.1971
N16 X44.8078 Y.25
........................
..........................
```

Fig. 7. Sequence of the NC Code program for the drilling operation

For the obtaining of the designed cassettes, a method of fabrication by adding material was used. These technologies have been used since they allow the creation of parts with extremely complex geometry in layers based on a 3D design. Also, cheap and affordable materials and installations are used [7]. Processes based on thermoplastic extrusion use a filament of materials such as ABS or PLA to heat it up to a temperature a few degrees below the melting temperature, then reduce its diameter to 0.15 mm by deposition on the work platform which moves in the xoy plane to make a section of the 3D model [5].

3 Experimental

The microorifices plates were executed on CNC machining center by milling and drilling (Fig. 8), based on the design presented.

Microorifices were obtained with good precision [5], and Fig. 9 shows a microscopic image of microholes with a diameter of 0.3 mm with a 50x magnification. It was realized plates with holes of 0.1, 0.3, 0.5, 0.7, 0.9 and 1.1 mm in diameter with different layout of the microorifices, which were mounted on the cassettes obtained by additive technologies.

Cassettes have been designed and constructed in two ways. The first way is to fix the plate to the cassette by means of a special adhesive, and the second method is based on mechanical fixing by screwing, the major advantage of this method is the possibility of interchangeability of the plates (Fig. 10 - upper green box). The assembly consisting of the microorifices plate and the fine bubble generator cassette was successfully tested in the experimental installation for water oxygenation, demonstrating the efficiency of the manufacturing methods (Fig. 11) [2, 3].

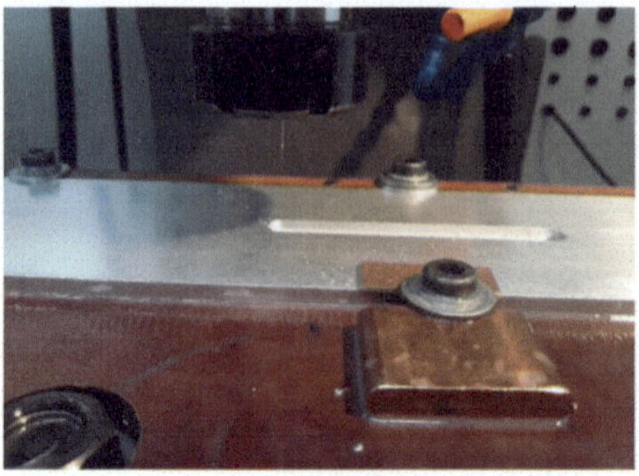

Fig. 8. Drilling process

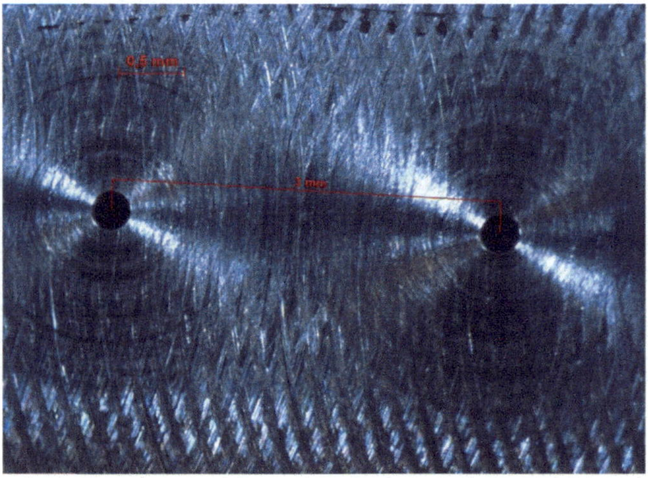

Fig. 9. Microscopic image of orifices with a diameter of 0.3 mm

A problem that may arise during the wastewater treatment process is the clogging of the microorifices. To solve this problem, we have proposed a mechatronic ultrasound system for microorifice cleaning [6].

Starting from the effects of ultrasound with a certain frequency and predetermined values of the amplitude of the ultrasonic vibrations in the presence of water, ultrasonic cavitation phenomena occur with acceleration values of several "g" (gravitational acceleration) that will favour and accelerate the cleaning of the microorifices of the diffusers through which the oxygen is blown [6].

Considering that the waste water to be treated contains organic impurities, grease and other residuals which in time obstruct oxygenation holes, cleaning is indispensable to ensure the continuity of the treatment process [6, 7]. The main advantage of the ultrasonic mechatronic system for cleaning of the microorifices of the diffusers used for wastewater treatment is that the microholes of the fine bubble generators are monitored

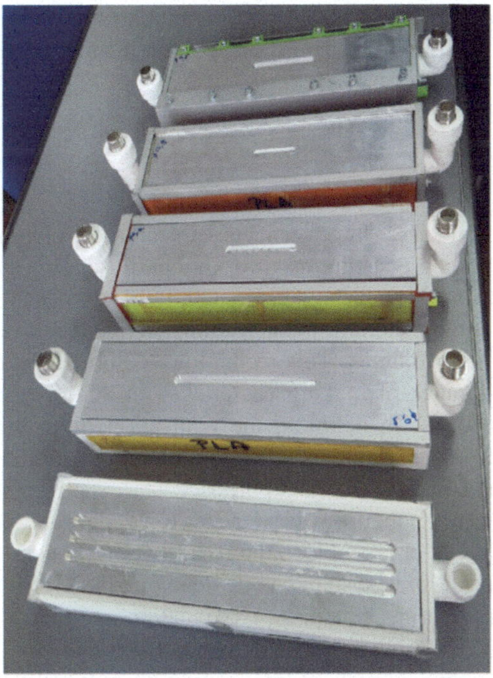

Fig. 10. Aeration systems for the ecological treatment of wastewater

Fig. 11. Testing of aeration system

online and when the sensors see the clogging of the holes, the mechatronic system automatically triggers the ultrasound emission and starts the cavitational process of the cleaning of microorifices.

The oxygen flow monitoring sensor sends the commands to the computation unit, which will trigger the start command of the mechatronic ultrasound system. In order to achieve a good ultrasound transmission performance in the diffuser, it is necessary to achieve a good rigid coupling between the ultrasonic concentrator and the plate to be ultrasonicated [6].

4 Conclusions

Wastewater treatment processes have seen a great deal of development over the last few years, with many technologies being used for this purpose. However, there are several problems related to wastewater oxygenation and we proposed an aeration system consisting of a fine bubble generator made by substractive technologies and its cassette obtained by additive technologies. This technology combination can provide the cost and time-saving advantages of making cassettes with additive technologies in comparation to metal cassettes. In addition, the realization of microorifices plates on a numerically controlled machining center offers a higher execution accuracy compared to other manufacturing methods for this purpose.

Biotreatment technology that uses wastewater oxygenation processes uses diffusers with microorifices that can clog and stop the oxygenation process. For this purpose, an ultrasonic assisted mechatronic system can be developed to clean the microorifices. By means of a sensory monitoring system for oxygenation parameters, the computing unit will start the ultrasonic system when the microorifices clogging is detected. In addition, we have also found an efficient method for cleaning ultrasonically the microorifices and designed a mechatronic system that monitors the efficient and continuous operation of the entire installation.

Acknowledgments. This work was supported by a grant of the Romanian National Authority for Scientific Research and Innovation, CCCDI – UEFISCDI, project number Manunet – MNET17/ENER2307 – CEBIOTREAT, within PNCDI III.

References

1. Robescu, D., Lanyi, S., Robescu, D., Constantinescu, D.: Tehnologii, instalatii si echipamente pentru epurarea apei. Editura Tehnica, Bucuresti (2000)
2. Dontu, O., Ioana, M.C., Tanase, B., Baran, N., Gheorghe, I.G., Moraru, E.: Researches regarding the use of additive technologies in the construction of water aeration elements. Int. J. Mechatron. Appl. Mech. **3**, 7–12 (2018)
3. Gheorghe, I.G., Dontu, O., Baran, N., Moga, C., Constantin, M., Tamasanu, E.: Researches on the measurement of the dissolved oxygen concentration in stationary waters. Int. J. Mechatron. Appl. Mech. **3**, 120–126 (2018)
4. Ghionea, I.G.: Proiectare Asistata In Catia V5. Elemente Teoretice Si APLICATII, Editura Bren, Bucuresti (2007)
5. Besnea, D., Gheorghe, I.G., Dontu, O., Moraru, E., Constantin, V., Moga, I.C.: Experimental researches regarding realization of wastewater treatment elements by means of modern technologies. Int. J. Mechatron. Appl. Mech. **4**, 61–65 (2018)

6. Dontu, O., Besnea, D., Moraru, E., Dontu, O., Besnea, D., Moraru, E.: Sistem Mecatronic Cu Ultrasunete Pentru Desfundarea – Decolmatarea Microorificiilor Difuzoarelor Pentru Tratarea Apelor Uzate, Cerere de brevet (Patent request) A/01057/05.12.2018
7. Berce, P., Balc, N., Caizar, C., Pacurar, R., Radu, A.S., Bratean, S., Fodorean, I.: Tehnologii de fabricatie prin adaugare de material si aplicatiile lor. Editura Academiei Romane, Bucuresti (2014)

Deep Learning Computer Vision for Sorting and Size Determination of Municipal Waste

Daniel Octavian Melinte[1], Dan Dumitriu[1(✉)], Mihai Mărgăritescu[2], and Paul-Nicolae Ancuţa[2]

[1] Institute of Solid Mechanics of the Romanian Academy, Bucharest, Romania
{octavian.melinte, dan.dumitriu}@imsar.ro
[2] National Institute of Research and Development for Mechatronics and Measurement Technique – INCDMTM, Bucharest, Romania

Abstract. This paper presents a mobile robotic system for picking and collecting waste and trash from the ground. The objects that the trash is made up are detected using a camera mounted on the robotic system and are processed for identification using computer vision and deep neural networks. In order to successfully pick waste and trash, the size and distance to the object are important. The proposed method takes into account that the focal length of the camera is constant, and consists of two phases: (a) from the first captured image one can measure the object dimensions l_1 and w_1 in pixels, (b) the camera moves towards the object by a distance of Δd = 100 mm capturing the second image, where l_2 and w_2 are the measured object dimensions in pixels. For better results, both captured images should be taken statically (the robot is not moving). The algorithm computes the distance to object as a function of Δd, l_1 and l_2, i.e., the dimensions in pixels measured on the two different/consecutive captured images (with $d_2 = d_1 - \Delta d$ [mm]). Then, the object dimensions are easily determined as a function of l_1 and l_2 (or w_1 and w_2), of d_1 and of the camera focal length f_{px} (in pixels). This determination of object size and distance to object is a simple but reliable method, showing good performance in practice. The image processing uses pre-trained deep convolutional networks with Single Shot Detectors (SSD) and MobileNetsV1. MobileNetsV1 architecture consists of 28 convolutional layers, one aggregation layer, and one fully connected layer. All 28 convolutional layers are followed by a non-linear function (RELU). The transition from linear convolutional to non-linear layers is done using a normalization function. After the convolutions are performed, the aggregation takes place, followed by the classification through a fully connected layer.

Keywords: Computer vision · Deep learning · Neural networks · Image processing · Object identification · Size determination · Waste management

1 Introduction

Municipal waste represent an major concern in cities across European Union and other parts of the world. In 2016, Europeans generated on average 480 kg of municipal waste per person. Municipal waste represents only around 10% of the total waste generated in

G. I. Gheorghe (Ed.): ICOMECYME 2019, LNNS 85, pp. 142–152, 2020.
https://doi.org/10.1007/978-3-030-26991-3_14

the EU, but it is one of the most complex streams to manage due to its diverse composition, its large amount of producers and fragmentation of responsibilities [1], also the waste disposal may be very diverse. By 2020, the preparing for re-use and the recycling of waste materials such as at least paper, metal, plastic and glass from households and possibly from other origins shall be increased to a minimum of overall 50% by weight. By 2035 the target is increased to 65% [2]. The latest EU report, from 2016, shows that the average of municipal waste that has been recycled was 45%.

This paper presents our approach for municipal waste sorting, especially plastic and glass bottles. We use deep learning algorithms (convolutional neural networks) and computer vision for waste identification/sorting and for a rough size determination of the identified object. The current distance between robotic gripper and detected object is also determined in this paper, being very important in the grasping process. The identification module is part of a bigger project which involves the development of a robotic system able to pick and sort municipal waste. The rough size of the object is important in order to calculate the stroke of the grasping mechanism which is placed on the robotic picking system.

The image processing module is made of Raspberry Pi controller and a camera module capable of capturing photos with a resolution of 3280×2464 (8 Megapixels). The Raspebery controller is used to run a pre-trained deep learning architecture and for basic image processing only. The convolutional neural networks (CNN) are pre-trained on other platforms.

2 Related Work

There are several researches in the area of waste sorting using deep leaning. In [3] the authors developed a database for municipal waste. The database contains 2400 images divided in six classes: paper, glass, plastic, metal, cardboard, trash. The classification has been made using two methods: Support vector machine (SVN) and CNN very similar to AlexNet model. In this case, SVN performed better, while the CNN hasn't been trained at its full capability. The results of this research stood as the basis for other papers like [4] and [5] where some adjustments of the hyperparameters, arhitecture or classification on the fully connected layers have been made. In [6] the classification of waste is improved by adding other classes that have a better degree of clarity, e.g. beverage and meal packages, cigarettes and derivatives, leaves, newspapers and papers, vegetable waste, etc. For training a OverFeat-GoogleLeNet architecture has been used at the beginning with some modification performed afterwards.

3 Methodology

3.1 CNN Architecture for Object Identification and Size Determination

In order to adapt the CNN to our needs we have tested different architectures such as: VGG, Inception, ResNet, DenseNet and MobileNet. The network that provided the best combination between the detection accuracy and processing time for our application is

MobileNet-Single Shot Detector (SSD) trained on COCO database and fine-tuned on VOC0712 and was trained by [7].

The first network layers are based on a standard CNN, MobileNetsV1 in our case. MobileNets uses depthwise separable convolutions to build light weight deep neural networks. This type of network will be used as the base model for the SSD framework. All layers are followed by a batchnorm and ReLU nonlinearity with the exception of the final fully connected layer which feeds into a softmax layer for classification. Down sampling is handled with strided convolution in the depthwise convolutions as well as in the first layer. A final average pooling reduces the spatial resolution to 1 before the fully connected layer. Counting depthwise and pointwise convolutions as separate layers, MobileNet has 28 layers [8].

In addition to MobileNet convolutional layers, the SSD architecture adds auxiliary structure to the network to produce multi-scale feature maps and convolutional predictors for detection. SSD networks discretizes the output space of bounding boxes into a set of default boxes over different aspect ratios and scales per feature map location. At prediction time, the network generates scores for the presence of each object category in each default box and produces adjustments to the box to better match the object shape. Additionally, the network combines predictions from multiple feature maps with different resolutions to naturally handle objects of various sizes [9].

In order to obtain high detection scores the database used for training is also important. There are three important databases: ImageNet, PascalVOC and COCO. The CNN base models analyzed in this paper were trained using ImageNet and the SSD architectures have been trained on COCO and some them fine-tuned on VOC. The Microsoft Common Objects in COntext (MS COCO) dataset contains 91 common object categories with 82 of them having more than 5,000 labeled instances. In total the dataset has 2,500,000 labeled instances in 328,000 images. In contrast to the popular ImageNet dataset which contains more than 14 million images and 20 k categories, COCO has fewer categories but more instances per category. This can aid in learning detailed object models capable of precise 2D localization. [10] The Pascal VOC, on the other hand, is made of 500,000 images, divided in 20 classes, that were retrieved from Flickr. The database and the labeled images are updated each year and contain significant variability in terms of object size, orientation, pose, illumination, position and occlusion.

The detection accuracy for each network was measured by averaging the confidence scores of each object identified in our image dataset. Although large networks perform better in term of detection scores, they tend to have a high computation time. Single shot detectors have an impressive frame per seconds using lower and medium resolution images and a good fps for higher resolution. For large objects, SSD performs better.

3.2 Proposed Technique for Object Size and Distance to Object Determinations

The proposed technique consists in acquiring two consecutive images: • the first one is acquired when the camera is placed at d_1 distance from the concerned object; • the second object image is acquired after the camera has moved/approached with

Δd towards the object, the current camera-to-object distance being thus $d_2 = d_1 - \Delta d$. The Δd displacement is robotically controlled, e.g., one can systematically consider $\Delta d = 100$ [mm]. Using only these two consecutive images, captured for two different distances between camera and object, the proposed technique simply computes the object size and the distance between camera and object, in [mm], based on the knowledge of: Δd, the constant focal length in [pixels], the object size/length $l_{1,obj,px}$ in [pixels] identified by image processing on the first image acquired for d_1 camera-to-object distance and also the object length $l_{2,obj,px}$ in [pixels] identified by image processing on the second image acquired for $d_2 = d_1 - \Delta d$ camera-to-object distance (after approaching with Δd).

The proposed technique is a very efficient one, for our particular application of determining the object size, as well as the distance between object and camera (mounted on the end-effector of the robotic manipulator for waste collection). The goal is to use a simple camera to determine these geometric data, very important for the grasping process. In general, the idea of exploiting the geometric information provided by available camera captures has already proved to be highly efficient, e.g., in the framework of the Perspective-n-Point (PnP) problem. More precisely, Alcantarilla et al. [11, 12] have implemented a visibility prediction algorithm for solving efficiently the PnP problem, using images provided by a stereo camera (which can provide higher distance ranges, depending on the stereo rig baseline). Thus, “visibility prediction exploits all the geometric relationships between camera poses and 3D map points in the prior 3D reconstruction”. More precisely. Alcantarilla et al. [1] use the following expression: $Z = f \times B/d_u$, where Z is the depth of the concerned 3D point, B is baseline of the rectified stereo rig, f the focal length and d_u is the horizontal pixel disparity.

Figure 1 illustrates our method for object size and distance to object determinations. A Raspberry Pi camera is used to acquire images of the object, e.g. a plastic bottle as in Fig. 1, with the real size to be identified denoted by l_{real_mm} (in [mm]).

Since the image acquisition and processing induce usual numerical errors, let us denote by l_{estim_mm} the estimated real distance (in [mm]), using the proposed method. Once the object is identified using the deep learning computer vision method presented in previous Sect. 3.1, the robotic system places the gripper and camera perpendicularly above the ground and the object.

While approaching towards the object, the first image is statically captured when the camera is placed at distance d_1 from the object. The developed image processing tool (deep learning computer vision, as already mentioned), identifies the object on this first captured image, computing also the object size $l_{1,obj,px}$ (in pixels). The constant focal length is also known, in pixels, being denoted by f_{px}. The following expression applies to our image size identification:

$$f_{px}[\text{pixels}] \times l_{estim\_mm}[\text{mm}] = l_{1,obj,px}[\text{pixels}] \times d_1[\text{mm}] = \text{constant} \quad (1)$$

After the first image is captured, the camera approaches the object on the direction of the perpendicular to the ground (preferably above the object), being driven by the controlled robotic manipulator. More precisely, the camera approaches with Δd (considered here as 100 mm), so the new camera-object distance is $d_2 = d_1 - \Delta d$ (with $\Delta d = 100$ mm), distance d_2 for which the camera captures statically the second image.

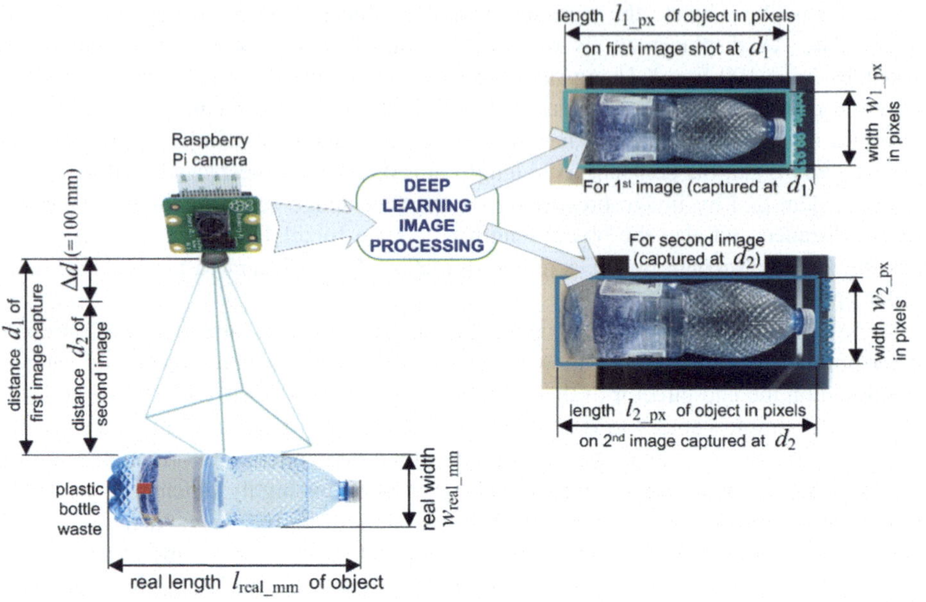

Fig. 1. Proposed technique for object size and distance to object determinations.

The same image processing is performed to this second image captured for d_2 [mm] camera-object distance, identifying the object-box and computing the object size $l_{2,obj,px}$ (in pixels). An expression similar with (1) applies to the second image size identification:

$$f_{px}[\text{pixels}] \times l_{estim\_mm}[\text{mm}] = l_{2,obj,px}[\text{pixels}] \times d_2[\text{mm}] = \text{constant} \tag{2}$$

From (1) and (2) and taking into account that $d_2 = d_1 - \Delta d$, it results easily:

$$d_1[\text{mm}] = \frac{\Delta d[\text{mm}] \cdot l_{2,obj,px}[\text{pixels}]}{(l_{2,obj,px}[\text{pixels}] - l_{1,obj,px}[\text{pixels}])} \tag{3}$$

then it follows:

$$l_{estim\_mm}[\text{mm}] = \frac{l_{1,obj,px}[\text{pixels}] \cdot d_1[\text{mm}]}{f_{px}[\text{pixels}]}$$

$$= \frac{\Delta d[\text{mm}] \cdot l_{1,obj,px}[\text{pixels}] \cdot l_{2,obj,px}[\text{pixels}]}{f_{px}[\text{pixels}] \cdot (l_{2,obj,px}[\text{pixels}] - l_{1,obj,px}[\text{pixels}])} \tag{4}$$

Equations (3) and (4) concern the length of the waste object, i.e., plastic bottle in the considered case. Similar equations hold for the estimated/computed width of the plastic bottle (waste object), denoted by w_{estim_mm} (in [mm]), while $w_{1,obj,px}$ and $w_{2,obj,px}$ are the width in pixels determined for the first image captured at d_1, the for the second image captured at d_2, as follows:

$$d_1[\text{mm}] = \frac{\Delta d[\text{mm}] \cdot w_{2,obj,px}[\text{pixels}]}{(w_{2,obj,px}[\text{pixels}] - w_{1,obj,px}[\text{pixels}])} \tag{5}$$

$$w_{estim\_mm}[\text{mm}] = \frac{\Delta d[\text{mm}] \cdot w_{1,obj,px}[\text{pixels}] \cdot w_{2,obj,px}[\text{pixels}]}{f_{px}[\text{pixels}] \cdot (w_{2,obj,px}[\text{pixels}] - w_{1,obj,px}[\text{pixels}])} \tag{6}$$

Equation (5) is an alternative to Eq. (3). In conclusion, the proposed technique for object size (l_{estim_mm} and w_{estim_mm}) and distance to object (d_1) determinations is simply based on two consecutive images, the first one statically captured at distance d_1 from the object, while the second one is statically captured after the camera approaches the object with $\Delta d = d_1 - d_2$ (e.g., = 100 mm), being robotically controlled by a precise manipulator mounted on mobile robotic system for picking and collecting waste and trash from the ground.

4 Results and Discussions

4.1 Object Detection

In order to test the accuracy of the CNN architecture, 25 images of different bottles with different shapes have been considered in this paper. Due to the fact that the detection of bottles in the municipal waste is a process that involves identification of objects that have different degrees of deformation, shapes, position, transparency, the dataset used for testing have taken into account these constraints. The straightforward identification of plastic or glass bottles provides more than 95% confidence, but our goal is to take into account as many scenarios as can possible.

The research in this paper focuses on the detection of one bottle in image but we tested the object identification for different degrees of clustering with good results as well.

In Fig. 2 the most important detections are presented. The bottles in the image are placed in different positions (including bottom up), are partially obturated, deformed, with different shapes, etc.

The deep learning detection algorithm has been tested on our collection of 25 images of plastic and glass bottles. The images are captured at different resolutions, with resolutions lower than the resolution of the camera which is placed on the robotic system.

The first tests have been made using different models of convolutional neural network trained on ImageNet dataset. In order to determine the average precision of each network the detection confidence for each image has been taken into account. Although the mAP and IoU of each CNN model is greater than the average confidence of our images, the results obtained are important considering that the bottles in our images are difficult to detect. The resulted bounding boxes in each image after detection

Fig. 2. Identification results for our collection on images.

are considered to be useful for further determination of object size although there are some offsets from the ground truth box. The rough size of the object is determined using the size of the bounding box and this information is sent to the object size determination algorithm.

Based on the detections scores for CNN models, presented in Table 1, we have observed that the best results for our test dataset are obtained using the NasNet and DenseNet. This type of CNN model have around 5-6 million parameters and need a good amount of processing time for training, classification or detection. Although they are more accurate, for real-time process this CNN's are slow. On the other hand, the Inception and MobileNet convolutional networks, have less parameter (around 3 million) and lack the same accuracy but are much faster. Based on this fact, we further used these models to increase the average confidence by implementing a Single Shot Detector network trained using COCO dataset.

The best results in this case were obtained for the Inception V2 + SSD + COCO architecture (59.1%). The last improvement that was tested for the MobileNet + SSD architecture was the training on MS-COCO and then fine-tuning on VOC0712. The average confidence for our collection of test images, in this case, was around 64% with better results when the images provided to the network where resized between 300 $\times$ 500 and 700 $\times$ 900 (Table 2).

The convolutional networks that uses VGG models did not detect anything regardless of the training dataset (COCO or ImageNet), test images or the CNN architecture(VGG models or SSD + VGG). The same thing occurred after testing our test images on ResNet and DenseNet121.

Table 1. CNN model.

CNN model	Average confidence tested images
VGG16 + ImageNet	0
VGG19 + ImageNet	0
Inception + ImageNet	47.45
Xception + ImageNet	44.21
ResNet + ImageNet	0
MobileNet + ImageNet	40.46
MobileNetV2 + ImageNet	29.65
NasNetMobile + ImageNet	45.5
NasNetLarge + ImageNet	62.07
DenseNet121 + ImageNet	0
DenseNet169 + ImageNet	54.84

Table 2. Single Shot Detectors + CNN model + Dataset.

Model	Average confidence tested images
MobileNet + SSD + VOC + COCO	63.97
MobileNet + SSD + COCO	27.06
MobileNetV2 + SSD + COCO	42.31
InceptionV2 + SSD + COCO	59.1
VGG + SSD + COCO	0

4.2 Size Determination

The next step after object identification is the determination of the bottle size. The determination is performed using the information from the identification algorithm. The identification algorithm provides the bounding box of the object and the size is determined based on the dimensions of the bounding box.

In order to perform the size determination three types of bottles are considered: (a) one 2L plastic bottle, (b) one 0.55L plastic bottle and (c) 0.75L glass bottle. The objects are placed at 40 cm and 50 cm. In Fig. 3 the resulted bounding boxes of the identified objects are presented.

4.3 Results Concerning the Determination of Object Size and Distance Between Camera and Object

The case study considers a real distance $d_{1\_real} = 500$ mm, which must be retrieved by the proposed technique of object size and distance between camera and object determinations. Obviously, $d_{2\_real} = d_{1\_real} - \Delta d = 500 - 100 = 400$ mm, with the already mentioned $\Delta d = 100$ mm. So, in reality, the first image is captured when the camera is distanced by 500 mm from the object, then the robotic system holding the gripper and the camera approaches with 100 mm towards the identified object, capturing statically the second image for a camera-object distance of 400 mm.

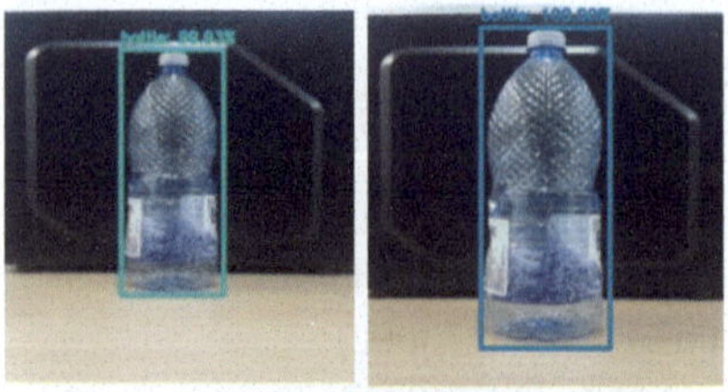

(a) 2L plastic bottle

(b) 0.55L plastic bottle

(c) 0.75 glass bottle

Fig. 3. Bounding boxes and identification confidence of tested objects for d_1 = 50 cm and d_2 = 40 cm

The constant focal length of the considered camera is f_{px} = 935 pixels. For the two captured images, the image processing tool presented above provides us the estimated lengths at widths of the observed objects: $l_{1,obj,px}$ and $w_{1,obj,px}$ for the first image captured at d_1, then $l_{1,obj,px}$ and $w_{1,obj,px}$ for the first image captured at d_2. Table 3 comprises these lengths and width in pixels (on the processed images), showing also the object sizes (length and width) and camera-object distance computed using the proposed technique, i.e., by Eqs. (3)–(6).

The d_1 distance can be computed either using Eq. (3), or using Eq. (5). For both of these estimated distances an error is computed with respect to the real distance d_{1_real} = 500 mm considered in this paper. The maximum error is 19% (computed in terms of d_1 distance), which is acceptable taking into account that this a rough estimation method. Reducing this error in estimating object size and camera-object distance will be the object of a further study, for example it might help considering d_{1_real} = 400 mm, i.e., capturing the first image at 400 mm, then the second image at 400 − 100 = 300 mm.

Table 3. Results concerning object sizes and distance to object determinations.

Waste object	Size at d_1 in px: $l_{1,obj,px} \times w_{1,obj,px}$ [px]	Size at d_2 in px: $l_{2,obj,px} \times w_{2,obj,px}$ [px]	Distance d_1 [mm] given by Eq. (3)	Distance d_1 [mm] given by Eq. (5)	Length l_{estim_mm} given by (4)/wrt real length	Width w_{estim_mm} given by (6)/ wrt real width	Relative error of computed d_1 [mm], w.r.t. real 500 mm [%]	
Small plastic bottle	282 × 152	374 × 183	**406**	**590**	**123**/95 (real)	**66**/70 (real)	**19**	**18**
Big (2l) plastic bottle	494 × 204	617 × 248	**502**	**563**	**265**/340(real)	**109**/100(real)	**0**	**13**
Glass bottle	464 × 152	573 × 197	**526**	**438**	**261**/310 (real)	**85**/74 (real)	**5**	**12**

5 Conclusion and Future Work

The proposed framework for object detection and size determination has a confidence score higher de 95% for common plastic and glass bottles and decreases to 64% for the test dataset that is made of bottles that are twisted, that have uncommon shapes, high degree of transparency, placed in uncommon position, etc. We consider that this score satisfies our system requirements.

SSD + MobileNets trained on COCO database and fine-tuned on VOC can even match other larger architecture's accuracies using better extractor. SSD + Mobilenets performs not as good for small objects comparing to other methods. Input image resolution impacts accuracy significantly. For large objects, SSD + MobileNets can outperform Faster R-CNN and R-FCN in accuracy with lighter and faster extractors.

Further we will use transfer learning for classification and detection in order to extend the number of classes using a dataset which will include images for other type of waste such as paper, cardboard, metal, etc. Other improvements to the network will focus on the hyper parameters tuning and the adjustments of the kernels.

As for the proposed method for determining/estimating the object size (length and width) and the distance between camera and object, further work will concern the reduction of the estimation error. More precisely, one may study which are the most suited camera-object distances d_{1_real} and d_{2_real} for statically capturing the first and the second images. Another improvement would be to be able to dynamically capture the two consecutive images, i.e., while the robotic manipulator is in motion, without being necessary to shortly stop for capturing images.

Acknowledgements. This work was supported by a grant of the Romanian Ministry of Research and Innovation, CCCDI – UEFISCDI, project number PN-III-P1-1.2-PCCDI-2017-0086/contract no. 22 PCCDI/2018, within PNCDI III.

References

1. European Commission: Report from the commission to the European parliament, the council, the European economic and social committee and the committee of the regions on the implementation of EU waste legislation, including the early warning report for Member States at risk of missing the 2020 preparation for re-use/recycling target on municipal waste, Brussels, 24 September 2018
2. The European Parliament and the Council of the European Union: DIRECTIVE 2008/98/EC OF THE EUROPEAN PARLIAMENT AND OF THE COUNCIL on waste and repealing certain Directives, In: Official Journal of the European Union, 19 November 2008
3. Yang, M., Thung, G.: Classification of trash for recyclability status. In: CS229 Project Report (2016)
4. Awe, O., Mengistu, R., Sreedhar, V.: Smart trash net: waste localization and classification. arXiv preprint (2017)
5. Rad, M.S., von Kaenel, A., Droux, A., Tieche, F., Ouerhani, N., Ekenel, H.K., Thiran, J.P.: A computer vision system to localize and classify wastes on the streets. In: International Conference on Computer Vision Systems, pp. 195–204. Springer, Cham, July 2017
6. Kennedy, T.: OscarNet: using transfer learning to classify disposable waste. In: CS230 Report: Deep Learning. Stanford University, CA, Winter 2018
7. https://github.com/chuanqi305/MobileNet-SSD. Accessed 15 May 2019
8. Liu, W., Anguelov, D., Erhan, D., Szegedy, C., Reed, S., Fu, C.Y., Berg, A.C.: SSD: single shot multibox detector. In: European Conference on Computer Vision, pp. 21–37. Springer, Cham (2016)
9. Howard, A.G., Zhu, M., Chen, B., Kalenichenko, D., Wang, W., Weyand, T., Andreetto, M., Adam, H.: Mobilenets: efficient convolutional neural networks for mobile vision applications. arXiv preprint arXiv:1704.04861 (2017)
10. Lin, T.Y., Maire, M., Belongie, S., Hays, J., Perona, P., Ramanan, D., Dollár, P., Zitnick, C. L.: Microsoft COCO: common objects in context. In: European Conference on Computer Vision, pp. 740–755. Springer, Cham, September 2014
11. Alcantarilla, P., Stasse, O., Druon, S., Bergasa, L.M., Dellaert, F.: How to localize humanoids with a single camera? Auton. Robot. **34**(1–2), 47–71 (2013)
12. Alcantarilla, P.F., Ni, K., Bergasa, L.M., Dellaert, F.: Visibility learning in large-scale urban environment. In: IEEE International Conference on Robotics and Automation (ICRA), Shanghai, China (2011)

The Non-linear Dynamic Response of Microstructures

M. Amin Changizi[1], Ion Stiharu[2(✉)], and D. Erdem Şahin[3]

[1] Mohawk Innovative, Albany, USA
[2] Concordia University, Montreal, Canada
istih@ensc.concordia.ca
[3] Bozok University, Yozgat, Turkey

Abstract. This paper presents the results of the solutions of the non-linear differential equation that models the dynamic performance of a microstructure such as a cantilever beam when is subjected to uniform electrostatic field. This situation is encountered in all capacitive MEMS sensors. The mass-damper-spring model is used to evaluate the critical pull-in voltage yield by the solution of the non-linear differential equation. This model is analyzed based on the adopted models in the literature dealing with the stiffness of the cantilever beam. The one degree of freedom non-linear differential equation used to model the dynamics of the cantilever subjected to electric field set close to pull-in is stiff and the only correct solution is yield by Isode algorithm. The equivalent stiffness of the model was considered based on four different models selected from the open literature. The validity of the solution was confirmed through experimental tests. The stiffness model corresponding to the best match for the deflection model is proved to be different from the one that yields the best match in the resonant frequency.

Keywords: Micro-cantilevers · MEMS · Non-linear dynamic

1 Introduction

Micro-cantilever beams are structures of great interest in MEMS due to simplicity to fabricate and predictability of their performance. Such structures are mainly used in inertial sensing. The interest in micro-cantilevers has driven investigations from various perspectives including the static and the dynamic performances under as the influence of the potential fields. Other aspects have been also investigated such as the thermal effect, the manufacturing influence, the effect of the inter-laminar stress, or the specific geometric configuration. The pull-in voltage is a topic of highest interest as most of micro-cantilever-like structures operate under electric fields. Pull-in voltage was investigated from the theoretical perspective as well experimentally. Although relatively low voltages are used for measurement purpose, the micro-structures including cantilever type may operate at frequencies significantly below their resonance [1], within the range that would yield instabilities at any frequency. For MEMS structures operating within electrostatic fields such a situation is very likely to occur close to the pull-in conditions. The objective of such studies which includes the present one is to

G. I. Gheorghe (Ed.): ICOMECYME 2019, LNNS 85, pp. 153–172, 2020.
https://doi.org/10.1007/978-3-030-26991-3_15

identify the limits of the potential for specific configurations. This purpose has been achieved through development of models that are using differential equation to model MEMS. For simplification purpose, the nonlinear mathematical formulations are generally converted in linear forms. In some investigations the deflection of the tip of the micro-cantilever is assumed very small [2]. Energy methods are often used to derive the governing equations which in many reports are numerically solved and information such as the pull-in voltage is found as a numerical value. Parametric studies revealed the correlation among the various involved parameters. Small deflections were also assumed in [3] and the results of the numerical solution for the pull-in voltage were compared to those retrieved from experiments. The results indicated that the longer the cantilever, higher is the error. However, the error would be significantly lower when the cantilever is subjected to potentials significantly below the pull-in voltage [4] even when assuming linearized models. Simplified non-linear models of the electrostatic force in conjunction with analytical solution expressed as Taylor series from where the pull-in voltage was found are presented in [5] and [6]. FEA method was used in [7] to calculate pull-in voltage and the results were compared with the experiments. A linear lump mass model was developed in [8], then the corresponding ODE was solved numerically to calculate the pull-in voltage. The effect of the AC and DC potential between micro-cantilever and substrate was investigated [9] by linearization of the non-linear part using the first terms in Taylor series to express the pull-in voltage. A continuum form of the ODE for small deflection of beams was considered in [10]. The system response to AC voltage was investigated but not for the pull-in voltage. Duffing equation with harmonic excitation was considered to analyze the vibration of the micro-cantilevers [11]. The authors used perturbation method to solve the non-linear equation and compare the results with the experiments for pull-in voltage. Continuum models for small deflection of micro-cantilever beams were also investigated [12] and Taylor series built while orthogonal functions were used to linearize the non-linear ODE. The validation was based on the results of the comparison of the experimental and analytical deflections of micro-cantilevers beam under electrostatic force field. The governing equation of the deflection for a micro-cantilever beam under electrostatic field was also derived by Hamiltonian method [13]. A linearized form of the nonlinear equation of a lump mass model was developed and the numerical results were compared with the pull-in voltage obtained from the experiments. In [14] the authors used the theory of the dimensionless continuous beam equation and solved the linearized form by Differential Quadratic Method (DQM). They studied the effect of the voltage on the natural frequency using the numerical approach. The results of the linear analysis for the pull-in voltage were compared with the experimental values in [15]. The influence of the width and thickness of the beam on the resonant frequency were both theoretically and experimentally studied in [16]. The work considered the lump mass model linearized through Taylor series. Deflection of micro-cantilever beams under electrostatic fields was studied using lump mass model and the constitutive ODE was found as a Duffing type equation [17]. A numerical solution of the ODE was found for the model. The literature includes a large number of models that although are formulated in the non-linear domain, the differential equation describing the loading condition is usually linearized such that the solution carries large errors and lays rather far away from the experiments. The main contribution of the present paper is that it formulates the non-linear model and it solves the non-linear differential equation of the perturbation while comparing the equivalent stiffness models previously presented in the open literature. The findings point towards selection of altered values for the correction factors for the effective width and area which yielded more accurate results than many reported so far [18, 19].

2 The Constative Model

2.1 The Vibration of a Micro Cantilever Beam Subjected to Uniform Electrostatic Field

Study of the dynamics of a micro-cantilever beam when is subjected to an electric field will be carried out using a mass lumped model as the one below.

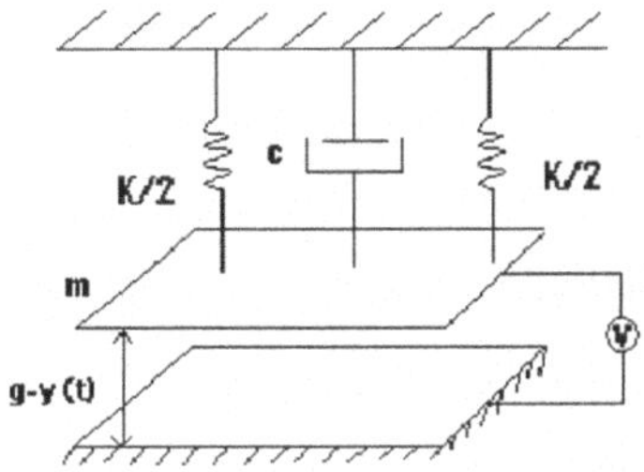

Fig. 1. The schematic of the equivalent mass-spring damper system to model dynamic performance of a micro-beam.

The model is developed to understand what it would happen to a capacitive micro-sensor made from an elastic beam that is mechanically excited and is subjected to an electrostatic filed due to the difference of potential between the elastic beam and the rigid substrate. The governing differential equation can be presented in a symbolic form as:

$$\frac{d^2y(t)}{dt^2} + 2\xi\omega_n\frac{dy(t)}{dt} + \omega_n^2 y(t) = \frac{\varepsilon_0 A V^2}{2m(g - y(t))^2} \tag{1}$$

Where all units are in [SI]:

m is the mass of plate [kg]

y(t) is deflection of mass center of gravity with respect to time [m]

ξ is damping factor where $\frac{c}{m} = 2\xi\omega_n$

ε_0 is absolute permittivity (8.854×10^{-12} F/m)

g is initial distance between plate and substrate [m]

A is area of plate [m^2]

V is the difference of potential between the microbeam and the substrate [V].

2.2 Effective Stiffness

The most common model the vibration of a cantilever beam is based on the one degree of freedom mass-damper-spring model as illustrated in Fig. 1. Under the circumstances, it is extremely important to correctly define the equivalent effective stiffness and damping for the model. In some prior work [18, 19] effective stiffness for a cantilever under the electrostatic force was formulated and numerically solved using FEA. In the current work, the effective stiffness is used to model the stiffness of the cantilever in deflection as in the mass-spring-damper model. The equivalent stiffness is used also to evaluate the dynamic performance of the micro-systems. The effective stiffness is defined in the above-mentioned works as:

$$K_{eff} = \frac{2}{3}\frac{E^{*}bh^{3}}{l^{3}} \tag{2}$$

Where:

$$E^{*} = \frac{E}{1 - \nu^{2}}, \tag{3}$$

E is Young's modulus [GPa]
ν is Poisson's ratio
b is width of beam [m]
h is thickness of beam [m]
l is length of beam [m]

Based on the effective stiffness, the pull-in voltage can be calculated in Eq. (4), using the effective stiffness values as in [18, 19]:

$$v = \sqrt{\frac{8k_{eff}d^{3}}{27\varepsilon_{0}A_{eff}}} \tag{4}$$

Where:

d is gap distance [m]
$\varepsilon 0$ is the absolute permittivity [8.854×10^{-12} F/m]
A_{eff} is the effective area [m^2]
From [18, 19]:

$$A_{eff} = \alpha b_{eff} l \tag{5}$$

With:

$$b_{eff} = b\left(1 + 0.65\frac{(1 - \beta)d}{b}\right) \tag{6}$$

The relationship between the two geometric parameters is:

$$\alpha = \frac{4}{\pi}\frac{1-\beta}{\sqrt{1-2\beta}}\arctan\left(\sqrt{1-2\beta}\right) \tag{7}$$

$$\frac{c}{m} = 2\xi\omega_n \tag{8}$$

Hence, the value of β is established to range between the two values in (8) as proposed in the open literature [18, 19].

3 Nonlinear Behavior and the Pull-in Voltage

The pull-in voltage was studied based on the linearization of the Eq. (1), whereas nonlinear pull-in voltage was calculated in this paper based on the solution of the same but non-linearized equation. The comparison of the results yield by the linear and nonlinear models shows significant difference between the values in the pull-in voltage yield by the solution of the two models. The difference increases with increasing the gap between the beam and the bottom electrode. The numerical values of the calculations are given in Table 1. In this table, for each gap distance the pull-in voltage was calculated by seven different accepted approaches which yield errors of up to 100% when compared with the results produced by the linear model. The methods in calculating the pull-in voltages are based on the equivalent geometry and stiffness of the micro-cantilever beam as available in the open literature and are shown below. The second column (linear) shows the pull-in voltage calculated based on equation which represents the linear approach expressed by Eq. (4). Third column (is called in this paper DCNF-Discontinuous-Cantilever-Natural-Frequency) shows the pull-in voltage value based on Eq. (4), while establish the theoretical stiffness of the beam as reference value such that $\omega_n = 3.515625\sqrt{\frac{3EI}{ml^3}}$ The fourth column (is called in this paper CCNF-Continues-Cantilever-Natural-Frequency) gives the pull-in voltage derived from the equation by assuming the resonant frequency as reference value as: $\omega_n = 3.515625\sqrt{\frac{3EI}{ml^3}}$. The fifth and the seventh columns represent the pull-in voltages derived from the same Eq. (4) where the two limits of the geometric parameters are assumed as recommended in prior publications as $\beta = 0.33$ and $\beta = 0.45$, respectively. The sixth and eighth columns provide the pull-in voltages from the nonlinear equation by correcting the effective area of the beam using Eq. (5) while keeping same limits for the geometric parameter β as in columns five and seven. As one can see from Table 1, DCNF in all cases yield results significantly different from the other models – results that are slightly below the one yield by the linear model. Nonlinear analyses of the CCNF model yield the maximum values for the pull-in voltage and the results come almost same as for the modified effective area for $\beta = 0.45$ from the Eq. (5). For the modified models, in any case the formulations give slightly higher value than the ones resulting from nonlinear analysis. The results reveal that the modified at $\beta = 0.45$ model yields the largest pull-in voltages. For small gaps, the CCNF model predicts pull-in voltage as low as the ones resulting from the modified β = 0.45 but while increasing the gap distance the pull-in voltage yield by the CCNF model will yield higher pull-in voltage than the one in which the geometry is modified by β = 0.45. Hence, one could conclude that the CCNF model yields the structure with the highest stiffness while DCNF produces the least stiff structure regardless the initial gap between the microbeam and the substrate. The modified β = 0.45 model yields a stiffer structure than the modified β = 0.33 model. The above finding is confirmed by the resonant frequencies predicted by the models when structures are subjected to low electrostatic forces. Thus, CCNF model exhibits highest natural frequency while DCNF shows the lowest natural frequency. Given the specific performance of each of the selected models, in the further analysis only four models will be considered as being relevant. They are: DCNF, CCNF, and the geometric modified models with β = 0.45 and β = 0.33.

The behavior of the system when the applied potential is close to the pull-in voltage represents another aspect of interest which is studied in this work. For the specific micro-cantilever model and gap distances, introduced in Table 2, in Figs. 2 four sets of results are shown for smallest gap distance. These four sets show: the time response of the system under the application of a step potential 99.99% of the pull-in voltage, the phase portrait for the applied potential at same voltage which is close to the pull-in voltage, deflection under various voltages for the different models and the resonant frequency variation produced by each of the four selected models with the variation of the potential. The reduction in resonant frequency makes the potential to be perceived as a factor that leads to the loss of stiffness of the structure under the application of the potential (weakening phenomena of the structure as detailed in reference [1]). The solutions of each of the four models were derived and plotted. Each graph has four sets of solutions and each set represents the solution of one of the above discussed models. The time domain solution of the ODE, Eq. (1), describing the dynamics of a microsystems as a micro-cantilever beam subjected to electrostatic forces reveal interesting trend that exhibit quite different aspect than the one reported in the open literature [1]. Figure 2a illustrates the time response of the four selected models for the gap of 2 μm: DCNF (7.1 V), CCNF (14.5 V), non-linear modified β = 0.33 (14.3 V) and non-linear modified β = 0.45 (14.6 V). It is seen that after the potential is applied, the system responds with a ramp followed by a flat which is same for all models (deflection representing 4/9 or 45% of the initial gap). It is important to mention here that the quasi-static analysis yield a unique limit position that corresponds to a deflection of 1/3 of the original gap. The duration of the flat which is a saddle point in the stability of the structure is dependent on the model. Thus, the duration of the marginal stability is inverse proportional with the instant stiffness. Lower stiffness yields longer duration to settlement. Once the system leaves the marginal stability seen in the phase portrait as the saddle point, the equilibrium position is attaining regardless the model in a position representing 2/9 of the initial gap or half of the marginal stability position. Although CCNF model proves to exhibit the lowest resonant frequency at the applied potential, this is due to the fact that the rate of decay of the resonant frequency of the CCNF model is higher than of the other models. It is important to mention that if the critical voltage is assumed in any of the models instead of close to the critical voltage, no solution is found.

Table 1. Pull-in voltage in linear and nonlinear analysis.

Gap distance	Linear model	DCNF	CCNF	Modified $\beta = 0.33$	Non-linear modified by $\beta = 0.33$	Modified $\beta = 0.45$	Non-linear modified by $\beta = 0.45$
2	7.52	7.17	14.57	14.36	13.69	15.38	14.66
4	21.27	20.28	41.22	39.80	37.95	42.78	40.79
6	39.80	37.26	75.73	71.70	68.36	77.32	73.72
8	60.17	57.37	116.59	108.33	103.28	117.16	111.71
10	84.09	80.18	162.94	148.66	141.74	161.24	153.73

Phase portraits in Fig. 2b show that the system described by the four models is still stable but it has developed sharp edge along the x-axis. These edges represent the saddle point that is seen in the time response as a flat at the peak deflection point or as previously described as the marginal stability duration.

Table 2. Ratio of distance and distance that pull-in happening.

	2 μm	4 μm	6 μm	8 μm	10 μm
Pull-in distance	0.9 μm	1.8 μm	2.7 μm	3.6 μm	4.5 μm
Ratio	45%	45%	45%	45%	45%

All phase diagrams regardless the model show the saddle point at the critical pull-in distance. Further, the plot indicates a convergence point which corresponds to the settling position as illustrated in Fig. 2a. The settlement distance represents 50% of the pull-in distance. The maximum velocity occurs in DCNF whereas minimum velocity is encountered in CCNF while velocities of the modified models are similar.

The microstructure deflects when subjected to a difference of potential, as illustrated in Fig. 2c. The four models yield different deflections for same potential difference. DCNF model exhibit the highest deflection which is consistent with the model that experience the lowest pull-in voltage. However, the pull-in voltage exhibited by this model is closer to the linear model and about half of any other three discussed models. The other three models (CCNF, modified $\beta = 0.33$ and modified $\beta = 0.45$) approach the pull-in voltage to similar values. The deflection increases with the applied potential. The least deflection is exhibited by the modified $\beta = 0.45$ and the most by the modified $\beta = 0.33$. The plot showing the resonant frequency of the system according to the formulations of the four models is illustrated in Fig. 2d. Here, CCNF model yields the highest resonant frequency, which comes consistent with the other findings. The modified models yield about same resonant frequency and similar pull-in voltage. The resonant frequency of the two models branches out when the structure is subjected to potentials closer to the pull-in voltage. The modified $\beta = 0.45$ model reduces less than modified $\beta = 0.33$ and the difference grows once the models are subjected to potentials closer to the pull-in voltage. The DCNF model exhibits evidently the lowest resonant frequency.

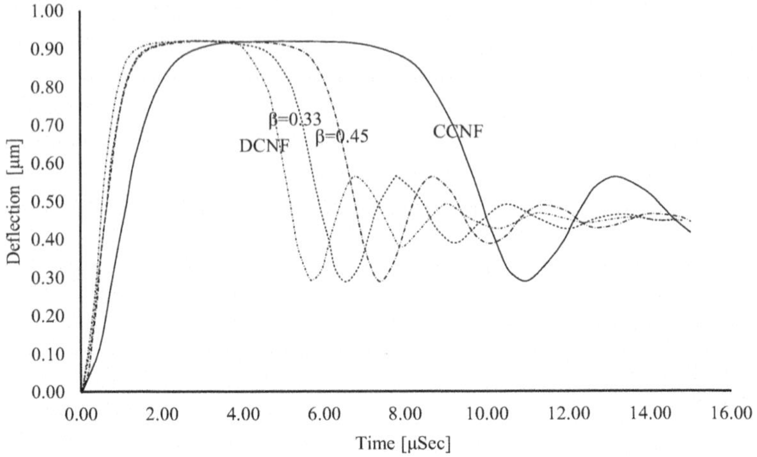

Fig. 2a. Time response of the system near the pull-in voltage for the four selected models (gap = 2 μm).

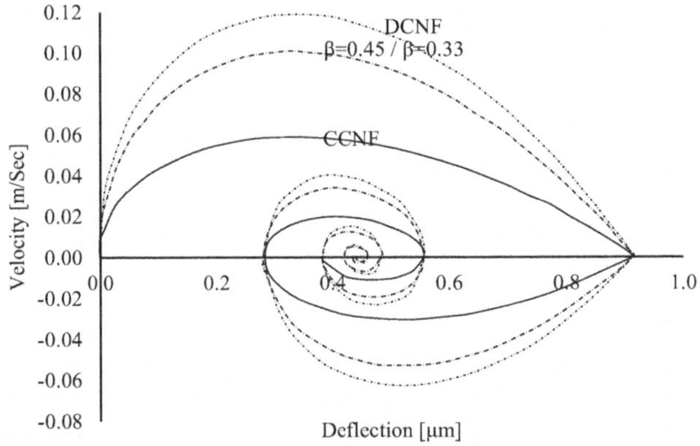

Fig. 2b. Phase portrait of the four models for potentials near the pull-in voltage (gap = 2 μm).

Further, Fig. 3a illustrate the same data as Fig. 3a but for an assumed gap of 10 μm. The pull-in voltage values for the four models and the assumed gap are as those given in Table 1. The pull-in voltages for DCNF and CCNF increase a little more than 11 times while for the two modified models with $\beta = 0.33$ and $\beta = 0.45$ a bit more than 10 times for the 5 times increase of the gap. Figure 3a also shows that the duration of the marginal stability of CCNF model significantly increases while for the other three models remains about same. This may be due to the fact that the assumed potential for the CCNF case (162.9 V) is closer to the pull-in voltage than the voltage assumed for 2 μm gap. The phase portrait illustrated in Fig. 3b is similar with the one plotted for 2 μm gap. It is important to mention that the dynamic behavior of the system regardless the model is same. Once the potential approaches the pull-in voltage, the structure

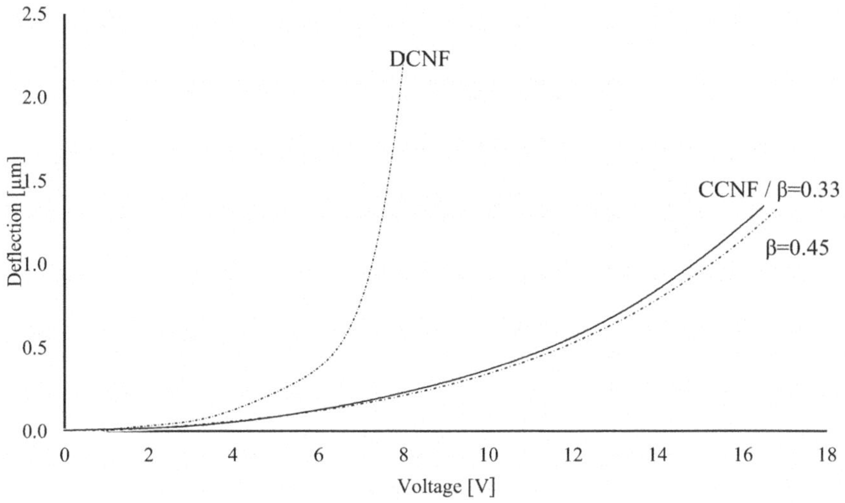

Fig. 2c. Dependency of the deflection with the applied voltage for the four different models (gap = 2 μm).

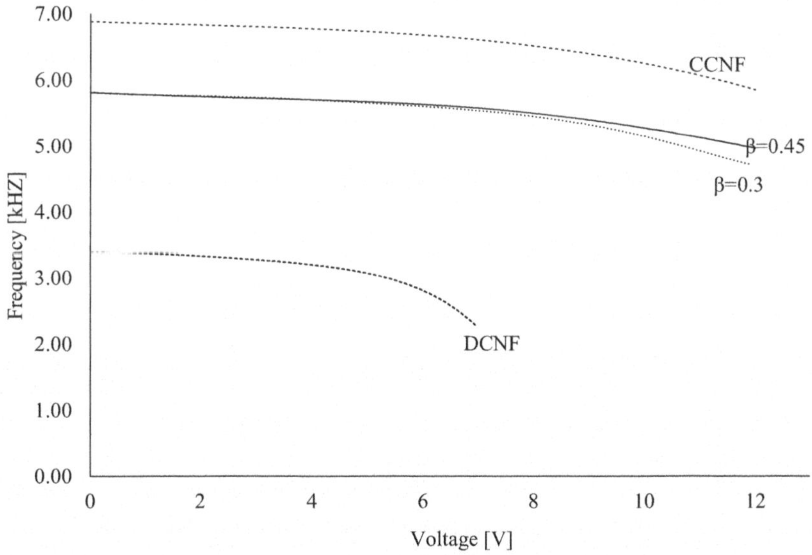

Fig. 2d. Variation of the resonant frequency of the system for the four models with the applied voltage (gap = 2 μm).

ramps to a deflection corresponding to 4/9 of the initial gap, and then settles in equilibrium position which corresponds to a deflection of 2/9 of the initial gap. The deflection vs. the applied potential illustrated in Fig. 4c is similar in trend with that of the model in which the gap was assumed at 2 μm except that the lowest deflection for

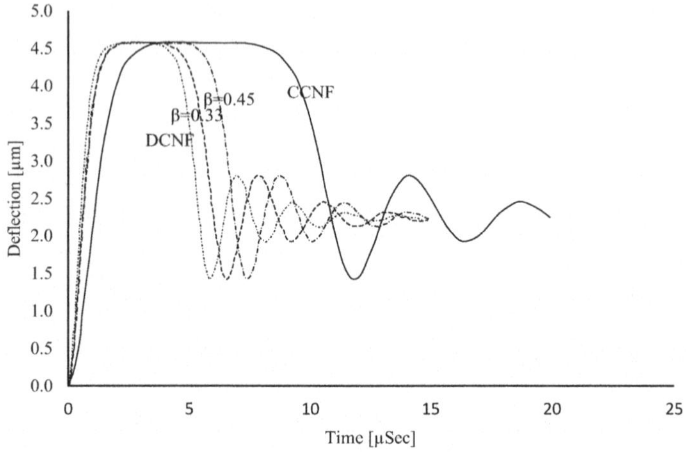

Fig. 3a. Time response of the system near the pull-in voltage for the four selected models (gap = 10 μm).

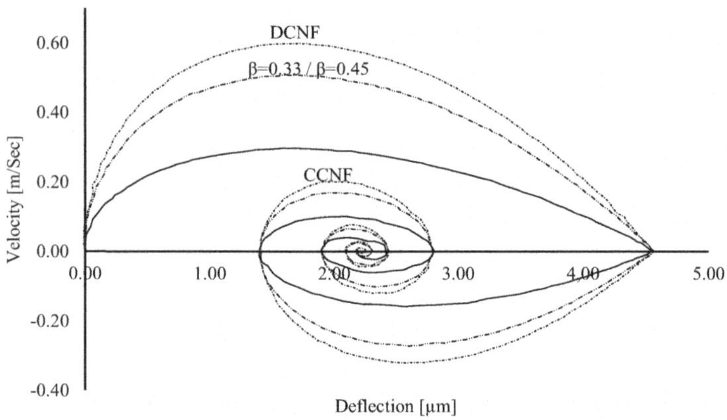

Fig. 3b. Phase portrait of the four models for potentials near the pull-in voltage (gap = 10 μm).

the same potential is performed by the CCNF model. The dependence of the natural frequency with the applied potential for 10 μm gap is illustrated in Fig. 3d similarly as in Fig. 2d, DCNF model yields the lowest resonant frequency while CCNF yields the highest.

A parametric study to investigate the influence of the gap on the time response, critical position and equilibrium position, deflection vs. applied potential and variation of the resonant frequency is carried out for each of the four considered models. However, the results for a single model are presented in Figs. 4. The general conclusion of the findings is that regardless the selected model and the gap choice, once the applied potential is close enough to the pull-in voltage in micro-system, the mobile restoring element will ramp, saddle always at the same position which is 4/9 of the

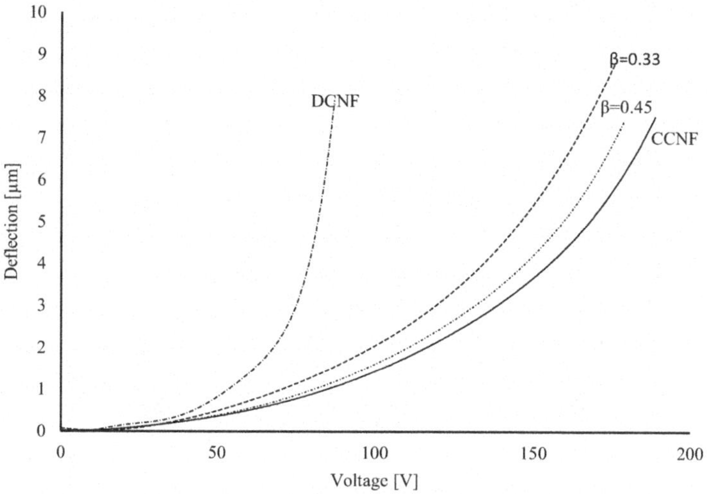

Fig. 3c. Dependency of the deflection with the applied voltage for the four different models (gap = 10 μm).

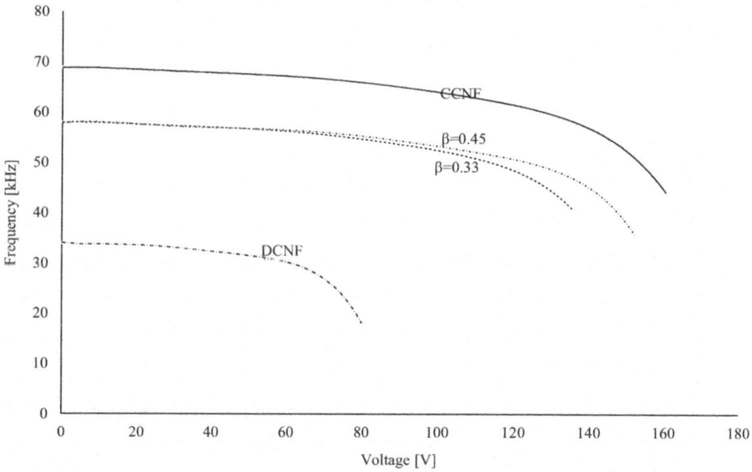

Fig. 3d. Variation of the resonant frequency of the system for the four models with the applied voltage (gap = 10 μm).

initial gap, and will settle in an equilibrium position which corresponds to a deflection of 2/9 of the initial gap which is always same, regardless the considered model. The duration of the marginal stability of the system is dependent on the effective stiffness of the structure which value si highly sensitive with the applied potential when approaching the pull-in voltage. Hence, the duration of the marginal equilibrium in Fig. 4a, is found to be irrelevant when a slight increase with the gap from 2 to 6 μm

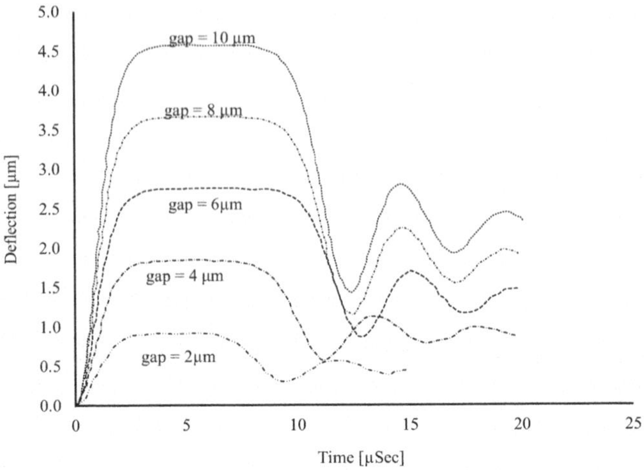

Fig. 4a. Pull-in voltage in different gaps

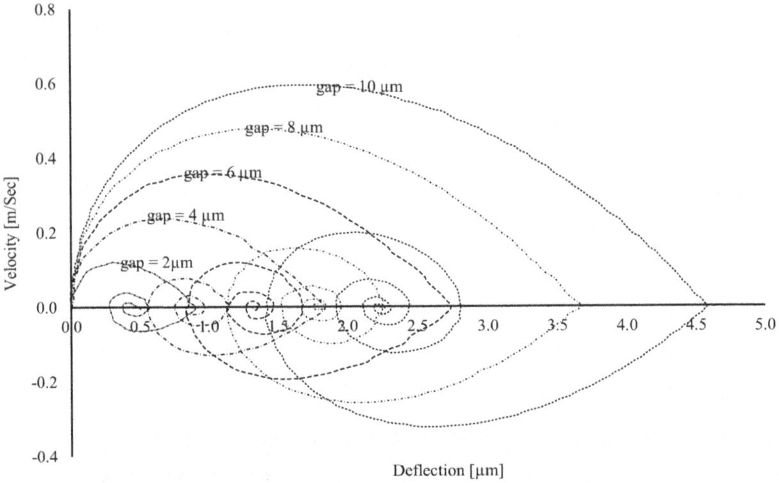

Fig. 4b. Phase diagram of pull-in voltage in different gaps

and a slight decrease back form 6 to 10 μm is observed. Closer is the applied voltage to the pull-in voltage, longer is the marginal stability duration is. The phase portrait for all assumed gap values provides same information as above and the results are illustrated in Fig. 4b. The variation of the deflection with the applied potential illustrated in Fig. 4c which shows a strong non-linear dependence of the deflection with the potential for any gap value. However, for low potential a linear dependency of the deflection with the voltage is noticed for a range of about 50% of the pull-in voltage. The resonant frequency of the system is dependent of the model and on the applied potential. For the

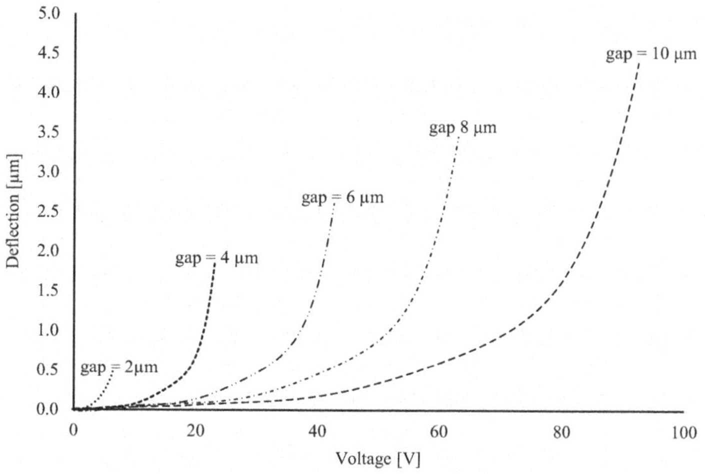

Fig. 4c. Effect of voltage on deflection of beams with different gaps

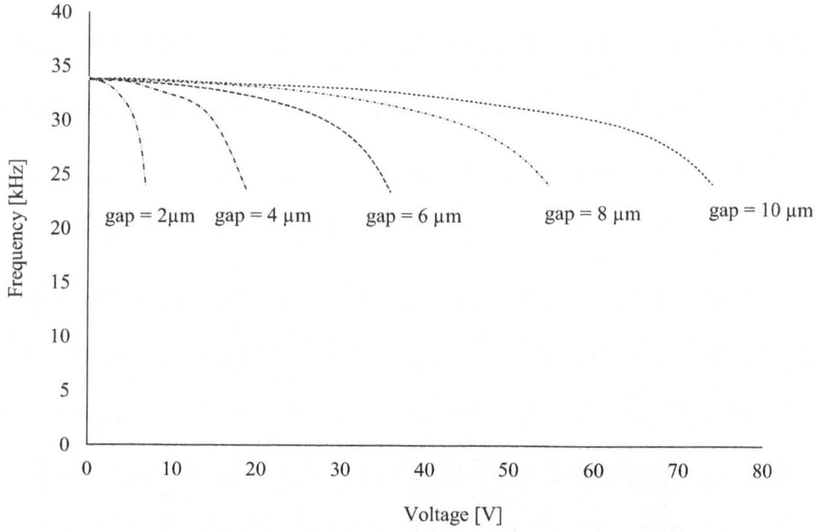

Fig. 4d. Effect of voltage on natural frequency of beam with different gaps

DCNF model, the resonant frequency of the system decays with the applied potential, faster for the lower gaps as illustrated in Fig. 4d. However, the frequency significantly drops when the applied potential is approaching the corresponding value of the pull-in voltage. The decrease is insignificant only for applied potentials that are remote from the pull-in voltage.

4 Experimental Validation

Sets of experiments to measure natural frequency and deflection of micro-cantilever beams subjected to electrostatic forces were carried out in CONCAVE laboratories. Figure 5a illustrates the setup of the experiment which includes a HeNe laser and photodiode detector in a laser vibrometer configuration, from Brüel & Kjær. The measurement system made use of an oscilloscope for the time-domain reference and a spectrum analyzer for frequency domain analysis. The micro-structures were excited with sweep harmonic oscillation produced by an audio speaker from the internal signal generator in the frequency analyzer. To measure the frequency response, a swept low amplitude harmonic excitation was used. The test was performed using non-contact measurement, by focusing the laser beam onto the substrate to extract the frequency response of the substrate. Then the laser beam was focused onto the micro cantilever while same swept frequency was applied. Finally, micro-cantilever response was calculated with respect to the base response. A DC power supply was used to obtain the deflection of micro cantilever and the static deflection was measured under an optical microscope. The deflection was measured from the overlapped pictures under microscope of the deflected cantilevers when specific voltages were applied, as illustrated in Fig. 5b. Two micro-cantilevers with dimensions as presented in Table 3 were studied.

Table 3. Dimensions of the selected beams

	Length [μm]	Width [μm]	Thickness [μm]	Young's modulus [MPa]
Beam I	351	34.5	0.94	169.5
Beam II	299	35	0.96	169.5

The resonance frequency of the beams subjected to the electrostatic forces was measured using the above-described procedure while a potential field was applied between the substrate and the beam and resonant frequency was determined for the various applied voltages.

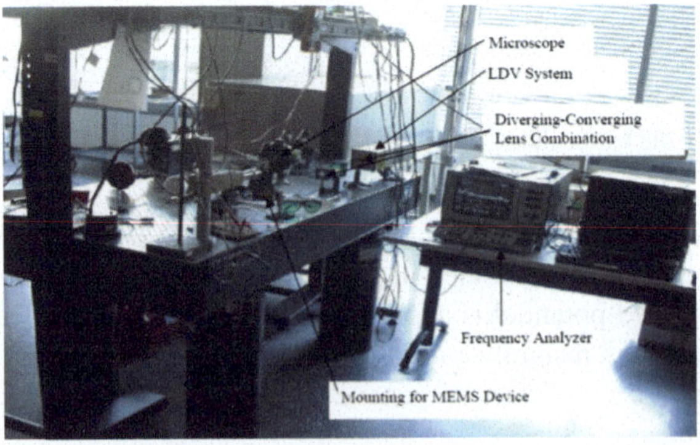

Fig. 5a. Experimental setup.

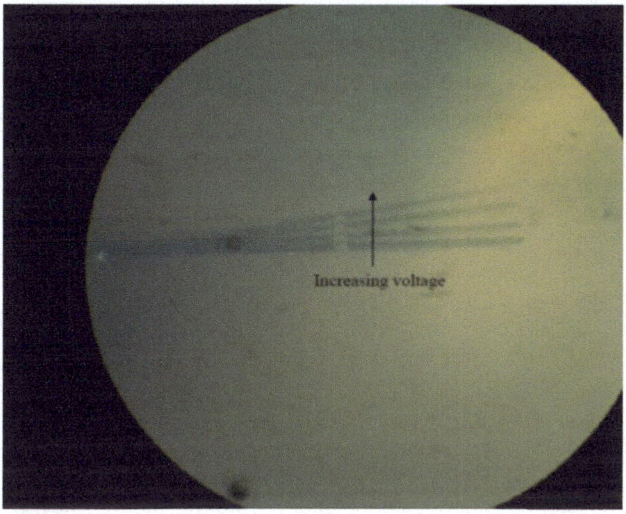

Fig. 5b. Positions of micro-cantilever under different applied potentials. The charged electrode (not seen in the picture) was positioned parallel to the micro-beam on the upper side of the picture.

The experimental and the calculated deflections under different voltages for two beams are illustrated (see Figs. 6 and 7). For both beams, maximum applied voltage for measuring the static deflection was 500 V. The four different theoretical models, as detailed in Sect. 4 were considered for validation. As illustrated, the best-fit curve for beam I is for $\beta = 0.775$ and while for beam II for $\beta = 0.73$. These values were calculated based on the error minimizing. The error was calculated as the relative error of the theoretical deflection with respect to the experimental deflection for each voltage. Summation of squares of these errors was considered as target value to minimize. The values of the effective stiffness as recommended in are significantly higher than reported values in [32, 33]. By increasing the value of β, the effective area decreases. If effective area decreases, the necessary applied force will decrease. It is clear that by decreasing the force, deflection will decrease. In this experiment, deflection is large so the area under the electrostatic force will be reduced; by reducing force deflection will also reduce. It will need to increase the value of β to match the experimental results with the model. According to the theoretical models presented and considering the above-mentioned values for β, the modified model has proved to yield closer results to experimental values. However, the matching value for β is considerably higher than the range recommended in [32]. The modified model proves to be the most suitable with the experiment for matching the deflection of the micro-cantilever beam.

The resonant frequency changes with the application of the potential electrostatic field. The theoretical and experimental findings are illustrated in Figs. 8. The four different models, which are CCNF, DCNF, Modified model with $\beta = 0.33$ and Modified model with $\beta = 0.45$ were considered and instead of using algebraic equations, differential equations, with considering the effect of these parameters, were solved. As

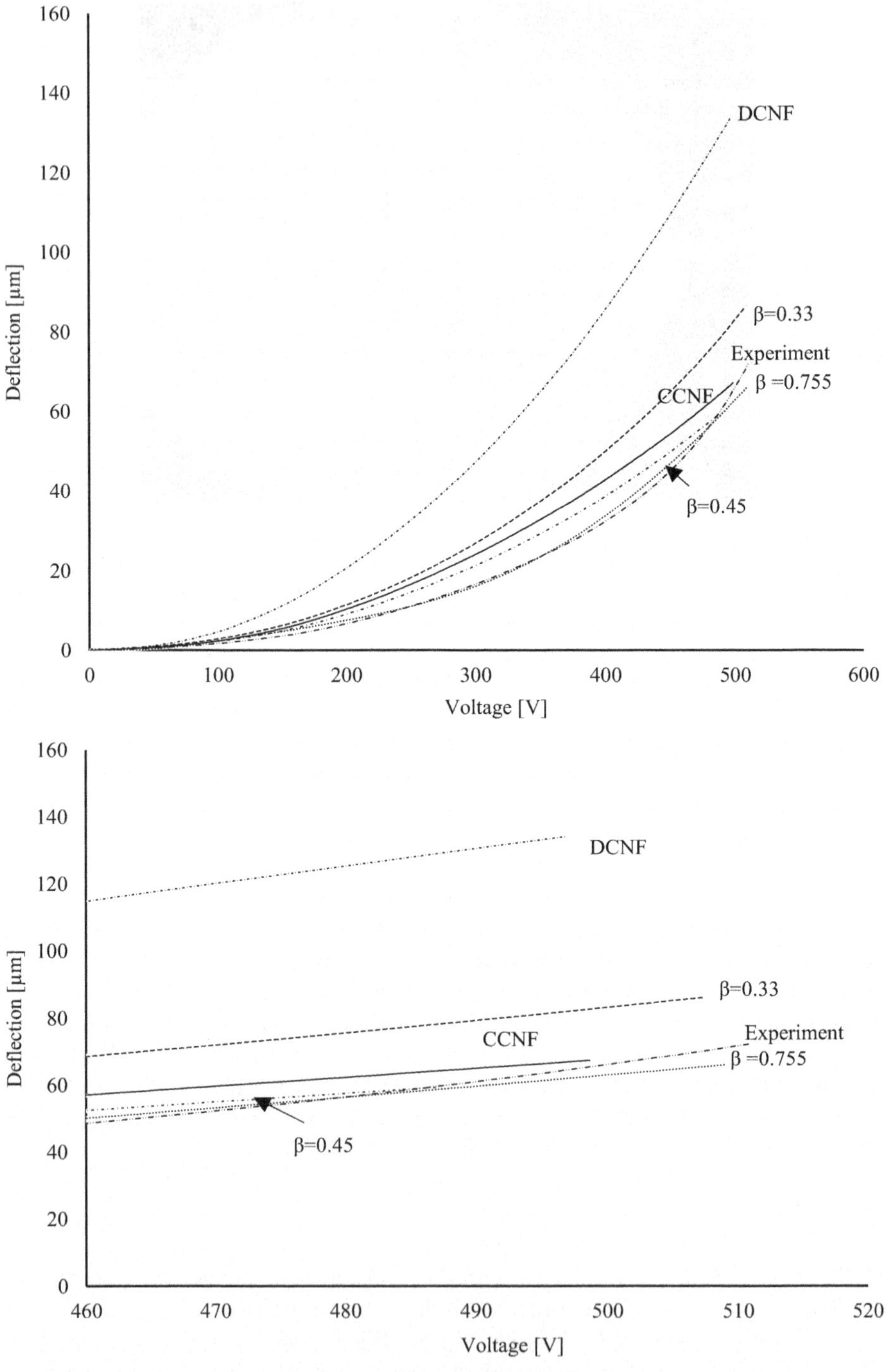

Fig. 6. (a) Deflection as a function of the applied potential; theoretical and experimental values of the deflection of the micro-cantilever beam I; (b) Zoom-in of the indicated zone.

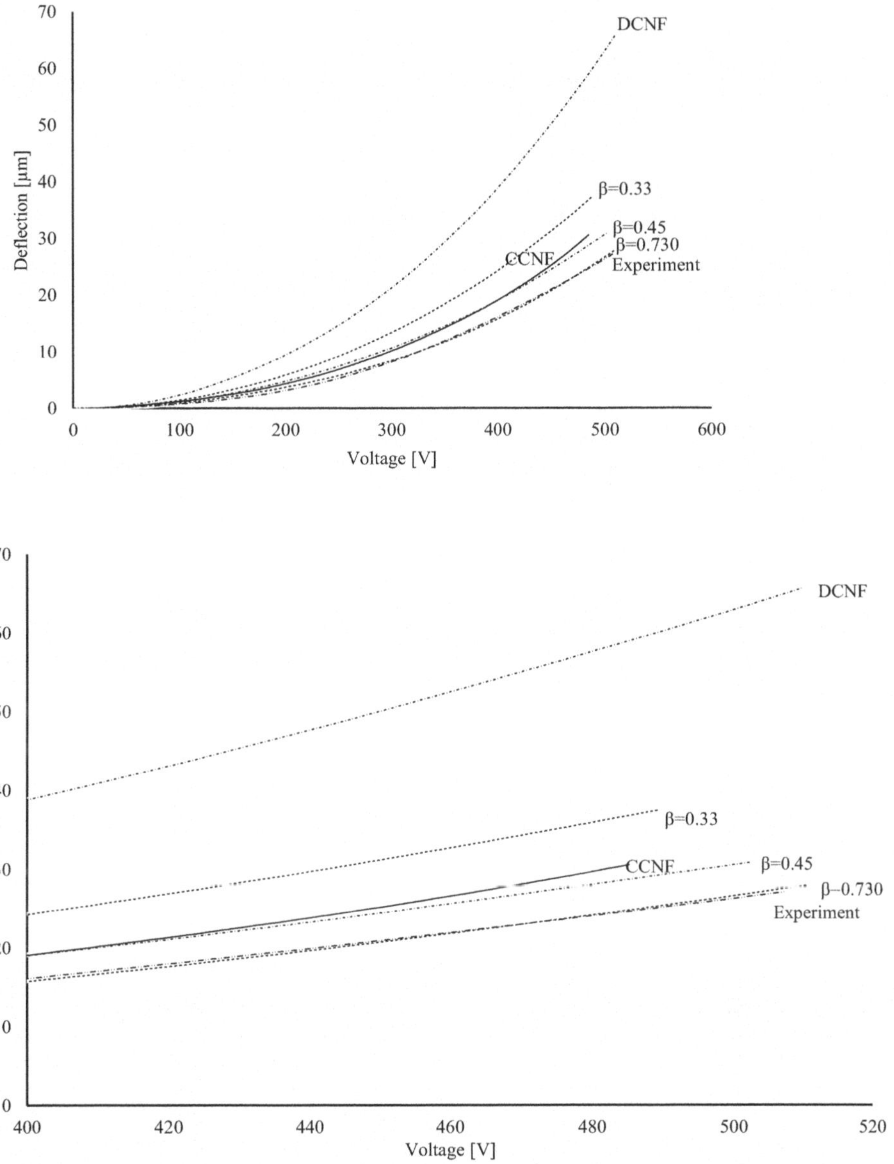

Fig. 7. (a) Deflection as a function of the applied potential; theoretical and experimental values of the deflection of the micro-cantilever beam II; (b) Zoom-in of the indicated zone.

one can see from the two figures, the best fit with experimental is not anymore, a modified model but the CCNF model. Modified stiffness selected within its range gives results with small divergence but, however, far from experiments. Both cases yield values of the resonant frequency below the experimental values. Both figures show that linear approach for the stiffness yield results that are significantly below the

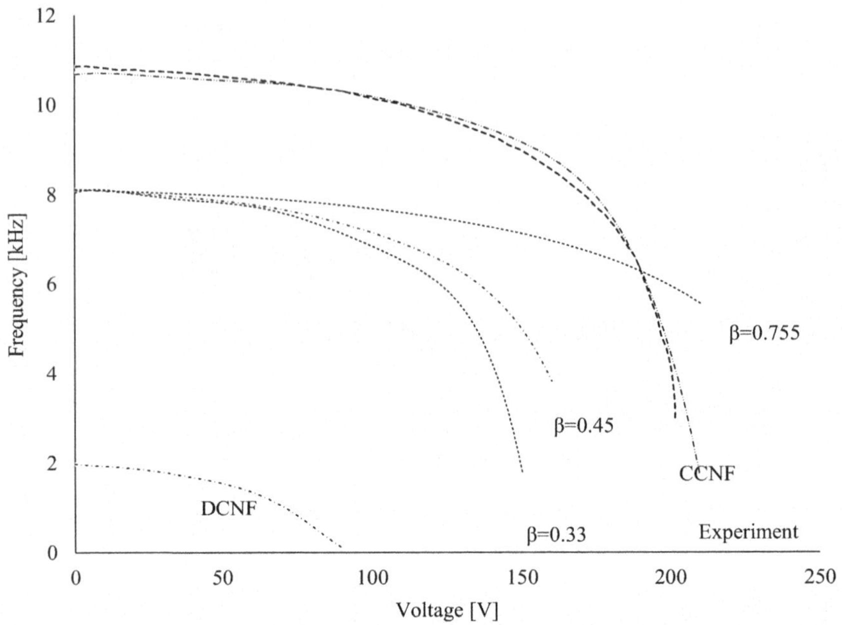

Fig. 8a. The frequency dependence on the applied voltage for beam I in comparison with the considered models.

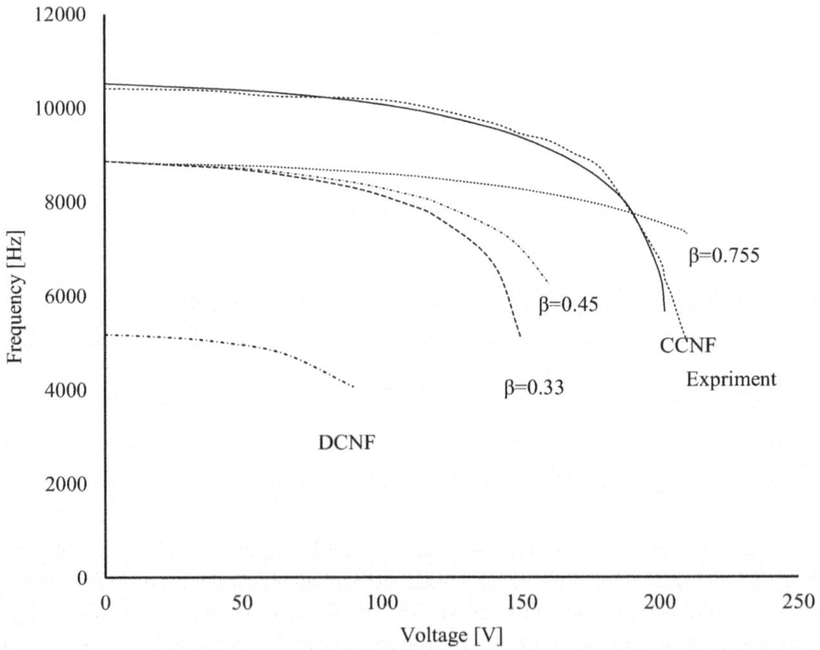

Fig. 8b. The frequency dependence on the applied voltage for beam II in comparison with the considered models.

experimental values, even less than the results produced by the modified models. In both vibration experiments, maximum applied voltage was 225 V. The error for modified stiffness and linear k ranges from about 35% and goes up to 200% with respect to the experimental recorded values.

The experimental results show that, although the modified model can well match the static deflection due to the electrostatic attraction in microstructures, the same model would not perform well for the dynamic response of the structure subjected to electrostatic field. The proposed CCNF can be a good matching model for dynamic model which instead, yields significantly large errors in the static deflection model.

5 Discussion

The analytical formulations in micro-systems have followed the formulations already available for the human-sized world. The simple explanation to this choice is that no other theories to be applied to micro-sized systems have been available. The quantum physics targets a much smaller size class. Still there is no known way to connect quantum mechanics to micro-mechanics. Most of the commonly used models of structures have been linearized for obvious reasons. The linearization would not necessary be suitable with the micro-sized systems and this is the motivation to re-consider the fundamental models while not neglecting potential non-linearities. The non-linear systems are represented either by non-linear ODE or PDE. Both type of equations rise significant challenges when required to be solved somehow by non-numerical methods. The sense of sensitivity of a certain parameter requires significant effort to be perceived. Usually, extensive parametric studies are required to gain such understanding. Otherwise, one could revisit mathematical methods to try solving the non-linear phenomenon-describing equations given the fact that an exact solution, if available, provides a wide perspective on the analysis of the behavior of the solution function. Under many circumstances, the popular routines for the well-known numerical methods which are embedded in multi-purpose software package such as Matlab, Maple, Mathematica, MathCAD or, Macsyma would yield no solution to stiff ODE. This is the case of the equation that describes the dynamic behavior of a micro-system subjected to an electrostatic field.

The time response of the system shows that a micro-system such as a micro-cantilever beam will respond to a voltage close to the pull-in value applied as a step input through a ramp that will hold in a marginal equilibrium position that occurs at a deflection that represents 4/9 of the initial gap. The system leaves the saddle point to settle in a position that corresponds to a deflection of 2/9 of the initial gap. Various models built based on the open literature show significant variance. The deflection and the resonant frequency of two micro-cantilever beams are experimentally measured and compared to the numerical methods for static and dynamic models. The best match model for deflection is different from the best match model for resonant frequency response when compared with experimental data. Thus, different models fit different experimental data.

The findings on this work bring more understanding on the phenomena at micro-size level.

Acknowledgment. NSERC Discovery Program is acknowledged to partial support this work.

References

1. Rinaldi, G., Packirisamy, M., Stiharu, I.: Dynamic synthesis of microsystems using the segment Rayleigh&-Ritz method. Microelectromech. Syst. **17**, 1468–1480 (2008)
2. Legtenberg, R., Gilbert, J., Senturia, S.D., Elwenspoek, M.: Electrostatic curved electrode actuators. Microelectromech. Syst. **6**, 257–265 (1997)
3. Schiele, I., Huber, J., Hillerich, B., Kozlowski, F.: Surface-micromachined electrostatic microrelay. Sens. Actuators A Phys. **66**, 345–354 (1998)
4. Hung, E.S., Senturia, S.D.: Extending the travel range of analog-tuned electrostatic actuators. Microelectromech. Syst. **8**, 497–505 (1999)
5. Castaner, I., Rodriguez, A., Pons, J., Senturia, S.D.: Pull-in time–energy product of electrostatic actuators: comparison of experiments with simulation. Sens. Actuators, A Phys. **33**, 263–269 (2000)
6. Chan, E.K., Dutton, R.W.: Electrostatic micromechanical actuator with extended range of travel. J. Microelectromech. Syst. **19**, 321–328 (2000)
7. Busta, H., Amantea, R., Furst, D., Chen, J.M., Turowski, M., Mueller, C.: A MEMS shield structure for controlling pull-in forces and obtaining increased pull-in voltages. J. Micromech. Microeng. **11**, 720–725 (2001)
8. Wu, J., Carley, L.R.: Table-based numerical macromodeling for MEMS devices, pp. 68–71 (2001)
9. Younes, M.I.: Investigation of the mechanical behaviors of microbeam-base MEMS devices. Virginia Polytechnic Institute and State University, Mechanical Engineering M.Sc Blacksburg (2001)
10. McCarthy, B., Adams, G.G., McGruer, N.E., Potter, D.: A dynamic model, including contact bounce, of an electrostatically actuated microswitch. Microelectromech. Syst. **11**, 276–283 (2002)
11. Zhang, W., Baskaran, R., Turner, K.L.: Effect of cubic nonlinearity on auto-parametrically amplified resonant MEMS mass sensor. Sens. Actuators **102**, 139–150 (2002)
12. Younis, M.I., Abdel-Rahman, E.M., Nayfeh, A.: A reduced-order model for electrically actuated microbeam-based MEMS. Microelectromech. Syst. **12**, 672–680 (2003)
13. Hu, Y.C., Chang, C.M., Huang, S.C.: Some design considerations on the electrostatically actuated microstructures. Sens. Actuators **112**, 155–161 (2004)
14. Kuang, J.H., Chen, C.J.: Dynamic characteristics of shaped micro-actuators solved using the differential quadrature method. J. Micromech. Microeng. **14**, 647–655 (2004)
15. Bruschi, P., Nannini, A., Paci, D., Pieri, F.: A method for cross-sensitivity and pull-in voltage measurement of MEMS two-axis accelerometers. Sens. Actuators **123–124**, 185–193 (2005)
16. Chowdhury, S.: A closed-form model for the pull-in voltage of electrostatically actuated cantilever beams. J. Micromech. Microeng. **15**, 756 (2005)
17. Zhang, W., Meng, G.: Nonlinear dynamical system of micro-cantilever under combined parametric and forcing excitations in MEMS. Sens. Actuators **119**, 291–299 (2005)
18. Pamidighantam, S., Puers, R., Baert, K., Tilmans, H.A.C.: Pull-in voltage analysis of electrostatically actuated beam structures with fixed-fixed and fixed-free end conditions. J. Micromech. Microeng. **12**, 458–464 (2002)
19. Zhang, W.M., Meng, G., Chen, D.: Stability, nonlinearity and reliability of electrostatically actuated MEMS devices. Sensors **17**, 760–796 (2007)

An Approach of Extracting Features for Fault Diagnosis in Bearings Using the Goertzel Algorithm

Daniel Cordoneanu(✉) and Constantin Nițu

Department of Mechatronics and Precision Mechanics,
University "Politehnica" of Bucharest,
Splaiul Independentei 313, 060042 Bucharest, Romania
daniel.cordoneanu@stud.mec.upb.ro

Abstract. Fault diagnosis has been a field of interest in the latest period, especially predictive maintenance, given the advances in artificial intelligence and state of the art machine learning algorithms available in a great number of libraries. In the industrial sector, fault diagnosis plays a very important role in order to avoid as much as possible downtime. Usually, rotating motors are involved in the actuation of the machines used in industry; therefore bearings are an important part of the kinematic chain. Given that faults in bearings can be detected in the frequency spectrum at frequencies that can be mathematically computed based on geometry, this paper proposes an approach to extract features for machine learning algorithms based on the computed frequencies and their harmonics. Since only a few frequencies are needed, the Goertzel algorithm can be used instead of the discrete Fourier transform to give a computational boost and have the feature extraction algorithm available on embedded systems.

Keywords: Goertzel · Bearings · Fault diagnosis · Vibration analysis · DFT · Features

1 Introduction

Fault diagnosis domain gets more and more attention nowadays as the hardware and software advancements provide more power and more efficient ways to assess whether a machine has a fault or not. Of course, fault diagnosis means, besides fault detection, an identification of the fault and the part of the machine that is broken. The importance of detecting an early fault can save a lot of money, as maintenance can be properly scheduled and the faulty part can be repaired before it collapses entirely and increases the downtime of a machine.

In [1], Natu describes the number of faults that can appear in a rotating machine per part: 40% bearing faults, 38% stator faults, 10% rotor faults, others 12%. Since almost half of the faults occur in bearings, it is of high importance to have these parts monitored, as an issue in these rolling elements is usually provoked by another component or suggests improper lubrication or wearing-off, in which case the bearing has to be repaired or replaced so that other components are not affected in time. In literature there are many algorithms described which usually use the spectrum analysis on which

G. I. Gheorghe (Ed.): ICOMECYME 2019, LNNS 85, pp. 173–183, 2020.
https://doi.org/10.1007/978-3-030-26991-3_16

different filter banks are applied, as in [2], where mel filter banks are used to extract features as mel cepstral coefficients that are later used in an algorithm that uses gaussian mixture models and kurtosis. Kurtosis is successfully used as well in [3] alongside envelope spectrum analysis. In [4], Nabhan et al. summarize fault detection techniques for ball bearings and shows that vibration measurements and spectrum analysis techniques are the most useful tools for fault diagnosis in rolling bearings. Since most of these techniques rely on the Fast Fourier Transform (FFT), which in acquisition and signal processing devices is done as the Discrete Fourier Transform (DFT), a more efficient algorithm can be used to extract specific frequencies information that doesn't have to deal with spectral leakage and is not dependent of the number of points of the DFT.

2 Bearing Fault Condition

2.1 Rolling Bearing Description

A rolling bearing is a rotating machine element that enables the movement by reducing friction and handles the stress coming from the linked components. In [5], bearing fault spectral analysis as well as other methods are described. In Sect. 4.3, it is explained that a fault can be indicated if there is a peak at the fundamental fault frequency and the 2nd harmonic of this frequency; also, if there is no peak at the fundamental fault frequency, but there are peaks in the next harmonics, then the fault is real. In Figs. 1 and 2, the geometric variables that help finding the fundamental fault frequencies are shown:

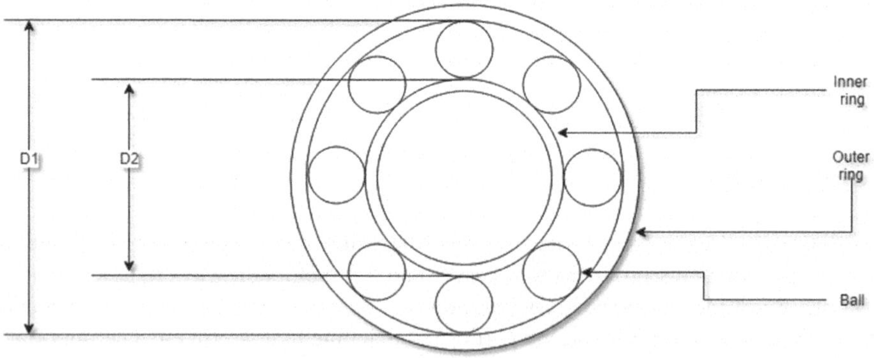

Fig. 1. Bearing elements

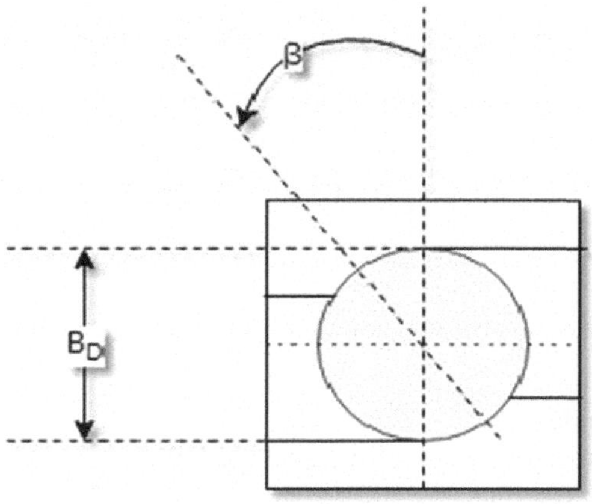

Fig. 2. Bearing ball

2.2 Types of Faults in Bearings

Given the geometries in the above figures, in [5] the causes of the bearing failures are described: bad lubrication, heavier load than expected, wear off because of time, misplacement of shaft, etc. Initially, bearing fatigue results in shear stresses below the load-carrying surface. In time, these shears turn to cracks in the exterior surface and as the load goes over them, fragments are ripped apart from the bearing. This type of fault is usually assessed through vibrations since mechanical waves that are produced by the cracks get higher amplitude as time goes. Surface distress is another type of fault that can produce cracks and it's usually provoked by improper lubrication or heavier load.

To diagnose properly a bearing fault, the following bearing frequencies can be calculated:

- F_{ord} = Frequency Outer Race Defect
- F_{ird} = Frequency Inner Race Defect
- F_{bd} = Frequency Ball Defect
- F_c = Frequency Cage

Based on Fig. 1, we can determine the following variable as the pitch diameter:

$$P_d = \frac{D_1 + D_2}{2} \tag{1}$$

The above described frequencies can be computed as follows:

$$F_{ord} = \frac{nf_{rotation}}{120}\left(1 - \frac{B_d \cos(\beta)}{P_d}\right) \tag{2}$$

$$F_{ird} = \frac{nf_{rotation}}{120}\left(1 + \frac{B_d \cos(\beta)}{P_d}\right) \tag{3}$$

$$F_{bd} = \frac{f_{rotation}}{60} * \frac{P_d}{B_d}\left[1 - \left(\frac{B_d \cos(\beta)}{P_d}\right)^2\right] \quad (4)$$

$$F_c = \frac{f_{rotation}}{120}\left(1 - \frac{B_d \cos(\beta)}{P_d}\right) \quad (5)$$

Where:

- n is the number of balls
- B_d is the ball diameter
- β is the contact angle
- $f_{rotation}$ is the rotation frequency of the shaft in RPM

Further, these base frequencies and their multiples will be of interest when computing the DFT magnitude using the Goertzel algorithm.

3 Goertzel Algorithm

3.1 Standard Goertzel Algorithm

As shown in [6] by Sysel and Rajmic, the original algorithm described by Goertzel in [7] that computes the DFT term of a signal x[n] with length N can be treated as a discrete linear convolution between the signal x[n] and $h_k[n]$ so that if the result of the convolution is $y_k[n]$, then:

$$y_k[m] = \sum_{n=0}^{N-1} x[n] e^{j2\pi k \frac{m-n}{N}} u[m-n] \quad (6)$$

Equation (6) is derived from the fact that h_k can be expressed as:

$$h_k[l] = e^{j2\pi k \frac{l}{N}} u[l] \quad (7)$$

As Sysel and Rajmic further show in their paper, Eq. (6) can be treated as an IIR linear system with the impulse response $h_k[n]$; the output sample N of this filter is the desired DFT component. Described as a second order IIR filter using differences, the equation of the filter can be written using state variables:

$$s[n] = x[n] + 2\cos\left(\frac{2\pi k}{N}\right) s[n-1] - s[n-2] \quad (8)$$

With the final output being:

$$y_k[n] = s[n] - e^{-j\frac{2\pi k}{N}} s[n-1] \quad (9)$$

As the author himself mentioned, only N multiplications and 2N additions are needed.

3.2 Goertzel Algorithm Compared to the DFT

There are multiple advantages of the Goertzel algorithm over the DFT under certain conditions. If $K < 4/7N$ [6], where K is the number of frequencies in which we're interested in, then the Goertzel algorithm is superior in computation speed to the DFT.

Another advantage of the Goertzel algorithm is that one can inspect a signal of length N without bothering that N is a power of 2 (case in which DFT is computationally fast).

Also, while inspecting the spectrum, based on the sampling frequency and the number of samples we don't always have the correct magnitude of a certain frequency since it can't be represented by the DFT due to spectral leakage (e.g. if we have a sampling frequency of 12 kHz with 1024 number of points, the frequency per bin of the DFT would be 12 kHz/1024 = 11.71 Hz per bin so that if we want to know the magnitude of the sinusoid with a period of 115 Hz, this would be leaked to the neighboring frequency bins of 117.1 Hz and 105.39 Hz).

3.3 Generalized Goertzel Algorithm

With a small computation expense, in [6] a generalized Goertzel algorithm is proposed that can use K as a real number, not only as an integer. This is helpful in bearing fault diagnosis case because the frequencies computed will most probably be real numbers since the cosine function is involved and also division to different numbers of different quantities.

4 Feature Extraction and Results

4.1 Proposed Algorithm

The proposed algorithm in this paper is a very simple one, yet it's one that can be computationally efficient and can be used to extract features for a machine learning algorithm, be it a multi-class identifier like a neural network or just an anomaly detection algorithm using gaussian distribution to assess whether the values recorded in the harmonics of the fault frequencies are part of the side probabilities of the normal distribution. The algorithm can be described in the following steps:

- Find the rotation frequency of the system that rotates the bearing
- Find the specifications of the bearing to be monitored
- Compute the frequencies described in (2), (3), (4), (5)
- Compute the harmonics of the frequencies up to 10 harmonics (as described [5], the magnitudes should rise in up to the 4th harmonic, but it's echoes should be visible in the higher spectrum as well)
- Apply the Goertzel algorithm on the computed frequencies to extract the DFT components and computer the squared length of the vector described by the components
- Split the computed harmonics per fault or use them all as features, depending on the chosen machine learning algorithm

In Fig. 3 we can see the flowchart for the algorithm.

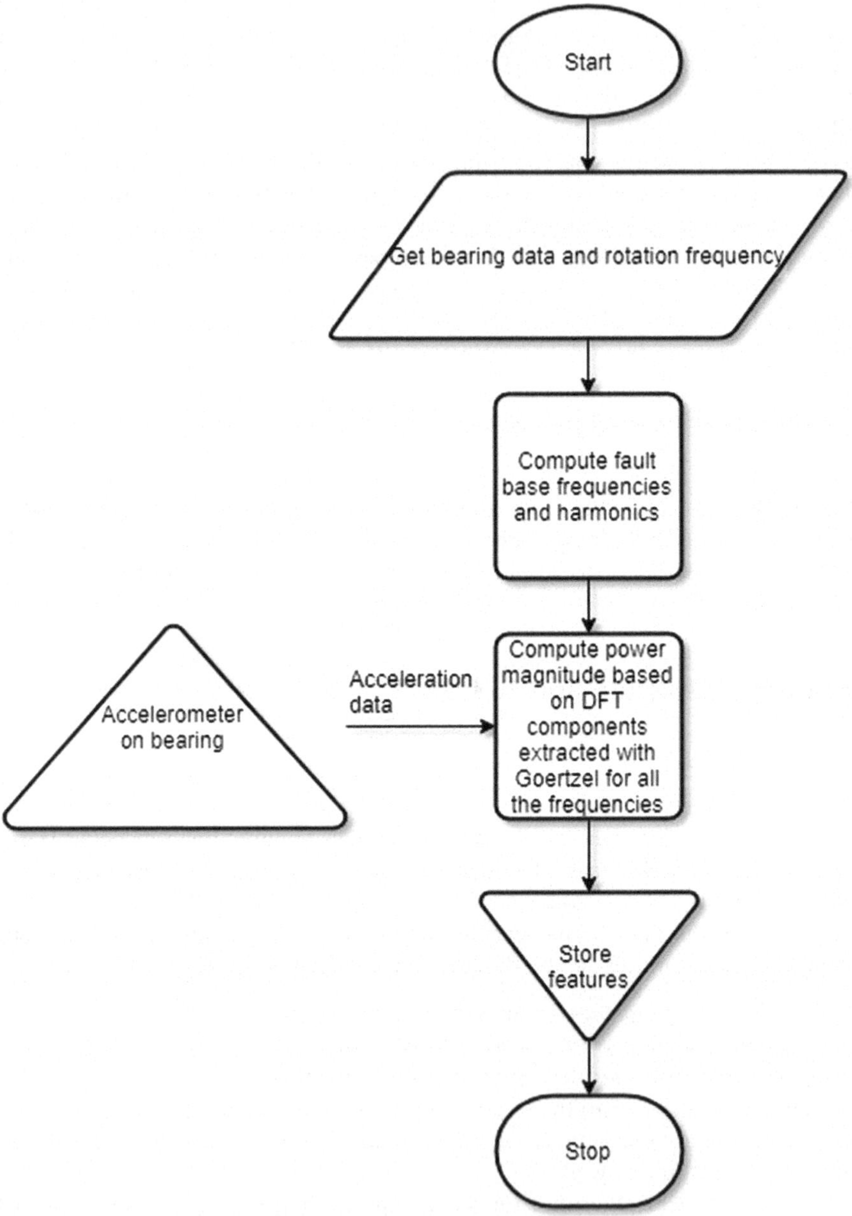

Fig. 3. Algorithm flowchart

4.2 Experimentation

To check the algorithm's results, the data obtained in the experimentation done by Lou and Loparo in [8] was used. The data comes from Rockwell Science Center and it's available on [9]. For the experiments, a 2 hp Reliance Electric motor was used and acceleration data was fetched from the motor bearings.

Faults were made to the motor bearings using electro-discharge machining (EDM). Faults diameters ranged from 0.007 to 0.040 in. and data acquisition was made at 12 kHz for bearings in normal conditions and faulty conditions at different speeds of the motor. Also, tables with fault frequencies are available directly on this resource so that the computations were easily made. Two bearings were used on the drive end and fan end.

The proposed algorithm will be applied on the drive end bearing (6205-2RS JEM SKF, deep groove ball bearing), for a motor speed of 1797 rpm.

In Table 1, the bearing's data is presented and in Table 2 the fault frequencies for 1797 rpm:

Table 1. Bearing data in mm

Inside diameter	Outside diameter	Thickness	Ball diameter	Pitch diameter
25	52	15	7.94	39.03

Table 2. Bearing fault frequencies (Hz) for 1797 rpm

F_{ord}	F_{ird}	F_{bd}	F_c
107.36	162.18	141.16	11.92

To be noted that in the data analysis, since there were no cage faults, the Fc was ignored.

4.3 Results

The goal in fault diagnosis is to detect the fault as early as possible, so the important data is the one for the faults with the diameter of 0.007 in. However, also the 0.014 faults were added for comparison.

In the following figures, ball fault, inner race fault and outer race fault data is shown in comparison with the normal data. The original data was split into batches of 5000 points each which is the equivalent of 0.41 ms. Here we can see that the Goertzel algorithm has no restriction regarding the number of points, nor the computed frequencies.

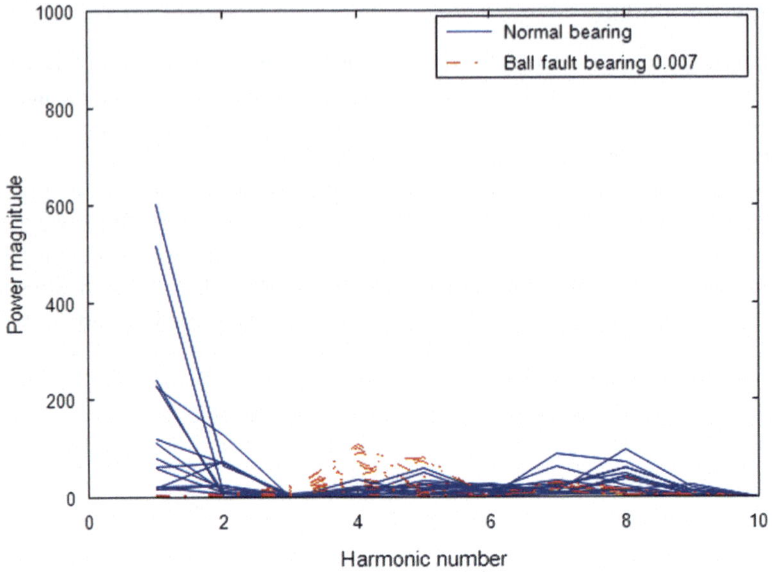

Fig. 4. Ball fault and normal data in 0.007 in. fault diameter

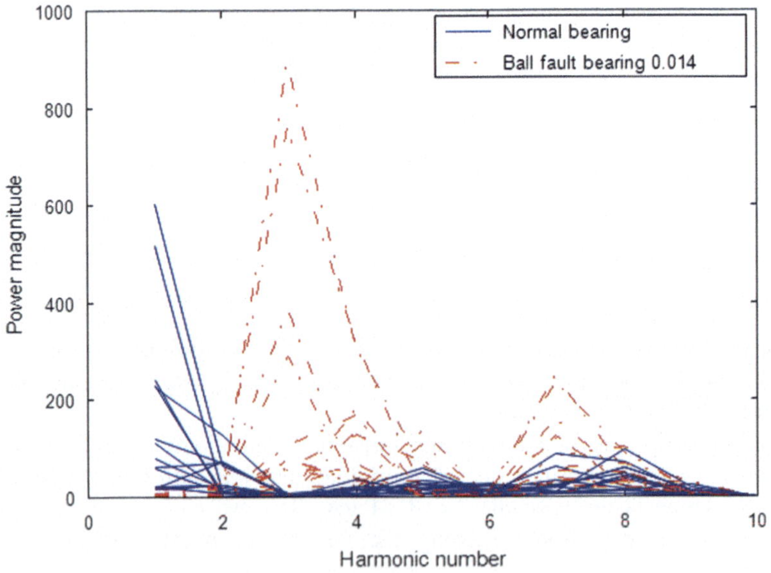

Fig. 5. Ball fault and normal data in 0.014 in. fault diameter

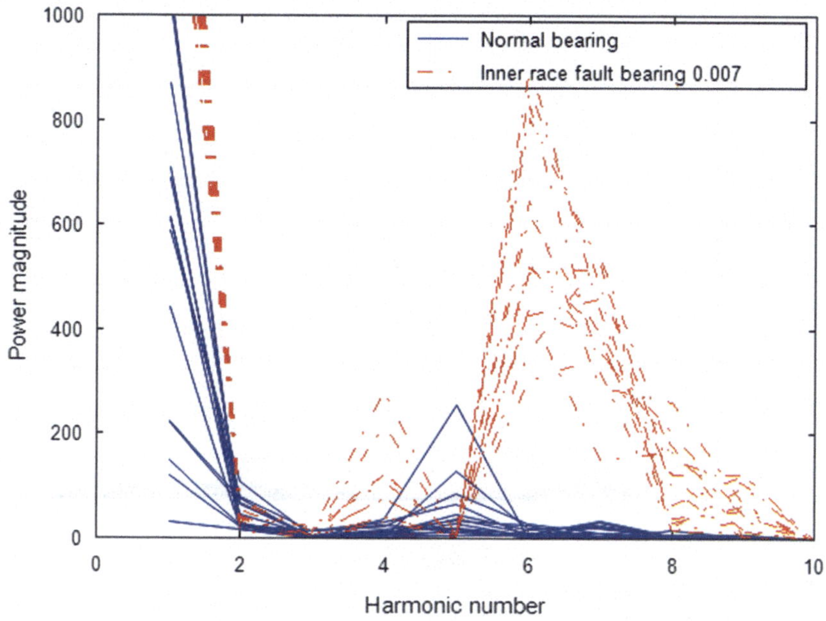

Fig. 6. Inner race fault and normal data in 0.007 in. fault diameter

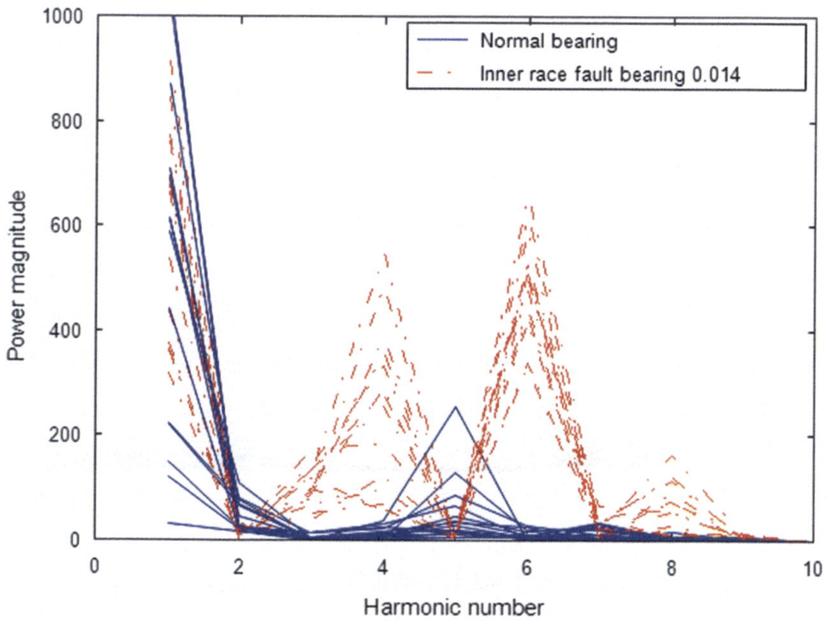

Fig. 7. Inner race fault and normal data in 0.014 in. fault diameter

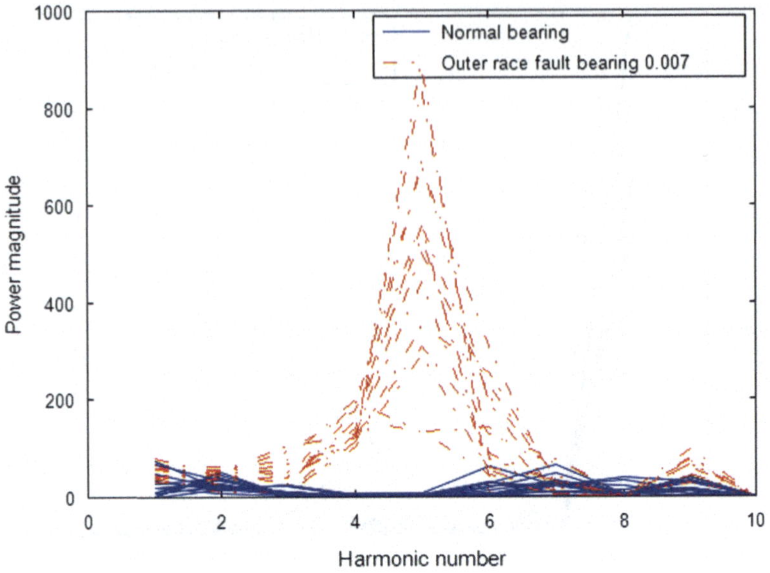

Fig. 8. Outer race fault and normal data in 0.007 in. fault diameter

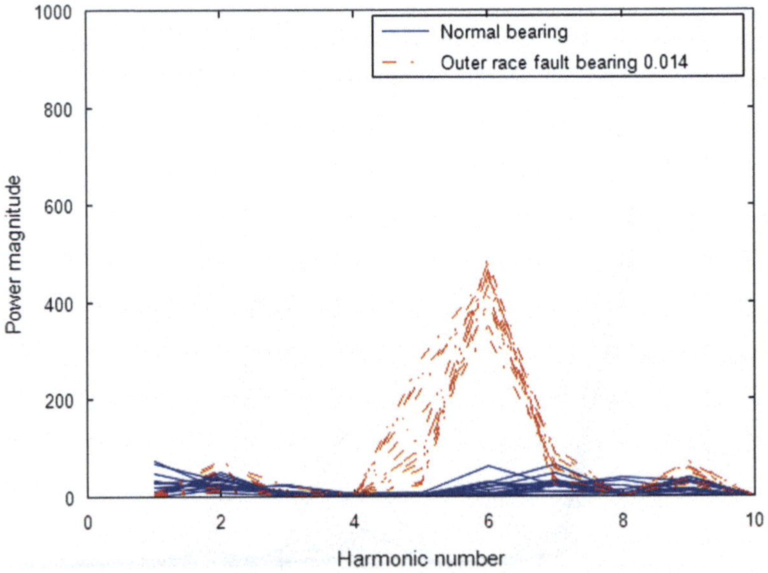

Fig. 9. Outer race fault and normal data in 0.014 in. fault diameter

5 Conclusions and Future Work

As it can be seen in Figs. 4 and 5 faults can be separated from the normal behavior in multiple training examples with the mention that for this type of fault (ball fault), as the fault is small in diameter, the magnitude is very small compared to the 0.014 in. fault but it can be separated from the normal data magnitudes. For the other type of faults (visible in Figs. 6, 7, 8 and 9), the faulty data can be easily separable from the normal data.

Therefore we can conclude that the proposed algorithm can be used to extract features for fault diagnosis in bearings without using the DFT, but a more computational efficient algorithm. The same approach can be used for other mechanical parts where fault frequencies can be computed.

In the future, this algorithm will be tested with different machine learning algorithms in order to assess its accuracy based on the way the features are separated.

References

1. Natu, M.: Bearing fault analysis using frequency and wavelet techniques. Int. J. Innov. Manag. Technol. **15**(6), 72–74 (2012)
2. Nelwamondo, F.V., Marwala, T.: Faults detection using gaussian mixture models, mel-frequency cepstral coefficients and kurtosis. In: Conference Proceedings - IEEE International Conference on Systems, Man and Cybernetics (2007)
3. Wan, S., Zhang, X., Dou, L.: Compound fault diagnosis of bearings using improved fast spectral kurtosis with VMD. J. Mech. Sci. Technol. **32**(11), 5189–5199 (2018)
4. Nabhan, A., Ghazaly, N.M., Samy, A., Mousa, M.O.: Bearing fault detection techniques - a review. Turkish J. Eng. Sci. Technol. (2015)
5. SKF, Bearings Spectrum Analysis (2012)
6. Sysel, P., Rajmic, P.: Goertzel algorithm generalized to non-integer multiples of fundamental frequency. EURASIP J. Adv. Signal Process. **2012**(1), 56 (2012)
7. Goertzel, G.: An algorithm for the evaluation of finite trigonometric series. Am. Math. Mon. **65**(1), 34–35 (1958)
8. Lou, X., Loparo, K.A.: Bearing fault diagnosis based on wavelet transform and fuzzy inference. Mech. Syst. Signal Process. **18**(5), 1077–1095 (2004)
9. Bearing Data Center. http://csegroups.case.edu/bearingdatacenter/home. Accessed 2 June 2019

Stand for Characterization of Shape Memory Wires

Emil Niță(✉) and Daniel Comeaga

University "Politehnica" of Bucharest, 060042 Bucharest, Romania
nitaemilionut@yahoo.com

Abstract. The paper presents a stand for tracing the voltage/displacement characteristic of actuators made by shape memory alloys such as: (Ni-Ti), cooper-zinc-aluminum (Cu-Zn-Al) and copper-aluminum-nickel (Cu-Al-Ni). The stand includes precision measuring equipment, electronic tensometer, signal acquisition equipment and dedicated software. In the paper is presented the working algorithm and the program designed to measure the hysteresis curve for different shape memory alloys at specific external loads.

Keywords: Hysteresis · Ni-Ti alloy · Shape memory wires

1 Introduction

Shape memory materials are materials that have a particular ability to respond to a series of external forces. This ability consists in changing the shape or properties under the action of the external constant mechanical loads due to an additional control signal and returning to the initial form even the external load remains constant after the external control signal, usually temperature, stops. For a constant control signal, the system reacts as a common elastic element under external mechanical load, with a linear or non-linear behavior. For example, change of shape caused by temperature variation is called the memory effect of the thermally induced form.

The "memory" of these alloys consists in the property of the material being under mechanical stress that leads to the deformation of the body made of this material, deformation which is removed only when the memory material is kept under a high recovery form temperature. Another outstanding feature of this material is the super-elasticity of shape memory alloys, which is successfully exploited especially in the construction of the springs [1]. This phenomena is explained by the presence of low symmetry martensite transformation phase for low temperatures or ambient temperature transformed in a phase with high crystallographic symmetry (austenite) with the heating of the material. This structural reorganization has macroscopic effects that lead to dimensional variations of alloys, which can be exploited in the construction of energy efficient actuators and in reducing the size of the devices.

The two above-mentioned properties: the shape memory effect and the pseudo-elasticity are due to structural changes, which consist in the molecular rearrangement of the alloy without phase changes (solid to liquid). The explanation consists in the modification of interatomic, intermolecular distances within certain limits, without the

G. I. Gheorghe (Ed.): ICOMECYME 2019, LNNS 85, pp. 184–193, 2020.
https://doi.org/10.1007/978-3-030-26991-3_17

destruction of these bonds. The material still remaining in a solid state. The two phases that allow changes in the physical characteristics of the alloy under specific thermodynamic conditions are called the Martensitic phase, and the Austenitic phase (Fig. 1).

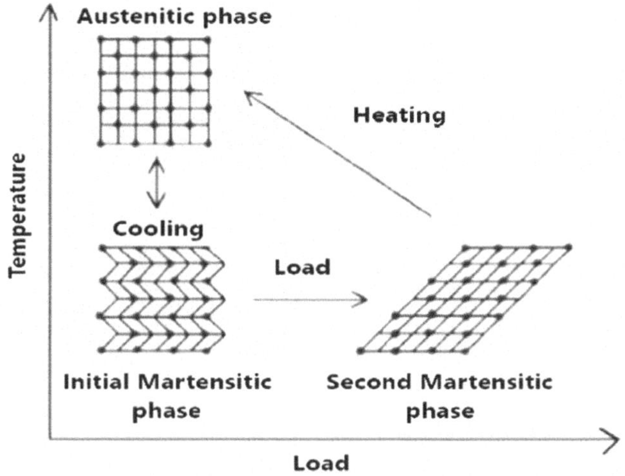

Fig. 1. Crystallographic transformation of the material [1]

Although there are at least 20 known alloy families that exhibit shape memory effect, currently only 3 have commercial importance: nickel-titanium (Ni-Ti), copper-zinc-aluminum (Cu-Zn-Al) and cooper-aluminum-nickel (Cu-Al-Ni). The development of these alloys was first made for low-temperature military applications, such as drainage connectors, low force insertion connectors, etc. using Ni-Ti or Cu-Zn-Al.

The amount of displacement that can be obtained from a shape memory component is limited by the total number of atoms that are reoriented at low temperature, deforming the cubic structure from the high temperature. The magnitude of movement is different from one alloy to another. For Ni-Ti, the maximum displacement is 8% (this means that 300 mm of nickel-titanium wire can be deformed with a 24 mm displacement). If the nickel-titanium shape memory element returns to free form at high temperature under external forces, this 8% displacement can be made constantly. If significant external forces are applied to the same element during form recovery, then the displacement obtained after repeated cycles will be less than 8% [2].

The most important phenomena of shape memory are:

1. Pseudo-elasticity effect (PSE);
2. Simple Shape Memory effect (EMF);
3. memory effect in both ways (EMFDS)
4. vibration damping effect.

The simple shape memory effect (EMF) is the unique and spontaneous recovery of "cold form" after heating the material.

The warm form is characteristic of the austenitic phase, and the cold form of the martensitic. The clearest evidence of this phenomenon is achieved by the variation of the elongation in relation to the voltage and temperature, as shown in Fig. 2.

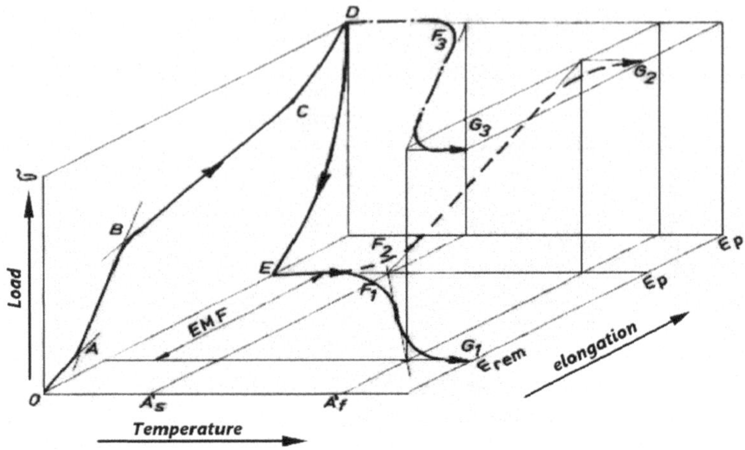

Fig. 2. Illustration of the simple shape memory effect (EMF) through the schematic curves in the voltage-deformation-temperature space [2]

In Fig. 2, there are several curves that highlight the simple effect of shape memory according to spatial restrictions and temperature, namely:

- Curve EF1G1 - free return EMF;
- Curve EF2G2 - EMF restricted return;
- Curve DF3G3 - EMF generator of mechanical forces;

2 The Current State of Achievements in the Field

Shape memory alloys are more and more present in modern medical applications such as: blade type implants, shape memory osteosynthesis plates for complex fracture reconstruction, Amphetz Septal Device (Ni-Ti Wire Mesh), etc. [3] In the mechanics industry, these alloys are used to drive mechanical arms as a mechanical work generator, for gripper mechanisms, for fastening and control mechanisms. For all this, it is necessary to determine a characteristic behavior of these alloys [4].

One way to trace the shape memory feature of these alloys is using the software simulation with the finite element method. Knowing the elasticity modules of the alloy materials and the weight of them inside the material a theoretical mathematical modeling and a simulation can be done. The process is complex and requires advanced knowledge of finite element analysis but has the advantage of flexibility. Material properties as well as alloy composition elements can easily be changed. The chemical alloy percentage is also essential and software can be changed as easily [7].

The major disadvantage of this method of testing and tracing the characteristics of materials is that simplifications are often used to reduce the number of mathematical equations that appear in the model. For example, all external actions on the tested element, such as the influence of the magnetic field in the environment, the influence of radio waves, etc. cannot be taken into account. This involves the occurrence of errors in the behavior of the physical element. It does not fully respect the theoretical mathematical model. A mathematical model close to reality requires staff with a higher education qualification able to make a synthesis of the studied phenomenon [4–6].

3 Experimental Stand for Shape Memory Wires Testing

Another method used to characterize shape memory alloys is using experimental laboratory stands. The test stand and the test algorithm with the corresponding program are present in the paper. The latter two were designed to test a Ni-Ti alloy shape memory wire, 0.15 mm thick and 120 mm long. The wire is kept under mechanical stress to elongate by attaching a mass to one of the ends.

To observe the changes in modulus of elasticity relative to mechanical stress, the Ni-Ti wire is stressed with three different masses.

3.1 Hardware Structure of the Stand

The experimental stand consists of the following equipment:

A. NI USB 6008 Acquisition Card (Fig. 3) used for signal acquisition and generation with the following features:
   - Analog inputs:
     - Differentials 4
     - Simple 8, selectable software
   - Input resolution:
     - 12 bits on differential inputs
     - 11 bits on simple inputs
   - Maximum sampling rate: 10 kS/s
   - Type of converter: successive approximations
   - AI FIFO: 512 bytes
   - Timeout resolution: 41.67 ns (24 MHz timebase)
   - Timing accuracy: 100 ppm of the current sampling rate
   - Field of entry:
     - Differential ±20 V, ±10 V, ±5 V, ±4 V, ±2.5 V, ±2 V, ±1.25 V, ±1 V
     - Simple ±10 V
   - Working voltage: ±10 V
   - Input impedance: 144 kO
   - Overvoltage protection: ±35 V
   - Trigger source: digital trigger, software or external trigger;

B. Shape memory wire and inductive differential transducer for measuring wire elongation (Fig. 3); the wire receives a current signal through the two terminals, one at each end of the wire. Due to the Joule effect, when the current passes through the wire, it heats up and the phenomenon of phase transformation occurs. The wire is strained by applying at the free end a weight. Unheated, the modulus of elasticity of the wire material is lower than in the heated state. As a consequence, in the cool state the wire is significantly deformed but upon its heating occurs the contraction. The lower end of the wire is rigidly secured by the movable metal core of the inductive differential transducer, so the wire shrinking will also move the core of the transducer. Movement of the core produces a variation in the inductance of both coils that compose the transducer (increasing the inductance of one of the coils and decreasing the inductance of the other coil) to be measured with the electronic tensometer.

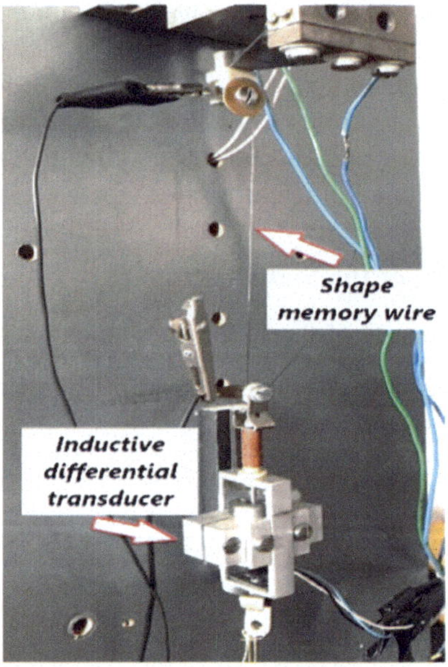

Fig. 3. Shape memory wire and inductive transducer

C. Electronic tensometer (model N2302) for measuring the inductance variation of the inductive differential transducer coils. It converts the inductance variation using a half Wheatstone bridge, where the two coils of the inductive transducers are connected, into elongation. The tensometer has six configurable bridges such as inductive half-bridge, resistive half-bridge or resistive full bridge. The inductive differential transducer connects to the channel 1 of the tensometer, configured as an inductive half-bridge. Because the coils have both electrical component, resistance and capacity, Wheatstone "parasite" bridges are formed, resistive and capacitive. The tensometer has correction elements to compensate for these additional bridges.

D. Circuit for converting and adapting the voltage signal generated by the NI USB-6008 DAQ device to the current control signal. The circuit is powered by an external 15 V source.

3.2 Test Algorithm

The algorithm for determining the characteristic is shown in Fig. 4 and consists of a "for loop" in which the output signal is incremented, equal with the voltage applied to the power amplifier, from zero to a maximum value programmed by the user.

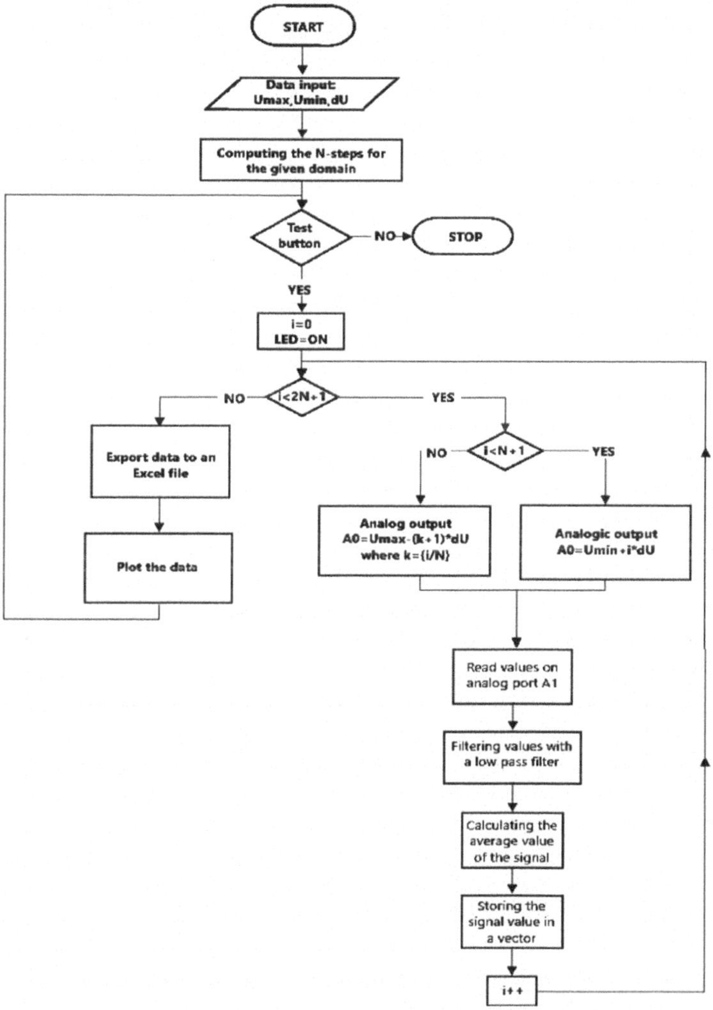

Fig. 4. Algorithm diagram

Incrementing is done in steps; the number of steps being indicated by the user. Upon completion of the increase and decrease of the voltage, the data is displayed and saved. The data acquisition system acquires a signal from the electronic tensometer coupled with the inductive transducer, indicating the deformation of the wire.

Based on the algorithm shown above, using the equipment of the stand, the program (Fig. 5) was developed in LabView software from National Instruments.

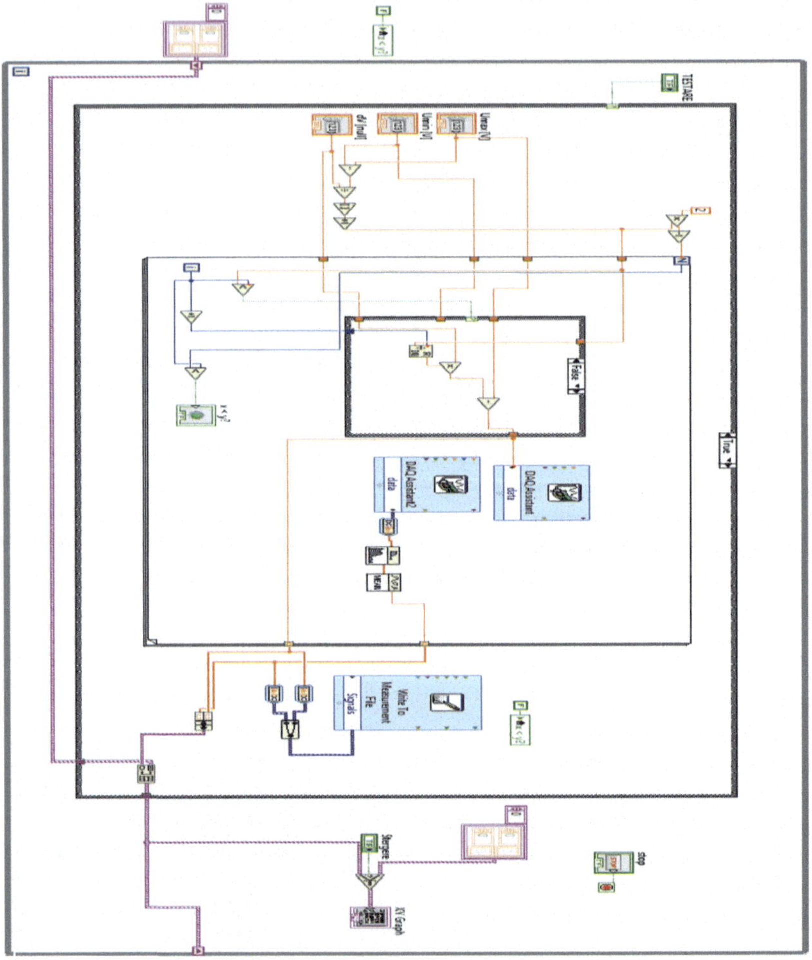

Fig. 5. Test program in LabView software

The main blocks used in the virtual instrument are the following:

- Numeric input blocks
- Repetitive function with finite number of steps FOR
- DAQ Assistant function for writing and reading signals

- Numeric blocks for arithmetic calculation (ADD, OR, Q & R, etc.)
- Decision blocks (GREATER, LESS, CASE, etc.)
- Low filter function
- The function of calculating the average value of the input signal string
- Bundle Block to generate the vector that contains the output/input signals
- XY GRAF block to display the vector
- Write to Measurement to export data

4 Hysteresis Characteristic of a Shape Memory Alloy Wire

After running the program, the graph in Fig. 6 result, which represents the electric voltage/elongation wire characteristic. We noticed the hysteresis behavior of the Ni-Ti alloy. The wire has been tested for a command voltage signal in the [0.3; 1.3 V] range. The graph shows how phase change of the material occurs around 1.03 V. This value corresponds to a current of 300 mA read on the power supply display.

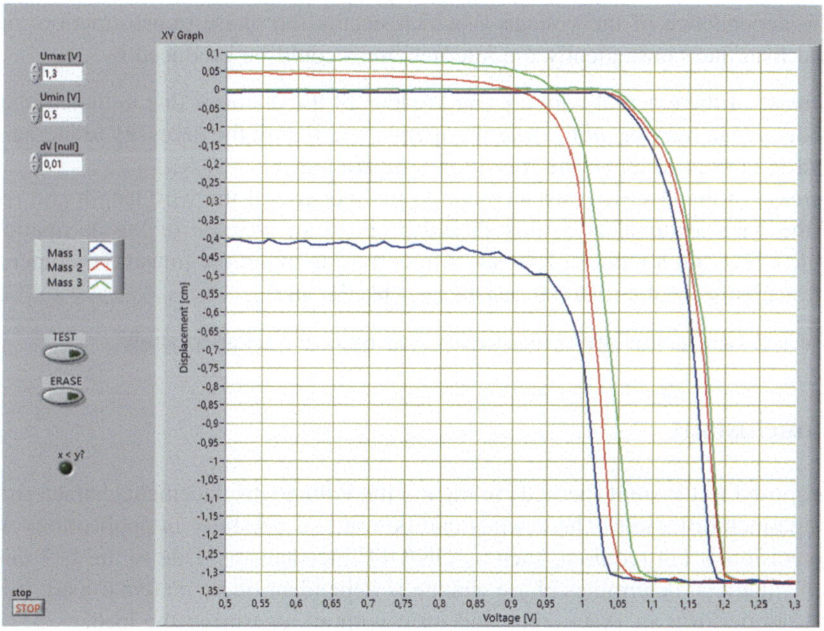

Fig. 6. Hysteresis characteristic for Ni-Ti wire

The blue chart corresponds to the first test. At the end of the wire is attached a weight of 100 g. After testing, the wire manages to overcome external stress, passes the martensitic crystalline state into austenitic state, a state with a higher modulus of elasticity, and contracts about 13.5 mm. On the unloading side, the phase shift begins around the same value of the command signal of 1.05 V. It is noted that the mass of

100 g is unable to reorganize the crystalline structure of the alloy and the wire fails to reach its original position, producing a 4 mm gap between the two positions.

The red graph corresponds to the external load generated by a mass of 200 g. On the second run of the program, respectively after the current command passes the wire, the change of martensitic – austenitic phase occurs. There is a shift towards a higher breaking value, which means that the wire has a slight dependence on the size of the stress it is subjected to. On the unloading side, this time, second weight manages to bring the wire back to its original position and even exceed that position. This suggests that at the unloading stage the wire reorganizes different than the initial crystalline structure and that the wire tends towards the plastic deformation limit.

The green graph corresponds to the last test, 300 g weight. There is still a shift to the right of the phase change starting point, which means that the wire requires more current to begin the reorientation of crystalline structure. On the unloading side, the alloy behaves as in the second case, exceeding initial position with 1 mm. The wire completely reorganizes its crystalline structure and is close to the plastic deformation.

The most important thing is that, regardless of the weight, the phase change from martensite to austenite and austenite to martensite occurs at the same voltage, but the disadvantage is that this voltage tends to vary according to the external load.

The dependence of the voltage at which occurs the phase transformation, on the external load and consequently the internal stress could be produced by:

- Increase of the external wire area and reduce of the diameter due to the elongation, producing an increase of the heat dissipation and in turn the necessity to increase the electric current to attain the transition temperature;
- Increase of the electric resistance due to the increase of the wire length and reduce of the wire sectional area; this should produce an increase of the dissipation by Joule effect and a reduce of the electric current to attain the transition temperature.
- The variation of the material resistivity with the stress.

The last two factors are contradictory and require a detailed study.

5 Conclusions

The proposed test system succeeds in tracing the voltage-displacement characteristic of the shape-memory wires. These wires can be used successfully in applications where high precision is not required, such as ON/OFF mechanisms. Due to the fact that the voltage value that determines phase change is dependent on the external load, it is not possible to perform an exact command on it without measuring the load.

The biggest advantage of these alloys is their high displacement. For a length of 120 mm Ni-Ti alloy, a displacement of approximately 13 mm was achieved but with a relatively low force, unlike other shape memory materials. This leads to a significant reduction in the gauge of the device being served.

In the future, it is desirable to improve the stand by introducing an electrical resistor to release a controlled temperature for tracing the temperature/displacement characteristic. Also, to optimize the control, the stand also contains a force transducer. The transducer has a triangular shaped cantilevered elastic element, with strain bridges glued on its faces and connected to another channel of the electronic tensometer in a complete Wheatstone bridges, together with two identically strain bridges glued on the cantilevering parts. The upper end of the wire is secured by the movable end of the spring. The force is used for optimizing the control voltage [4].

References

1. Kennedy, S.P.: Material characterization of nitinol wires for the design of actuation systems. Degree Master of Science in Mechanical Engineering, Faculty of California Polytechnic State University, San Luis Obispo, August 2013
2. Bujoreanu, L.G.: Materiale inteligente. Ed. Junimea, Iasi (2002)
3. El Feninat, F., Laroche, G., Fiset, M., Mantovan, D.: Shape memory materials for biomedical applications. Adv. Eng. Mater. **4**(3), 91–104 (2002)
4. Huang, W.M., Ding, Z., Wang, C.C., Wei, J., Zhao, Y., Purnawali, H.: Shape Memory Materials. School of Mechanical and Aerospace Engineering, Nanyang Technological University, 50 Nanyang Avenue, Singapore 639798
5. Jani, J.M, Leary, M., Subic, A., Gibson, M.A.: A review of shape memory alloy research, applications and opportunities. Mater. Des. (1980–2015) **56**, 1078–1113 (2014)
6. Silva, J.D., Resende, P.D., Garcia, P.R., Azevedo Lopes, N.I., Arruda Santos, L., Buono, V.T. L.: Fatigue resistance of dual-phase NiTi wires at different maximum strain amplitudes. Int. J. Fatigue **125**, 97–100 (2019). https://doi.org/10.1016/j.ijfatigue.2019.03.040
7. Montalvao, D., Shengwen, Q., Freitas, M.: A study on the influence on Ni-Ti M-Wire in the flexural fatigue life of endodontic rotary files using finite element analysis. School of Engineering and Technology, University of Hertfordshire, College Lame Campus, Hatfield, Herts AL10 9AB, UK

Analysis of the Static and Dynamic Mechanical Behavior of a Tibial Bone-Knee Implant Assembly Without a Tibial Extension

Mihai-Constantin Balaşa[1,2] and Viviana Filip[2(✉)]

[1] Macartney Hydraulics A/S, 7620 Lemvig, Denmark
[2] Doctoral School of Engineering Sciences,
Valahia University of Targoviste, 130105 Targoviste, Romania
v_filip@yahoo.com

Abstract. The ever-growing number of total knee replacement (TKR) surgical procedures has brought a proportional increase in the number of knee replacement revision operations. Among the causes of the early failure of a knee implant are the weakening and displacement of the tibial component. To avoid revision surgery and extend the lifespan of the implant, it is really useful to perform a finite element analysis in order to identify the areas that may be prone to cause the weakening of the tibial component.

The purpose of this paper is the virtual modeling of the tibia-knee implant assembly and the analysis of its mechanical behavior both in the static and the dynamic regime. For this purpose, we will look at the case of an implant without tibial extension and analyze the healthy bone in comparison with the osteoporotic bone. The results from the static stress, fatigue, linear and nonlinear dynamic stress tests confirm that the healthy bone has a higher mechanical strength and a lower degree of damage than the affected bone, under the same stress conditions.

Keywords: CAD techniques · Tibial bone · Knee implant · Finite element analysis

1 Introduction

Currently, the choice of the implant type for a knee replacement (Fig. 1) is made after medical imaging procedures. A radiography machine provides 2D images (with the final result consisting in a negative photograph) that make it possible to establish the diagnosis and choose the most appropriate implant that will be used to replace the knee joint. The factors the implant choice depend on and the surgeon will have to consider before scheduling the procedure are related to the patient's age, weight, activity, and health condition. Just as the wear occurring in the natural joint leads to the necessity to replace the joint, so the wear of the implant may require possibly a second procedure, called revision surgery. To extend the lifespan of the primary implant by choosing the best option available and in order to avoid or postpone as much as possible the revision surgery [1], it would be really useful for the surgeon to know, before the primary

G. I. Gheorghe (Ed.): ICOMECYME 2019, LNNS 85, pp. 194–205, 2020.
https://doi.org/10.1007/978-3-030-26991-3_18

surgery [2, 3], how the implanted bone behaves in time. This type of analysis is possible with the help of CAD techniques and is the subject of this paper.

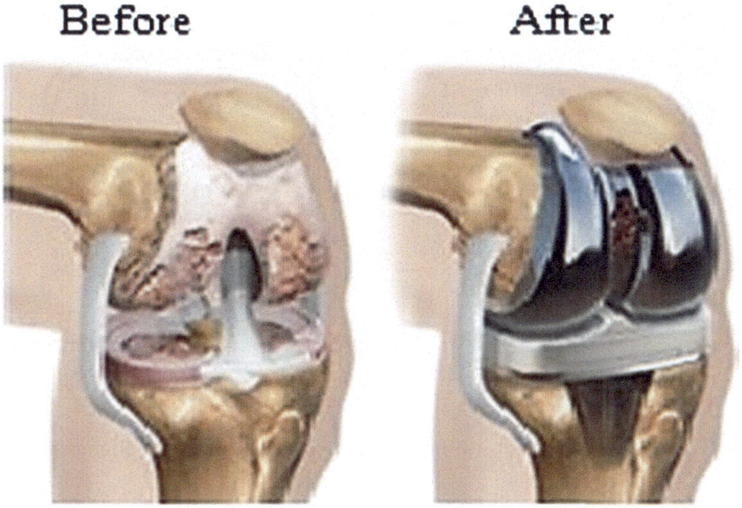

Fig. 1. Knee prosthesis [4]

In order to study the static and dynamic mechanical behavior of the tibia-knee implant assembly, we used the SolidWorks software to create the virtual model of a 320 mm-long tibial bone (based on an existing physical model which was previously 3D scanned) with the thickness of the cortical bone (i.e., the compact outer surface) varying between 0.8 mm (at the extremities, in the knee and the ankle areas) and 4 mm (in the middle area) [5]. The same software was used to create the components of the implant (based on an existing physical model of a widely-used standard prosthesis), including the cement used for fixation [6]. Then we used the finite element method to study the mechanical behavior of the bone-implant assembly.

2 Working Method and Resources Used

After creating the virtual model of the tibia and knee implant, we produced the tibial bone-implant assembly (Fig. 2-a), seeking the correct positioning of the implant on the tibial plateau. The tibial component of the implant is aligned correctly by dividing hypothetically the tuberosity into three parts (Fig. 2-b): the medial part (the inside of the bone, i.e. the side facing the other leg), the central part (the middle) and the lateral part (the outward side). A line is drawn that separates the medial part from the central part, then the tibial metal component of the implant (i.e. the tibial plateau) is positioned on the borderline.

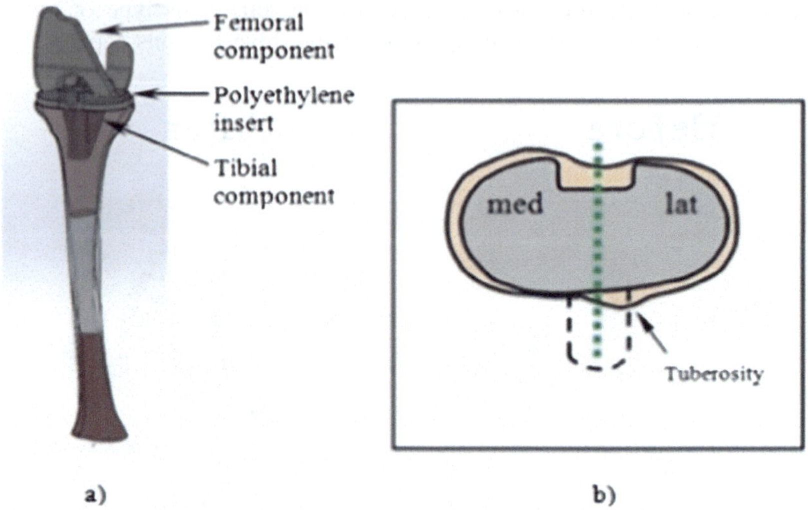

Fig. 2. Knee implant, (a) Implant components, (b) Tibial alignment

Next, we assigned material properties to each of the assembly's components: cortical bone, spongy bone, polyethylene insert, tibial metal component, and fixation cement– as follows:

For the healthy cortical bone: Young's modulus E = 1700 N/mm^2, Poisson's ratio = 0.2, density = 800 kg/m^3, Tensile strength 1.2 N/mm^2, Compressive strength 1.9 N/mm^2.

For the healthy spongy bone: E = 700 N/mm^2, Poisson's ratio = 0.2, density = 600 kg/m^3, Tensile strength 1.3 N/mm^2, Compressive strength 1.8 N/mm^2.

For the polyethylene insert we selected the High-Density Polyethylene option from the list of materials and assigned the following default parameters: Young's modulus E = 1070 N/mm^2, Poisson's ratio = 0.41, density = 952 kg/m^3.

For the tibial component we selected the Ti-6Al-4V alloy from the list of materials and assigned the following default parameter: Young's modulus E = 104800 N/mm^2, Poisson's ratio = 0.31, density = 4428 kg/m^3.

For the fixation cement we assigned the following parameters: Young's modulus E = 2150 N/mm^2, Poisson's ratio = 0.48, density = 1100 kg/m^3.

We discretized the entire virtual model into nodes and elements, we applied a 2100 N load force (3 times the average weight of a patient) on the surfaces of the polyethylene insert and a gravitational acceleration of 9.81 m/s^2 along the longitudinal axis of the of the tibial bone.

We analyzed the mechanical behavior of both assemblies (healthy tibia-knee implant, respectively impaired tibia-knee implant) under static stress, linear and non-linear dynamic stress and fatigue testing conditions.

3 Results of the Static Stress Analysis for the Healthy Tibial Bone-Knee Implant Assembly

The highest stress (von Mises) values obtained under static load conditions are 21.9 MPa and were measured on the polyethylene insert, which indicates that this component, like the lateral meniscus, absorbs part of the stresses, as can be seen in Fig. 3.

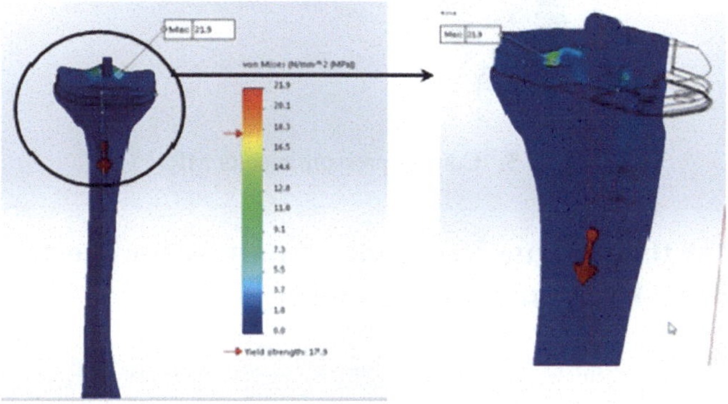

Fig. 3. Von Mises stress values (MPa). Detail – section through the assembly, with the corresponding maximum stress values

The largest displacement values measured 0.117 mm and were found on the polyethylene insert. These displacements propagate progressively through the tibial metal component (Fig. 4). On the Z-axis, the largest displacements are up to 0.098 mm.

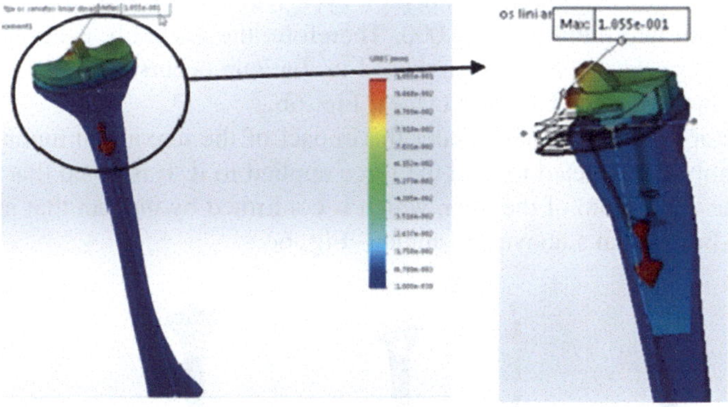

Fig. 4. URES displacements (mm). Sectional detail

As for the contact pressure between the tibial component, the cement and the proximal part of the tibial bone (near the implant), a high pressure was measured between these components, which indicates a high pressure resistance of 13.1 MPa, Fig. 5.

Fig. 5. Contact pressure results MPa

4 Results of the Fatigue Resistance Analysis for the Healthy Tibial Bone-Knee Implant Assembly

After assigning the initial conditions, we used the software to add an "event" which enabled us to analyze the resistance of the implanted bone when going through 700,000 functioning cycles.

After the event simulation, we obtained the following results:

- Damage percentage – the percent of the bone assembly that is damaged after going through the selected number of cycles, seen as a whole. In our case, the maximum damage percentage was 41.7% and was measured on the polyethylene insert, Fig. 6a.
- Total Life (cycle) – the total number of cycles the assembly withstands. In the case under study, the maximum number of cycles that the tibial bone-knee implant assembly can withstand is 1.000.000. Therefore, the assembly is able to withstand the 700,000 cycles it has been subjected to. Fatigue occurs in the back area of the polyethylene insert after 83,500 cycles, Fig. 6b.
- Load Factor – the load factor under the impact of the maximum number of cycles the assembly is subjected to, and the force applied to it. It is noted that the red area is found at the bottom of the tibia, which is confirmed by the fact that in most cases the tibial bone breaks above the ankle – Fig. 6c.

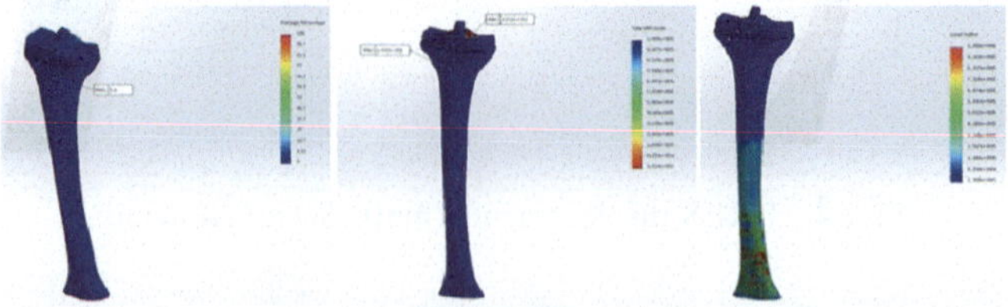

a) b) c)

Fig. 6. Fatigue analysis results (a) Damage percentage (b) Total life (maximum number of cycles) (c) Load factor

5 Results of the Linear Dynamic Stress Analysis for the Healthy Tibial Bone-Knee Implant Assembly

For the purpose of this analysis, we will apply a 214 kgf (2100 N) inertial mass distributed along the longitudinal axis of the tibia. This force will be applied to the assembly for one second, Fig. 7.

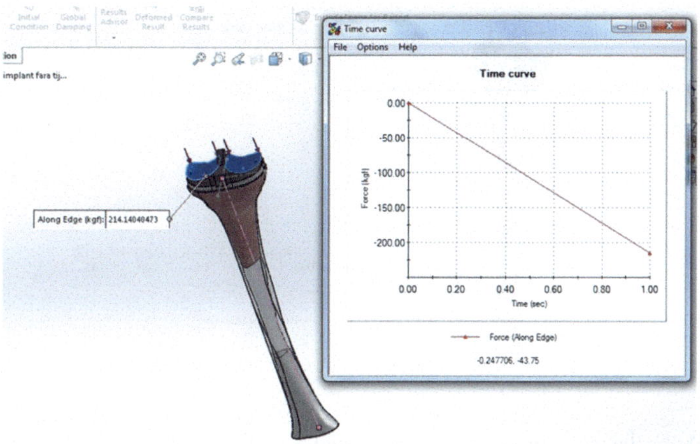

Fig. 7. Applying the force. Graph showing the force applied to the assembly for a period of one second

The highest von Mises stress values measure 43.2 MPa and are found at the contact area between the metallic tibial component and the proximal part of the tibial bone, Fig. 8.

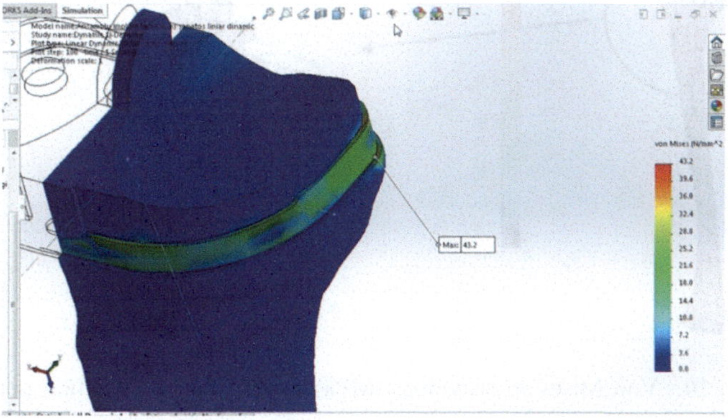

Fig. 8. Von Mises stress values (MPa)

The largest displacements are found on the Z-axis, on the polyethylene insert, and measure 0.165 mm.

6 Results of the Nonlinear Dynamic Stress Analysis for the Healthy Tibial Bone-Knee Implant Assembly

The force applied is 2100 N on the surfaces of the polyethylene insert, along the longitudinal axis, Fig. 9a. The curve showing the force variation for a period of one second is nonlinear, as can be seen in Fig. 9b.

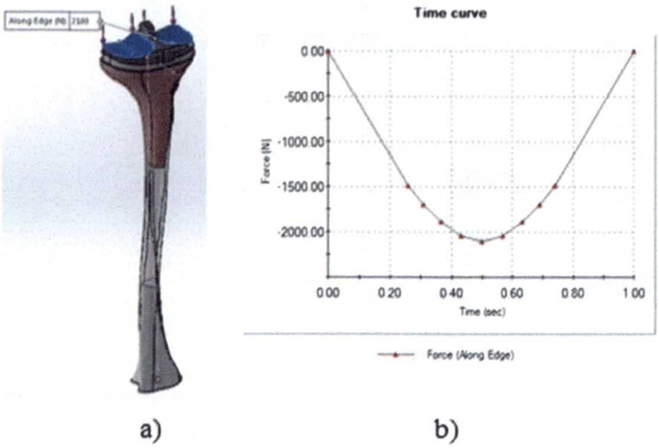

Fig. 9. Applying the load force: (a) force direction – along the longitudinal axis (b) graph showing the force variation with time

The highest stress values are found on the edge of the tibial component, at the contact with the polyethylene insert, and measure 84.4 MPa, Fig. 10.

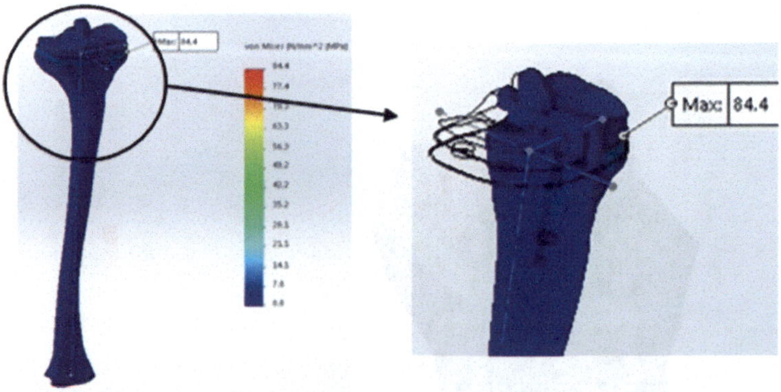

Fig. 10. Von Mises stress values (MPa), von Mises stress (MPa) detail

In this case also, the displacements are found on the polyethylene insert and measure up to 0.151 mm. The displacement values decrease gradually towards the tibial component, Fig. 11. The detail below shows a section through the assembly and the displacements measured on the polyethylene insert.

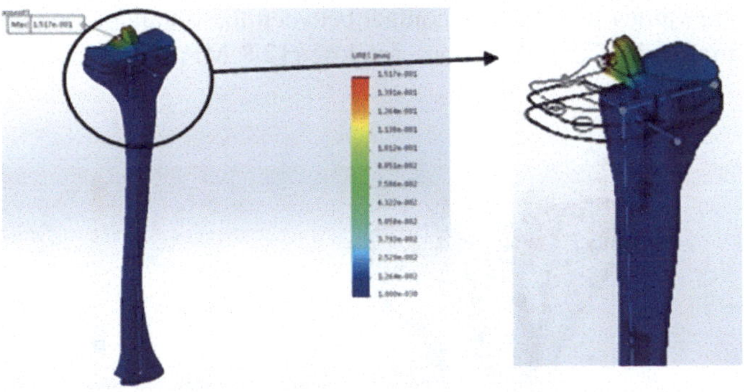

Fig. 11. URES Displacement (mm) Result, sectional detail of the assembly

The equivalent displacements are very small, up to 5.93×10^{-6} mm.

The next step was to analyze the osteoporotic bone under the same stress conditions, in order to determine its behavior in comparison with that of the healthy bone.

For this purpose, we altered the mechanical properties of the healthy spongy bone the implant is fitted into and we assigned the following material properties to the osteoporotic spongy bone:

Young's modulus E = 1000 N/mm^2, Poisson's ratio $\upsilon = 0.2$, density $\rho = 400$ kg/m^3, Tensile strength 1.1 N/mm^2, Compressive strength 1.7 N/mm^2.

7 Results of the Static Stress Analysis for the Impaired Tibial Bone-Knee Implant Assembly

The highest stress values were found on the polyethylene insert and measure 21.3 MPa (Fig. 12). As can be noticed from the detail showing a section through the bone, these stresses propagate also along the tibial component.

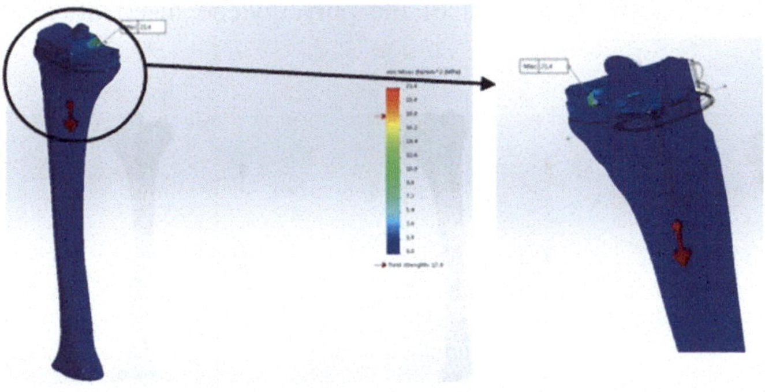

Fig. 12. Von Mises stress values (MPa), sectional detail of the bone

The pressure values found at the contact between the tibial component, the cement and the proximal part of the tibial bone are low (12.8 MPa, Fig. 13).

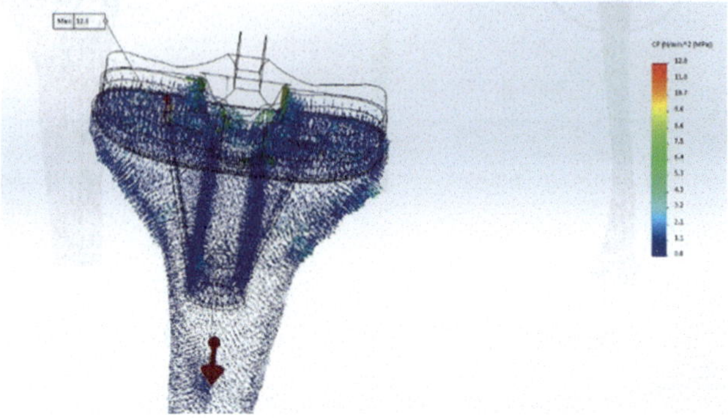

Fig. 13. Contact pressure results (MPa)

The highest displacement values are found on the polyethylene insert, they measure 0.117 mm and are transferred progressively to the other components. The displacements on the Z-axis are lower than the overall values and measure 0.97 mm.

8 Results of the Fatigue Resistance Analysis for the Impaired Tibial Bone-Knee Implant Assembly

After assigning the initial conditions, we used the software to add an "event" which enabled us to analyze the resistance of the impaired bone when going through 700,000 cycles. We configured the Damage percentage parameter for the entire model and we found that our assembly is damaged to a proportion of minimum 70% and maximum 91.7% in the area of the polyethylene insert, Fig. 14a.

We also found that the maximum number of cycles the impaired bone-knee implant assembly can withstand is 1,000,000 cycles.

Fatigue occurs in the back area of the polyethylene insert after 60,441 cycles, Fig. 14b.

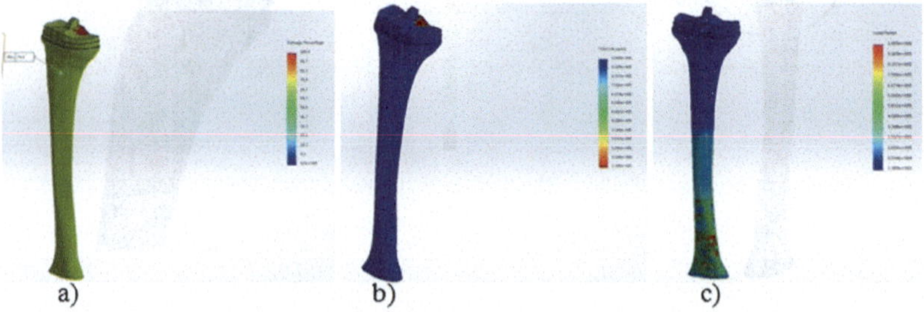

Fig. 14. Fatigue analysis results. (a) damage percentage (b) maximum number of cycles (total life) (c) load factor

The load factor reaches its maximum values at the bottom of the tibial bone, which confirms that in most of the cases the tibial bone breaks above the ankle, Fig. 14c.

9 Results of the Linear Dynamic Stress Analysis for the Impaired Tibial Bone-Knee Implant Assembly

The inertial mass applied is 214 kgf (2100 N) distributed along the longitudinal axis of the tibia, increasing linearly, for a period of one second, Fig. 15.

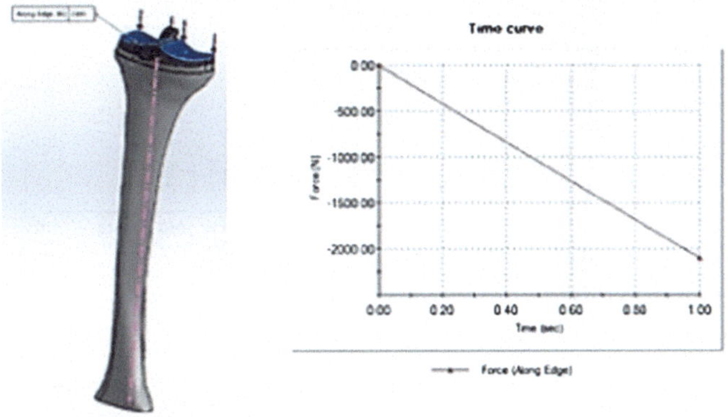

Fig. 15. Left: applying the force; Right: time-force graph for a force applied for one second

The highest von Mises stress values, measuring 28.4 MPa, are found on the edge of the tibial component, as seen in Fig. 16a, and on the polyethylene insert as well.

In this case also, the largest displacement values are found on the polyethylene insert and measure 0.154 mm, as seen in Fig. 16b. These displacements are transferred to the tibial component as well. On the Z-axis the displacements measure maximum 0.150 mm, as seen in Fig. 16c. The displacements are found on the polyethylene insert and at the contact with the tibial component.

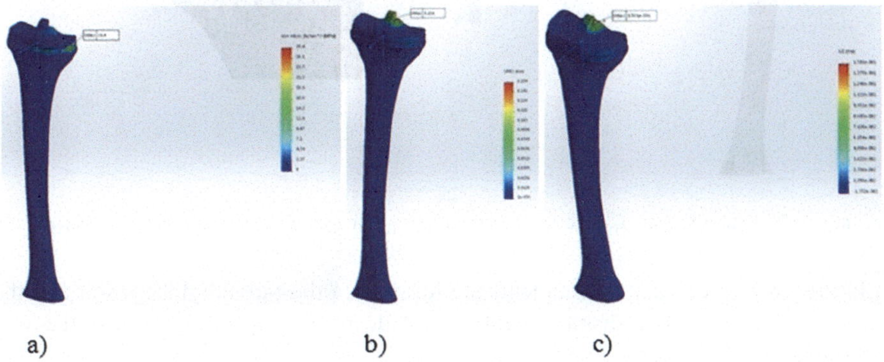

Fig. 16. (a) Von Misses stress values (MPa), (b) URES displacements (mm), (c) URES displacements on the Z-axis (mm)

10 Results of the Nonlinear Dynamic Stress Analysis for the Impaired Tibial Bone-Knee Implant Assembly

The 2100 N force is applied along the longitudinal axis of the tibial bone, on the selected surfaces of the polyethylene insert. The force varies in a nonlinear manner, and the graph representing the force applied for a period of one second is shown in Fig. 17.

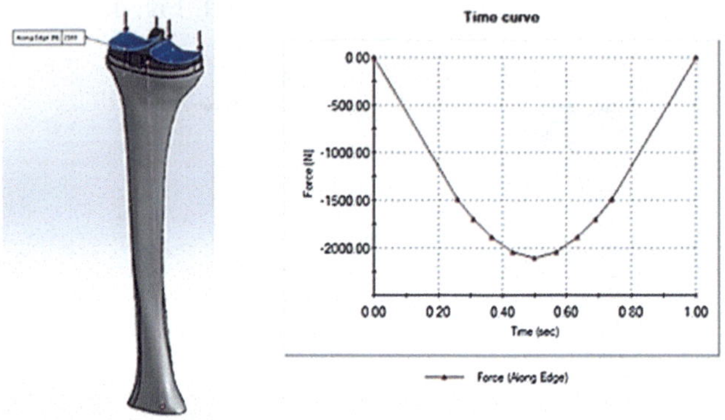

Fig. 17. Applying the force. The time-force graph showing the force variation over one second

The highest stress values, measuring 32 MPa, are found on the edges of the tibial component and of the cortical bone, but are also transferred through the polyethylene insert, as seen in Fig. 18.

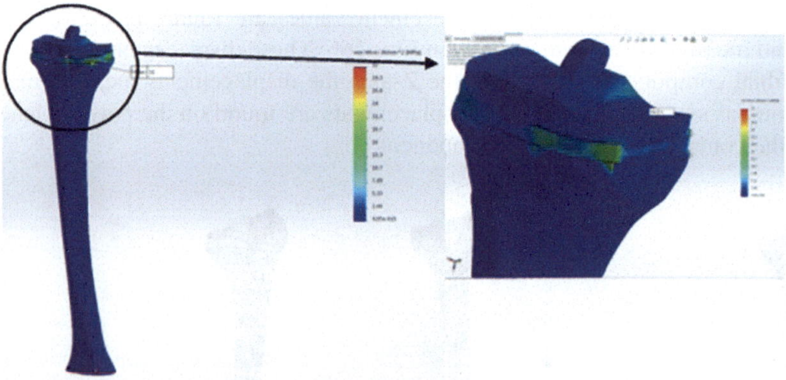

Fig. 18. Von Misses stress values (MPa), detail of von Misses stress results (MPa)

In this case also, the displacements are located on the polyethylene insert and they measure 0.068 mm. The displacement values decrease gradually towards the tibial component.

The maximum equivalent displacement values measure 1.139×10^{-5} mm.

11 Conclusions. Future Research Directions

The results obtained from the analysis performed on a healthy bone and a diseased bone subjected to static stress, fatigue, and linear and nonlinear dynamic stress, confirm that the healthy bone has a higher resistance than the diseased one, under the same stress conditions. These results allow us to substantiate, based on the bone density values, the predictions as to the long-term behavior of an implanted bone. Predictive biomechanical studies based on the finite element method may provide surgeons with valuable information that can help them choose the most appropriate type of implant.

As a future research direction, it will be necessary to perform the finite element analysis of an impaired bone-knee implant assembly with a tibial augmentation, to compare the results of such analysis to the those obtained from the analysis performed on the impaired bone-knee implant assembly without a tibial extension (from the current paper) and to determine which option is best suited to be used from the very beginning (at the moment of the primary surgery): the implant that includes the extension or the implant without the extension. If further to the comparative analysis it turns out that using the implant without an extension will result, after a certain number of functioning cycles, in a high degree of damage, which exceeds the damage that occurs when using an implant with extension, then the conclusion would be that it is advisable to choose the implant with the tibial extension at the moment of the primary surgical procedure. Otherwise, the revision surgery required to add the tibial extension will come soon after the primary surgery.

References

1. Cristea, Şt., Prundeanu, A., Groseanu, Fl., Gârtone, D.: The role of arthroscopy in mini-invasive treatment of Tibial Plateau fractures. In: Modern Arthroscopy, pp. 225–236 (2011)
2. Cofaru, I.I.: Summary of the Ph.D. thesis, Researches regarding the biomechanics of the axial deviations of the human lower member and the development of the correspondent surgical devices (2013)
3. Nicolescu, C.M., Bumbac, M., Mihai, S., Gheboianu, A.I., Balasa, M.C., Filip, V., Cuculici, S., Cristea, S., Pantu, C.: X-RAY diffraction and nanoindentation characterization of bone tissue affected by severe osteoarthritis. J. Sci. Arts **1**(42), 265–274 (2018). ISSN: 1844–9581
4. LNCS Homepage. http://mihairascu.ro/proteza-totala-a-genunchiului/. Accessed 29 May 2019
5. Balaşa, M.C., Mihai, S., Filip, V., Negrea, A.D., Tomescu, G.: Modelling the tibial bone using CAD techniques, starting from the 3D scan model. Int. J. Mechatron. Appl. Mech. **3**, 217–223 (2018)
6. Balaşa, M.C., Cuculici, Ş., Panţu, C., Mihai, S., Negrea, A.D., Zdrafcu, M.O., Leţ, D.D., Filip, V., Cristea, Ş.: Using 3D scanning techniques in orthopedic systems modeling. Sci. Bull. Valahia Univ. Mater. Mech. **15**(13), 41–47 (2017)

Processing of Captured Digital Images for Measuring the Optometric Parameters Required in the Construction of Ultra-personalized Special Lenses

George Baboianu, Constantin Nitu, and Constantin Daniel Comeaga(✉)

University Politehnica of Bucharest, Bucharest, Romania
comeaga_daniel@yahoo.com

Abstract. Devices for processing and capturing images have grown a lot. Softwares are increasingly complex, making processing with great accuracy being used in many areas of activity. Database acquisition systems are designed to store information of any kind (signal or image). In the field of Ophthalmology, these image capturing and processing devices help determine the patient's individual parameters for glasses' construction in order to provide comfort and clarity of vision. Investigating the functioning of the visual system, analyzing of the results, highlighting the problems, recommending the methods and means of improving the functioning, in order to achieve visual comfort in relation to the needs of the subject represents the science called Ophthalmic Optometry. In recent years, optometry has developed more, considering the individual as an integral part of the living environment, analyzing the performance and visual problems daily, depending on the work developed during a day. The maximum visual performance achieved by a patient is due to the measuring methods and means of the optometric parameters. Existing devices help both in detecting eye diseases, but also in measuring individual parameters necessary for a clear and comfortable view. All of these things depend on capturing and processing images with special devices and software. The paper presents such a software – Eye Fit, which is able to run the received data and have all measurements of patient's parameters and frame chosen in order to build an ultra-personalized progressive lens after measurements are performed with a tablet device that includes this specially designed software.

Keywords: Eye Fit software · Optometric parameters · Personalized lenses

1 Introduction

Image processing and analysis (often referred to as image processing only) was born due to the need of replacing the human observer through a device. It is important to note that image analysis went further than simply replacing the human observer because innovative solutions for problems which he had not previously encountered - as with non-visible images (acoustic, ultrasonic, radar) - appeared. As noted in [Jain A. K - 1989], image processing includes the possibility of developing a vision machine capable of

G. I. Gheorghe (Ed.): ICOMECYME 2019, LNNS 85, pp. 206–217, 2020.
https://doi.org/10.1007/978-3-030-26991-3_19

performing the visual functions of any living entity (of course, after the achievement of important theoretical and technological developments) [1].

"Image processing holds the possibility of developing the ultimate machine that could perform the visual functions of all living beings".

The term digital means: "the representation of discrete information within the computers".

So as long as we accept the idea that the computer is working in the image processing and it is also digital, processing is also digital, as a particular case of any numerical processing. Of course there are also image processing that are analogical - as are all the processing that take place within the standard television transmission and reception chain.

Ophthalmic optometry is the science that deals with the investigation of the human visual system activity, analyzing the results, highlighting the problems, recommending methods and means of improving the functioning to achieve visual comfort in relation to the needs of the subject, but without medical treatment, if possible. Functional optometry is a new trend in ophthalmic optometry. It considers the individual as part of his environment and systematizes from this point of view the analysis of the performances and the synthesis of the visual problems [1].

On functional level, the eye is ready to see despite of no light. It is possible that the intra-uterine life even in the absence of light to register sensations of light, thus allowing the initiation of the maturation of nerve paths with reflexes indispensable for the subsequent functioning.

These "visual" stimulations would be especially of a tactile nature due to pressure variations of the osmotic environment. This functioning start of the retina favors from birth all the adaptive responses to the shocks of light such as the pupillary reflex. Precocity functional maturation of the visual system is confirmed in premature infants. A prematurely reaching the age of nine months is ahead of a baby born in due time. At birth, the neural activity is largely under subcortical control, which explains the driving reflex responses to sensations [1].

2 Acquisition of Images to Measure the Individual Parameters Required for Building Ultra-personalized Lenses

If your eyes are no longer emmetropic and need correction, this is done with eyeglass lenses. It is very important to choose the frame that meets the comfort and positioning conditions for the patient's activities and to respect the size required for each type of mount so that it is technically suitable for the lens to be mounted and which helps the clarity of image. The new eye and frame system must be adjusted so that the patient has maximum comfort and clarity of the image. For people who have facial asymmetries coming out of the "standard" area, adjustments should be made to both the eyewear frame and the lens construction. Therefore, the correct calculation and determination of patient's individual parameters and frame parameters should be done with special equipment. All these determined parameters will be used to build individual lenses so that the patient benefits from maximum image clarity (Fig. 1).

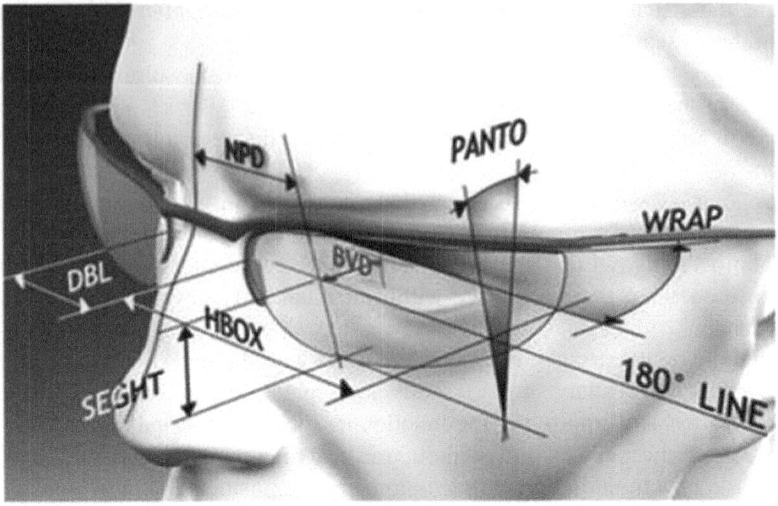

Fig. 1. Individual parameters of the patient and the selected frame [2].

All these individual parameters are used in the construction of the progressive lens in order to personalize it:

- interpupillary distance in the range
- the vertex distance
- pantoscopic angle
- the convergence of each eye
- radius of curvature of the frame
- the height of installation
- nasal bridge

The measured parameters are based on the physiognomy of each patient, the frame must be adjusted prior to image capture and it must be well positioned on the nose and ears.

2.1 Eye Fit Device - Special Software for Measurement of Optometric Parameters

Measurements are performed with a tablet device that includes specially designed software [2] (Fig. 2).

Very important for making the measurements is the frame clamping device, which has three markers positioned so that all measurements can be calculated in relation to their arrangement. The device is made of carbon and has 9 grams, and it's easy not to change the position of the eyeglass frame on which it snaps. All dimensional, constructional and guiding angles of the device are known and will be part of the calculation of the measurements run by the software. It is very important to grip and position

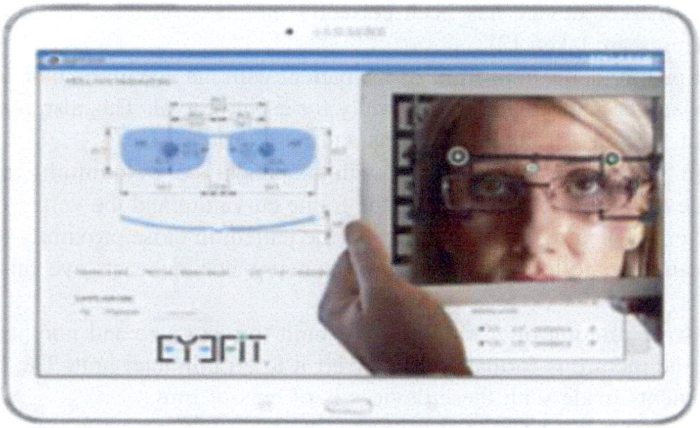

Fig. 2. EYE FIT mobile device to determine the individual parameters [2].

the device with the three guides, over the patient's selected glasses frame. Positioning is done taking into account that the center marker is exactly half the face and frame chosen [2] (Fig. 3).

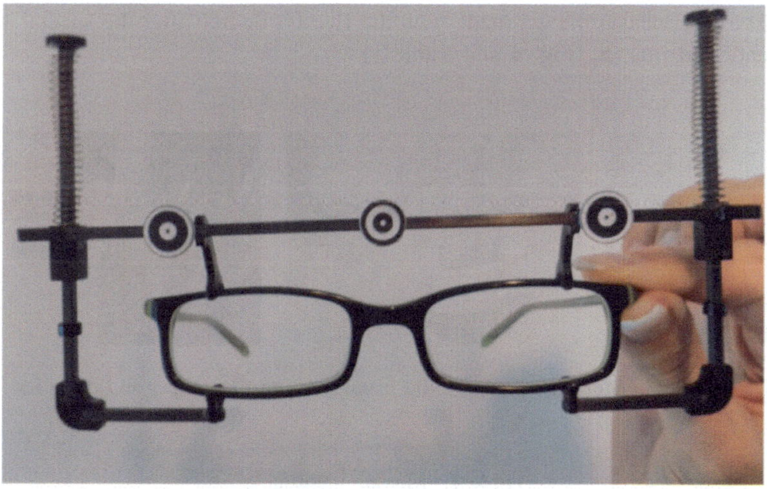

Fig. 3. The frame clamping device with three markers, [2]

The device is required to construct a correct position of the coordinate system in which the measurements are made and to establish image correction parameters. Without markers, it is not possible to correctly determine the coefficients of the passage matrices from units in pixels to units in mm in space.

After the marker device has been correctly mounted on the eyeglass frame, the patient's pictures are taken [2].

The first picture is taken in front of the patient with his eye in the tabletop room so he can calculate the pupil distance to infinity for each interval. This also results in lens mount height, eyeglass frame dimensions and nasal bridge size.

There are two pictures in the profile, with the patient's eyes to infinity, pictures that determine the frame pantoscopic angle, the frame curvature and the vertex distance for each eye. The last picture will be taken with the patient in close proximity to the tablet cell to measure the following parameters: convergence on each eye and lens read position.

The software running on such a tablet is built to recognize and number the pixels from which the picture is formatted to convert it to a millimeter unit. The accuracy of the measurements made with these devices is of tens of mm.

After the patient has been photographed, the EYE FIT program runs the received data, and as a result we have all measurements of patient's parameters and frame chosen in order to build an ultra-personalized progressive lens [2].

2.2 Taking Pictures with Eye Fit and the Accuracy of the Measurements Made

To begin with, a mannequin was used with a frame was placed and the individual parameters were measured with the caliper. The results of these measurements were compared to the measurements made with the Eye Fit device after the special software ran the captured images (Fig. 4 and Table 1).

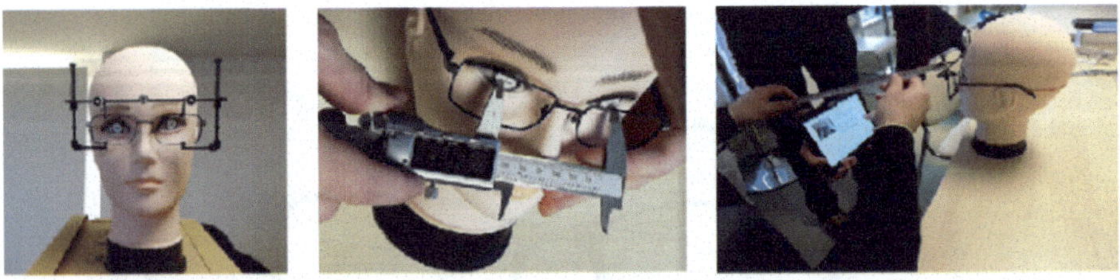

Fig. 4. Measurements made on the manikin using the caliper and the ruler

Table 1. Measurement table

Parameters	Measuring [mm]
Pupilar distance [PD]	64.77
Vertex distance	R: 17/L: 16.8
Reading distance	36.5
Nasal bridge	19.16

It is noted that the measurements made with Eye Fit are very accurate, the software being able to measure with an accuracy of tenths of a millimeter. Image processing is very important in making measurements of these individual parameters. Photos must be made very clear to be able to be processed by the software, image clarity influences the result of the measurements [2] (Figs. 5 and 6).

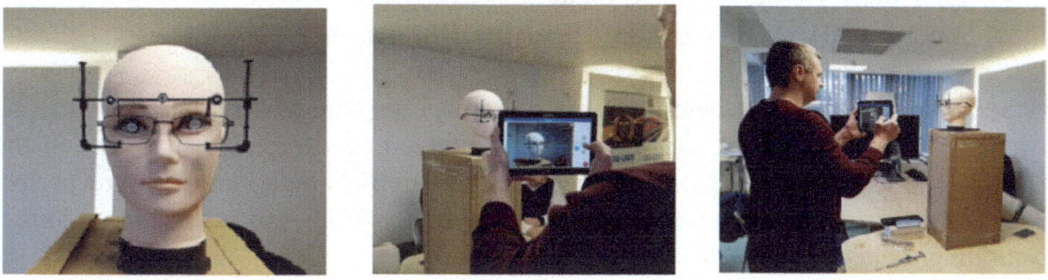

Fig. 5. Images captured with the Eye Fit device [2]

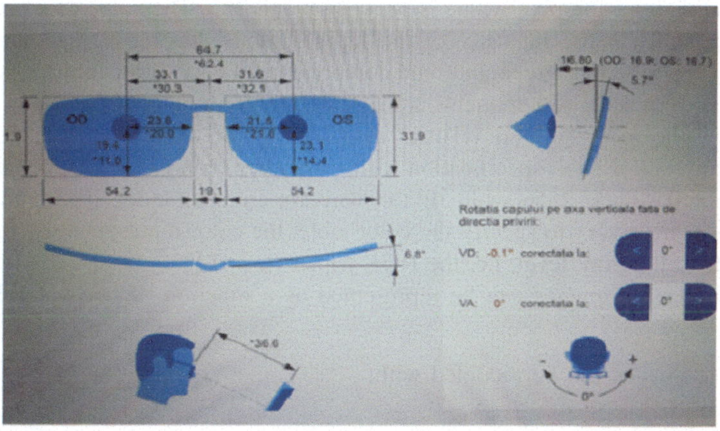

						* Dimensiuni pentru vederea la aproape		
Dip:	64.7	mm	BH:	31.9	mm	*Dip:	62.4	mm
MpD:	33.1	mm	BL:	54.2	mm	*MpD:	30.3	mm
MpS:	31.6	mm	Br:	19.1	mm	*MpS:	32.1	mm
PpD:	23.6	mm	Vx:	16.8	mm	*PpD:	20.0	mm
PpS:	21.8	mm	VxOD:	16.9	mm	*PpS:	21.0	mm
HD:	19.4	mm	VxOS:	16.7	mm	*HD:	11.0	mm
HS:	23.1	mm	UP:	5.7	°	*HS:	14.4	mm
			UC:	6.8	°	Distanta la citire: *36.5 cm		

Fig. 6. Measurement results with Eye Fit [2]

Dip = interpupillary distance; MpD = Interpupillary distance for right eye; MpS = interpupillary distance for the left eye; Vx = vertex distance (from cornea to lens); HD, HS = mounting height; BH = the height of the frame; BL = frame length; Br = nasal bridge; UP = pantograph; UC = frame angle of curvature [2].

3 Filters and Image Processing Used by Eye Fit Software to Accurately Measure Individual Parameter

For accurate measurement determination, the image capture to be processed must be extremely clear. Parameters are measured by converting the pixels from which the image is formed into a unit of millimeter length [mm]. You cannot capture perfect images due to external factors that influence the clarity of the images - light, incorrect positioning, patient collaboration, device camera, etc.

Eye Fit software is built to apply some filters to the captured image so it can accurately determine the measurements. Below I will describe some of the processing and filters applied to the images for their improvement.

Pixel is the smallest element of an image. Each pixel corresponds to any value. In an 8-bit scale image, the pixel value is between 0 and 255. The value of a pixel at any point corresponds to the intensity of the light photons that hit at that point. For a grayscale image, the red, green and blue components of each color in the color palette are the same. If the color components are specified by 8-bit numbers (so between 0 and 255, that is, the most commonly used case), the color table will have 256 colors (different grayscale). Their specification will be between 0 and 255, assigned according to convention 0 - black, 255 - white. In this way, for an index indexed with gray levels, it is no longer necessary to specify the color table; the color represented by the index i corresponds to the gray level, i.e. the RGB triplet (i, i, i) [3].

Mathematically, images can be represented as a function of two variables in L2 (R2), as follows:

- grayscale images can be modeled with:

$$f(x;y) : R2 \rightarrow R, \tag{1}$$

in which case the values of the function f represent the values of the luminance of the objects in the image, in the points (x; y) of the space.

- color images can be modeled with:

$$f(x;y) : R2 \rightarrow R3, \tag{2}$$

in which case the f function values represent 3-vector components of a color space.

For example, it may be the three components of the RGB model.

The discrete image model is the model used in practice. The function f is discrete value, also found on a range of discrete values, namely:

$$f(k;l) : Z^2 \rightarrow Z^+ \text{ sau } f(k;l) : Z^2 \rightarrow Z^{+3} \quad (3)$$

Switching from the continuous field to the discreet field is done by sampling and quantization.

3.1 Histogram of an Image and Histogram Equalization

Histograms have many uses in image processing. The first use is image analysis. We can predict about an image just by looking at its histogram. Another use of the histogram is to improve the brightness of the image. Histograms are also used to adjust contrast and equalize an image. The histogram of an image shows the frequency of the pixel intensity values. In a histogram of the image, the x-axis indicates the intensity of the gray level and the y-axis indicates the frequency of these intensities [3] (Fig. 7).

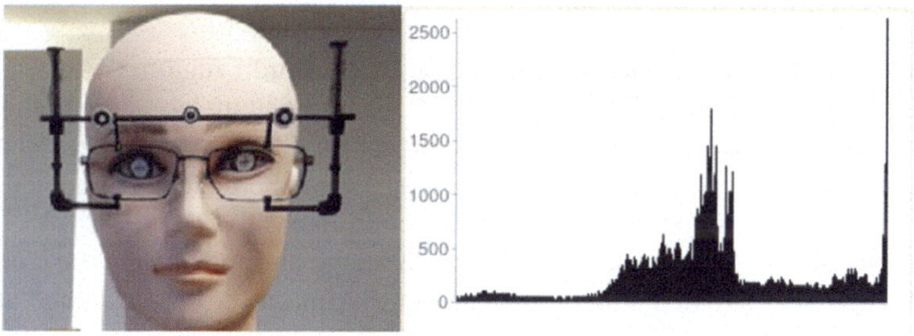

Fig. 7. Image capture and image histogram [4]

The x-axis of the histogram shows the range of pixel values. From an 8 bpp image, it means that it has 256 levels of gray or shades of gray in it. For this reason, the x-axis range starts at 0 and ends at 255 with a gap of 50. While the y-axis is the number of these intensities. As you can see from the graph, most of the bars that have a high frequency are in the portion of the first half of the section, which is the darkest. This means that the image we have is darker [3].

Histogram alignment is used to enhance contrast. First, we need to calculate the PMF (probability mass function) of all the pixels in this image. The next step involves calculating CDF (cumulative distributive function). We apply this technique to our original image.

After applying, we have the following image and the following histogram (Fig. 8):

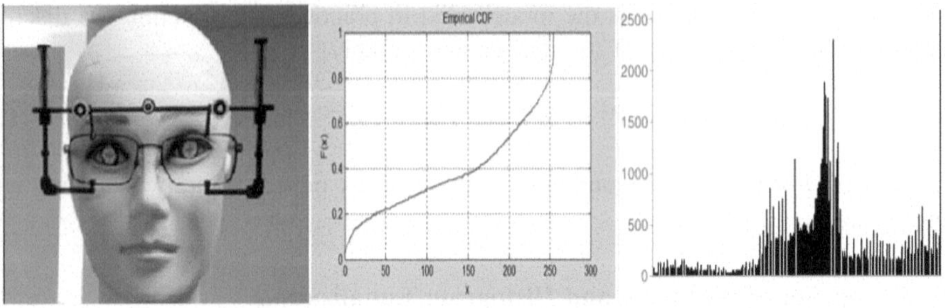

Fig. 8. Image - cumulative distributive function of this image, - equalization histogram [4]

As you can clearly see in Fig. 9, the new image contrast has been improved and its histogram also has been aligned. There is also an important thing to note - during histogram equalization, the general histogram pattern changes, the histogram stretches, but the overall shape of the histogram remains the same [3].

3.2 Filters Applied to an Image

The human system tends to deepen the profile of transition regions between uniform regions. The study of the physiology of the visual system has shown that this is accomplished by derivative processing occurring in the various stages that visual information goes through. The global effect can be mathematically modeled by subtracting from the original signal a weighted second derivative [3].

Filters can also be used to detect edges in an image and to enhance the clarity of an image. We can also say that sudden changes in discontinuities in an image are called margins.

Generally, the edges are of three types:

- Horizontal edges
- Vertical edges
- Diagonal edges

Most information about the shape of an image is enclosed in the margins. So, first we detect these edges in an image by using these filters and then by improving those areas of the image that contain the edges, the image clarity will increase and the image will become clearer [3].

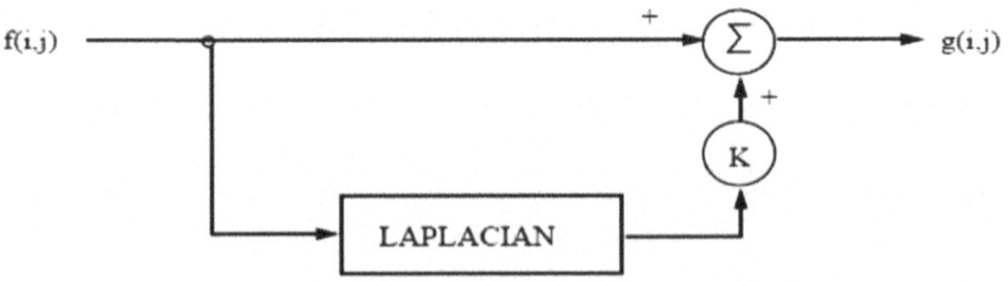

Fig. 9. Highlight filter [3]

Operator Laplacian. Laplacian is a derivative operator; its applications highlights the gray level discontinuities in the image and try to highlight regions with different gray levels. This operation in the result produces such images that have lines of gray edges and other discontinuities in a dark background. It produces inner and outer edges in an image. The important thing is how to apply these filters to the image. Note that we cannot apply both the Laplacian positive and negative operator on the same image, we only need to apply one, but we have to remember that applying a positive Laplacian operator to the image subtracts the resulting image from the original image to get the sharp image. Similarly, if we apply a Laplacian negative operator, we must add the resulting image to the original image to get the sharp image [3].

Operator Prewitt. The Prewitt operator is used to detect horizontal and vertical margins [3].

Operator Sobel. The Sobel operator is very similar to the Prewitt operator. It is also a derived mask and is used for edge detection. It also calculates the edges in the horizontal and vertical directions [3].

For example, we started from a captured image with the Eye Fit, which has the three markers on the special frame as datum points for parameter computation. Once the markers have been detected accurately - using special filters, the software can calculate the pixels from which the image is formed. For the accuracy of the measurements, a series of filters are used to help detect pixels, margins vertically and horizontally (Fig. 10).

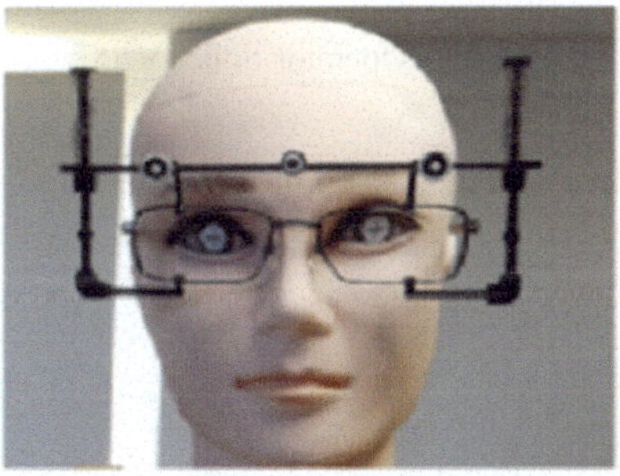

Fig. 10. Eye Fit captured image [5]

Fig. 11. Operator Sobel H [5]

To apply the Sobel filter to detect gray levels so we can better 'read' the image, we apply the following software script where we highlight the grayscale horizontally. Example (Fig. 11):

if(grupFiltre.selectedData=="sobleH")

{var sobleH:ConvolutionFilter = new ConvolutionFilter(3,3,[1,0,-1, 2,0,-2,1,0,1], 1,127);

bmp_temp_modificat.bitmapData.applyFilter(bmp_temp_modificat.bitmapData, bmp_temp_modificat.bitmapData.rect,new Point(0,0),sobleH); [5]

Apply Sobel filter and detect vertical grayscale (Fig. 12):

Fig. 12. Operator Sobel V [5]

As an example, we have in Fig. 13, applied in the same way the Prewit filter.

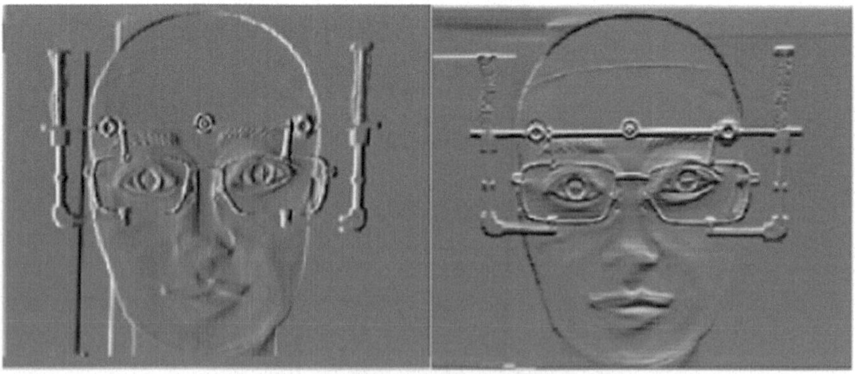

Fig. 13. Operator Prewit H; Operator Prewit V [5]

After applying the image filter set, the Eye Fit software can accurately calculate the parameters according to the three markers that are attached to the special frame. Once computed, the parameters are introduced into the calculation and processing of the ultra-sophisticated progressive lenses, helping the patient to enjoy maximum comfort while wearing a very clear view due to the individual construction.

4 Conclusions

Image capture and processing is used in a wide range of domains. Image processing, especially in the medical field, helps a lot in prescribing a correct diagnosis without exposing the patient to stress. The medical devices for investigation have developed a lot lately, managing enviable performances. MRIs and ultrasounds are currently highly accurate, detecting pathologies that would otherwise remain unknown. In the field of ophthalmology, existing devices help both in detecting eye diseases, but also in measuring individual parameters necessary for a clear and comfortable view. All of these things depend on capturing and processing images with special devices and software. Technology is constantly evolving. The help that man receives from advanced technology is substantial. Technological evolution shows that new knowledge and technological advances in the field of image acquisition, computing power in portable devices, the explosive development of mathematical fundamentals and image processing algorithms, and non-contact spatial measurements have allowed the transition to manual measurements affected by numerous errors to automatic, performance measurements.

It should be noted, however, that there are still major issues related to patient collaboration, uneven illumination, the need to achieve more successive pictures and their processing.

Consequently, advanced studies are needed to improve the method, to reduce the number of shots (by improving algorithms to be more robust in relation to error sources and to making an optical system with multiple micro-cameras to allow simultaneous more photos).

References

1. Dumitrescu, N.: Bazele opticii fiziologice. Politehnica, Bucureşti (1991)
2. Centwins SRL: dezvoltator soft OPTIVIO – EYE FIT – Prezentare aplicatie, capturi imagini masuratori, Bucuresti (2019). https://www.optivio.ro
3. Buzuloiu, V.: Prelucrarea imaginilor: Note de Curs, Universitatea “Politehnica” Bucuresti (1998)
4. Tutorial: Procesare de imagini. https://www.dcode.fr/image-histogram
5. Centwins SRL: dezvoltator soft OPTIVIO – EYE FIT – Prelucrare imagini - scriere soft, Bucuresti (2019). https://www.optivio.ro

Clamping Mechanisms of an Inspection Robot Working on External Pipe Surface

Bogdan Grămescu(✉), Laurențiu Adrian Cartal, Ahmed Sachit Hashim, and Constantin Nițu

University POLITEHNICA of Bucharest, 313, Splaiul Independenţei, 060042 Bucharest, Romania
bogdan.gramescu@upb.ro

Abstract. The paper presents the clamping mechanism for an external pipe inspection robot developed for use in oil and gas industry. During regular motion, the robot resembles to a crank-slider mechanism with equal lengths of the crank and rod, while for stepping over flanges it becomes an open chain with double actuated leverage. In order to keep the contact with the pipe, the robot needs two clamping mechanisms. The proposed solution analyzed for a small-scale demonstrator is presented.

Keywords: Mobile robot · Inchworm locomotion · External pipe inspection · Clamping mechanism

1 Introduction

It is well known the importance of oil and gas industry for the world economy. The climate conditions of the areas where these resources are mainly exploited, and the personnel costs are reasons for use of robotics and automation equipment within this industry [1]. Robotics is expected to help for inspection and detection of pipelines defects: cracks, corrosion, deposition and leakage, or for maintenance operations, like cleaning. Achieving these goals requires movement along the pipe.

Most of the developed robots do this inside the pipe [2–4]. The inner surface of the pipe is used for different types of locomotion: wheels, inchworm or screw techniques. The main advantage of these developments is the adjustment to variable pipe diameters and detection/cleaning of the internal depositions, while the main disadvantage is the need for pipe dismantling in order to perform the robot tasks.

The alternative is to use the outer surface of the pipe for locomotion and inspection tasks [5, 6], for which the inchworm technique or actuated wheels are regular solutions. In these circumstances, the locomotion principle and mechanism should be also adequate to avoiding obstacles like flanges, elbows turning left/right or up/down, etc. That means the existence of at least two grasping mechanism, with controlled action.

G. I. Gheorghe (Ed.): ICOMECYME 2019, LNNS 85, pp. 218–230, 2020.
https://doi.org/10.1007/978-3-030-26991-3_20

2 Clamping Mechanism Structure

The proposed clamping solution was designed for use with a robot developed [7] to travel along the pipelines, carrying sensorial equipment and communication devices, in order to perform the inspection of the pipes. The entire system consists in a specific mechanism – Electro-mechanical unit, which is the plant of the automatic system, integrated with sensors and actuators, two Power supply units of different nominal voltages, with step down circuits, a Control unit and a Communication unit. The above presented structure is valid both for a small-scale demonstrator and for a real scale robot, for which the adequate components (mechanical structure, power supply and actuators) must be chosen when it is built.

The most important function of the robot structure is the locomotion, which is, in this case, of inchworm type (see Fig. 1) It is called so, due to similarity of its locomotion mode with the living being having the same name. This specific motion consists of alternating clamping of the worm's body ends, followed by body extensions and contractions alternatively. This technique is often used by people to climb on a rope.

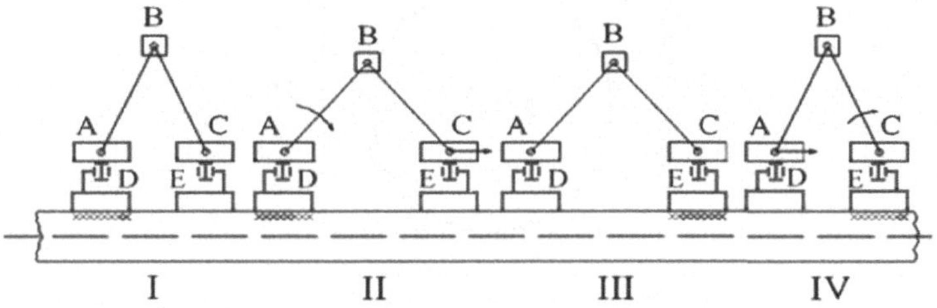

Fig. 1. Inchworm locomotion of the inspection robot along a pipe

One of the functions of the robot kinematic chain is clamping on the pipe. For clamping the ends of the developed robot body, if taken into consideration that its pathway is mostly a cylindrical surface (external surface of the pipe), a grasping mechanism (gripper) seems to be the most appropriate solution. These devices are useful, during operation, mainly for two tasks: to ascertain their adequate position, in order to be aligned to the pipeline axis; to fasten the robot's ends on the pipe.

There are several classifications depending on different criteria, but the shape and dimensions of the grasped objects and the nature of the grasping force are the most important.

In this case, the grasping mechanism should fasten the robot on the cylindrical pipe, without request for adapting to different diameters. It is to remind that the actual research aims to development of a small-scale demonstrator, able to travel along a pipeline of 50 mm diameter, and to avoid obstacles like elbows, flanges, etc.

Among different grasping devices, such as mechanical, magnetic, with vacuum or adaptable to shape and dimensions of the grasped object, a mechanical one, electrically actuated is the adequate solution of the problem.

Some design decisions must be made from the very beginning, based on functional considerations, which can optimize the constructions in connection with particular criteria. For example, a rotational motor would diminish the size of the clamping mechanism, while letting fixed one of its jaws and the other mobile will lead to a cheaper design. It is obvious that only a theoretical and experimental evaluation of the fixation grade will be the criterion for final decision making.

Following the idea with one fixed jaw, it is to be noticed that this one will constitute an alignment basis for the robot during locomotion, but it will not be allowed to have a V-shape.

A V-shape of the jaw increase the friction between the jaw and pipe and, therefore improve the clamping force, but this type of jaw requests an angular mobility, in order to obtain an adequate opening, which is over the pipeline diameter. So, this makes possible to detach the grasping mechanism from the pipe, when an obstacle is met. A sketch of the jaws and clamping forces is shown in Fig. 2.

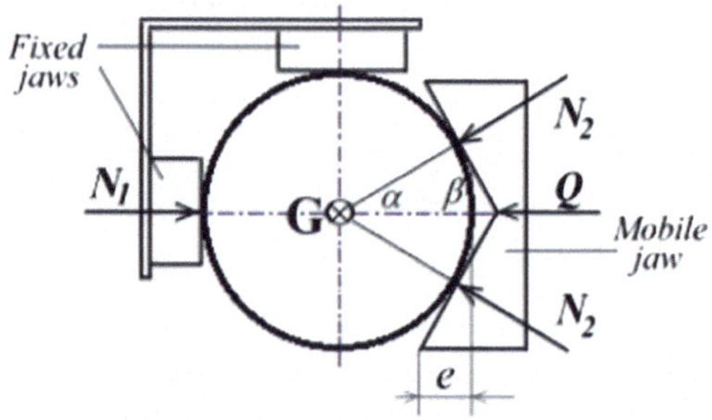

Fig. 2. Sketch of the jaws, pipe and their forces

For the proposed solution, the force Q is generated by the torque of a RC servo motor and amplified by a four-bar linkage mechanism presented in Fig. 3. This kind of motor offers a angular rotation, not a clamping force, but, using a compliant jaw, the angular position is converted into a force at the pipe level.

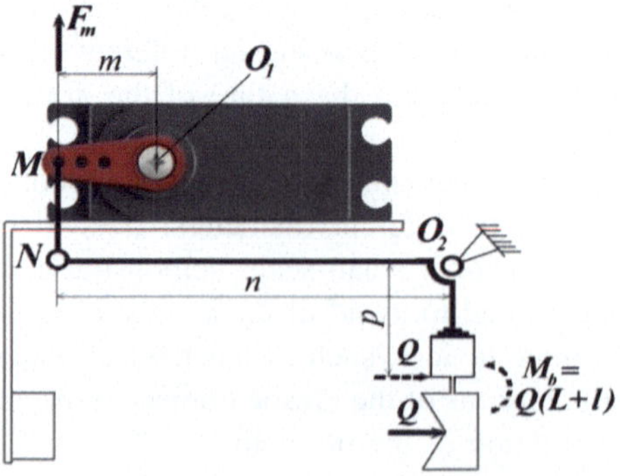

Fig. 3. The clamping mechanism

3 Clamping Mechanism Calculations

It is proposed an analyze of the mechanism in order to determine the relations between the motor moving angle, geometrical dimensions of the mechanism and the clamping force generated by the compliant jaw.

The balance of the forces acting on the pipe is:

$$N_1 = 2N_2 cos\alpha \tag{1}$$

The forces acting on the V-shaped jaw are also in equilibrium, resulting:

$$2N_2 cos\alpha = Q = N_1 \tag{2}$$

If N_1 cannot be amplified as concerns Q, N_2 depends on the V-shaped jaw angle, $\beta = 90° - \alpha$. In this case:

$$N_2 = \frac{Q}{2cos\alpha} = \frac{Q}{2sin\beta} \tag{3}$$

Equation (3) shows that the smaller is the angle β, the bigger is the reaction force, N_2. It is necessary to have a compromise between the increase of the clamping force when β is small and the necessary rotation angle of the jaw, which increases when β decreases. The rotation angle must allow the detaching of the mechanism for obstacle avoidance.

The obstacles, listed among operational objectives, can be elbows between horizontal and vertical segments of the pipe, when the robot has to climb. It is the most disadvantageous position for the required clamping force, Q, because the entire weight of the robot is supported only by the friction forces, associated with the reactions N_1 and N_2. Assuming that both jaws are made of the same material at the contact area (same friction coefficient with the pipe), the forces balance is described by:

$$\mu(N_1 + 2N_2) = kG \tag{4}$$

where: μ – friction coefficient between jaws and pipe; G – robot's weight; k – safety factor. By substituting (1) and (3) into (4), it will be obtained:

$$\frac{1 + sin\beta}{sin\beta} \cdot Q = G(\beta) \cdot Q = \frac{kG}{\mu} \tag{5}$$

The Eq. (5) describes the dependence of the clamping force on the half-angle of the V-shaped jaw. As it could be noticed, there is a gain provided by the V-shape of one jaw, a factor expressed as a mathematical function of β. Even this one could present a significant value variation, there are geometrical restrictions based on constructive reasons, that limits its range. For example, the most used value of the V-shape half angle is β = 30°. Smaller values than 30° will increase the jaw size, while a limit of β = 90° transforms the V-shaped jaw into a plane one. So, the investigation of the G(β) factor variation could help to optimize the grasping mechanism.

Using the above relations, it could be observed that, when β takes values within the range [30°…90°], the gain factor, G(β), varies from 3 to 2. This result shows the capital influence of the V-shape jaws, with the best value for the regular value of β = 30°. Someone should remember that only one jaw is V-shaped in the proposed clamping mechanism, so, if both jaws would be V-shaped with the half angle, β = 30°, the gain factor, G(β) = 6. As usual, a compromise is made between the grasping force amplification and the necessary opening of the jaws. For only one V-shaped jaw, it results that G(β) = 3, if β = 30°, G(β) = 2.42, if β = 45° and G(β) = 2.15 if β = 60°. The conclusion is that the slope of the gain factor variation is much lower after β = 45° and the optimal value is for β < 45°, but the jaw width influences the necessary opening for avoiding the pipe touch when the clamping device is retracted.

In the same time, the clamping force, Q is another variable, depending on G(β), which can influence the design, especially the chosen servo motor providing the clamping/unclamping movement and force. According to (4):

$$Q = \frac{kG}{\mu G(\beta)} \tag{6}$$

If the robot weight is G = 20 N, and the safety factor is chosen to be, k = 1.2, it is necessary to evaluate the friction coefficient between the jaws and the pipe. The jaws will be manufactured by 3D printing, with Nylon 645 wire (Taulman 3D) and the pipe for testing the small-scale robot is a commercial one, made of PVC. Most producers and handbooks recommend using, for the contact between these materials or similar, a value of the friction coefficient within the range [0.15…0.25].

With these values, the necessary clamping force is Q = 75…56 N, for β = 60° and Q = 66…50 N, for β = 45°. This one is provided by a RC servo and four bar linkages mechanism, which amplify the grasping force. An RC servo is a closed loop control system, which accomplish a position control of the output shaft, while for clamping it is necessary to have a force/torque control. This can be obtained, without hacking the RC servo, if a compliant part is introduced in the position-controlled mechanism. The most appropriate part to be compliant is the moving jaw. For this purpose, a flexure will be added between the V-shaped block and the fixing end (Fig. 4).

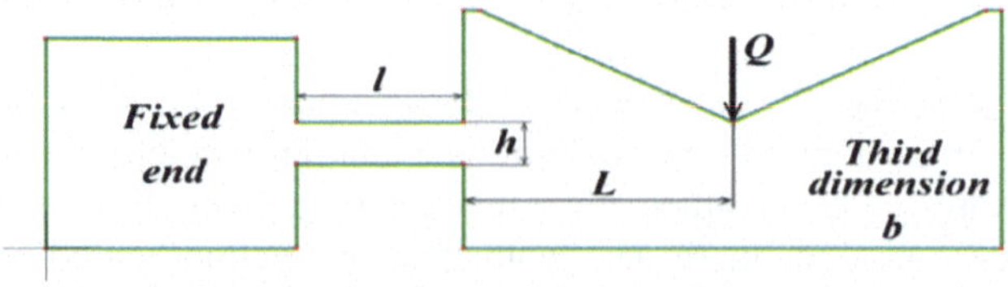

Fig. 4. Sketch of the compliant jaw

The flexure is of rectangular type, to be easier for accurate manufacturing, with a less known material. Moreover, the design is more versatile as concerns changing the dimensions, while a test trial for a notch type flexure has failed by cracking.

The rectangular type flexure has also the advantage to be more accurately investigated with analytical methods, than the notch type one. Anyway, a finite element analysis should be performed in both cases.

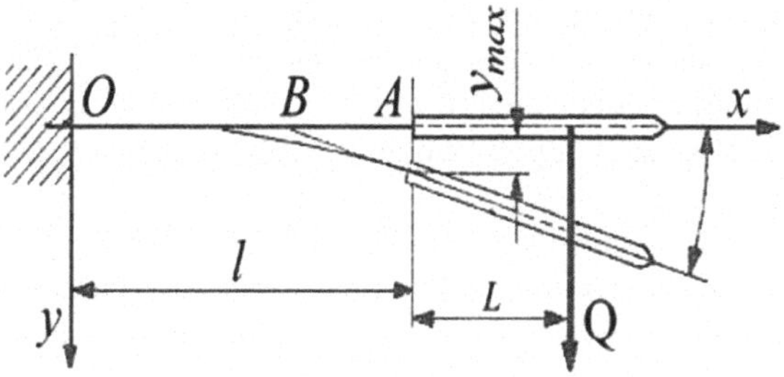

Fig. 5. Structural model of the mobile jaw

As concerns the analytical method, the structural model of the mobile jaw in Fig. 4 is shown in Fig. 5. It consists of a deformable beam fixed at one end (length, l), and a rigid one (the V-shaped part of the jaw) – beam of length, L, up to the application point of the clamping force, Q. The action of this one upon the flexure (deformable beam of the model) is equivalent with the force, Q, moved at the mobile end of the flexure and the moment Q.L, concentrated along the deformable beam. By assuming the stress-strain state is in the linear domain of Hooke's law, the superposition of effects is applicable. So, the linear deflection of the flexure free end is:

$$y_{max} = \frac{Ql^3}{3EI_z} + \frac{QLl^2}{2EI_z} = \frac{Ql^2}{6EI_z}(2l + 3L) \tag{7}$$

Where: E – Young's modulus of the flexure material; $I_z = bh^3/12$ – cross section moment of inertia of the flexure beam; b – beam width; h – beam thickness. Similarly, the angle of rotation of the flexure free end is:

$$\theta_{max} = \frac{Ql^2}{2EI_z} + \frac{QLl}{EI_z} = \frac{Ql}{2EI_z}(l + 2L) \tag{8}$$

The center of rotation is approximately the point B in Fig. 5, and its position can be found as:

$$\overline{AB} = y_{max} \cdot cot\theta_{max} = \frac{y_{max}}{\theta_{max}} = \frac{l}{3} \cdot \frac{2l + 3L}{l + 2L} \tag{9}$$

The maximum bending stress is at the fixed end of the flexure beam:

$$\sigma_{max} = \frac{6Q(L+l)}{bh^2} \tag{10}$$

If the values of the clamping force, Q, were calculated for the most disadvantageous situation (e.g. minimum friction coefficient and clamping on a vertical pipe), the influence of the V-shape angle can be another criterion for design, because the overall size of the clamping mechanism issues new constraints. These dimensional constraints limit the dimensions of the jaw (L, l), while the clamping force is related to the V-shape angle, β and also to the dimension, L. For calculations it is used the dimension l = 10 mm, for both V-shaped angles, while L = 16 mm for $\beta = 60°$ and L = 20 mm for $\beta = 45°$.

A tensile strength within the flexure is established as $\sigma_{max} = 40$ MPa < 47 MPa = σ_{all}, which is the maximum allowable stress for Nylon 645. The producer of the material also specifies the value of the Young's modulus as being E = 1940 MPa.

With these data, the cross section dimensions of the flexures (b, h) are determined so to fulfil the Eq. (10). Practically, the width, b, is calculated from (10), for different thickness, h within the range [2…3]. The maximum deflection, y_{max} and the rotation angle of the flexure moving end, θ_{max} are also calculated from Eqs. (7) and (8). The optimal values are Q = 75 N, L = 16 mm, h = 3 mm, b = 32.5 mm, $y_{max} = 0.6$ mm, $\theta_{max} = 0.111$ rad = 6.36°.

When analyzing these data, someone should compare the values of the maximum deflections, y_{max}, and the resulting width, b, of the jaws. Even the loads differ with about 14%, the table presents very closed results for both cases (jaw with $\beta = 60°$, and $\beta = 45°$). They can be explained by the maximum bending moment on the flexure strip, which is 1950 N.mm for $\beta = 60°$, and 1980 N.mm for $\beta = 45°$. When a similar design of the mobile jaw is adopted for both cases, the different location of the contact points of the V-shaped jaws with the cylindrical pipe is the cause of lengthening the 45° jaw (Fig. 6).

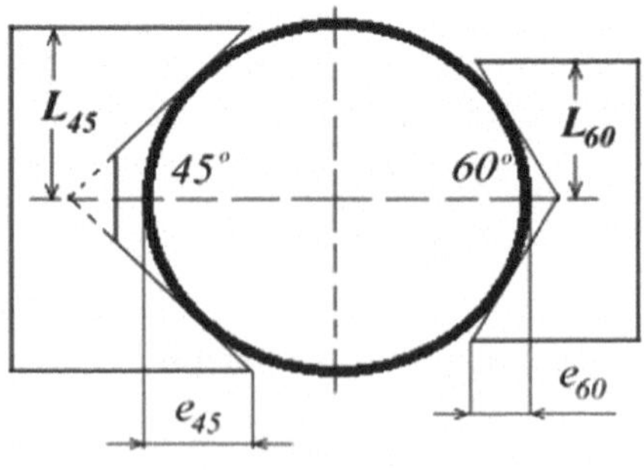

Fig. 6. Comparison between V-shaped jaws with 45° and 60° respectively

It is to be noticed in Fig. 6, that the 60° jaw is also smaller on the transversal direction, while the minimum necessary opening, e_{60}, for completely gripper detaching from the pipe, is also smaller than e_{45}.

The last comparison between the use of β = 60° V-shape jaw and the one with β = 45° is based on the influence upon the necessary motor torque.

Particular position of the mechanism in Fig. 3, $(\overline{O_1M}||\overline{O_2N})$ corresponds to the pipe clamping, when the force Q should be maximum. It is easy to see that the tangential force, developed by the motor torque, F_m, is best exploited when the angles $\measuredangle O_1MN = \measuredangle MNO_2 = 90°$.

In this case, the force vector Q, which acts on the jaw, has to be translated to the fixing point of the bending flexure, with the distance, L + l. The result is that a bending moment is added to act in this point, without influence upon the force balance of the four-bar linkage mechanism, but being balanced by the elastic reaction of the deformed flexure. Therefore:

$$F_m \cdot n = Q \cdot p \tag{11}$$

Where: n – length of the rocker bar, in Fig. 3; p – length of the rigid part of the mobile jaw. The necessary motor torque, T_m, to develop the clamping force, Q, is:

$$T_m = F_m \cdot m = \frac{Q \cdot m \cdot p}{n} \tag{12}$$

The bar lengths have a little value range, due to their correlation with the motor and pipe sizes. The proposed motor is a RC servo, which is very often used for developing prototypes of mobile robots. This is a closed loop positioning system, consisting of a dc motor with gearbox, a potentiometer for position feedback and a controller. This one interprets the input signals as position reference, converts them into voltage for comparison with the output feedback, in order to generate the error signal. Then, the error signal is amplified and converted into the adequate PWM command of the dc motor. A high ratio gearbox multiplies the output torque, while the angular velocity is diminished with the same factor. For the actual clamping mechanism, a Power HD product was chosen (HD 1501 MG), with a high output torque and gear ratio (supply voltage 6 V dc, stall torque 1666 N.mm, gear ratio 298 and speed 0.14 s/60°."The opening parenthesis does not have a corresponding closing parenthesis in sentence starting with "For the actual clamping mechanism… (supply voltage 6 V dc" under Sect. 3".Please insert the quotes in the appropriate position." –>

These data allow to determine the mechanical characteristics of the dc motor with gearbox, even in an approximate way, and to evaluate if this one is able to develop the clamping force. If the duration of 60° angular stroke is known, the stationary value of no-load angular velocity is the ratio between the angular stroke and this duration. For 6 V supply voltage, the no load angular velocity is 7.48 rad/s, while for 4.8 V supply voltage, this one is 6.545 rad/s.

Recalling the result from Eq. (12), the necessary motor torque is strongly dependent on the dimensions of the four-bar linkage mechanism, m, n, and p. It is obviously that m and p small, while n large will reduce the necessary torque, but there are space limitations which cannot be exceeded, for a thorough design. For example, n is limited to 55 mm, due to the side length of the angle profile (60 mm), and m can be chosen 10 mm, depending on the servo horns, which are delivered together with the RC servo (not longer than 20 mm). The length p depends on the jaw size and can be the same for both designs, even it could equalize the necessary torque, with a smaller value for larger Q. After all, the minimum same value was chosen for both jaws, p = 16 mm. It follows from Eq. (12) that the necessary motor torque for a clamping force, Q = 75 N, is T_m75 = 218.2 N.mm, while for a clamping force Q = 66 N, is T_m66 = 192 N.mm. These values represent approximately 13% of stall torques at 6 V and 4.8 V, respectively, which means the RC servo is able to provide the clamping forces. Anyway, the control system has to limit the motor current to values corresponding to the clamping torque, when stopped.

The other function of the clamping mechanism is to detach from the pipe, when the other end of the robot is clamped (fixed). For this purpose, the four-bar linkage mechanism has to pull out the jaw from the pipe, with a minimum distance, measured at the end of the elastic jaw. This one should be larger than the sum between the maximum deflection of the compliant jaw and the dimensions e_{60} or e_{45}, in Fig. 6, which are the minimum necessary openings, for completely gripper detaching from the pipe.

For performing the gripper opening, the four bar linkage mechanism (Fig. 7) starts from the particular position, shown in figure, when the angular position of the crank, θ_c, is equal to the one of the rocker, $\varphi_c(\varphi = \varphi_1 + \varphi_2)$, and the bar MN is perpendicular to the crank and rocker. This particularity allows to establish a mathematical relationship between the bar lengths, m, n, q, and a:

$$a = \sqrt{(n - m)^2 + q^2} \tag{13}$$

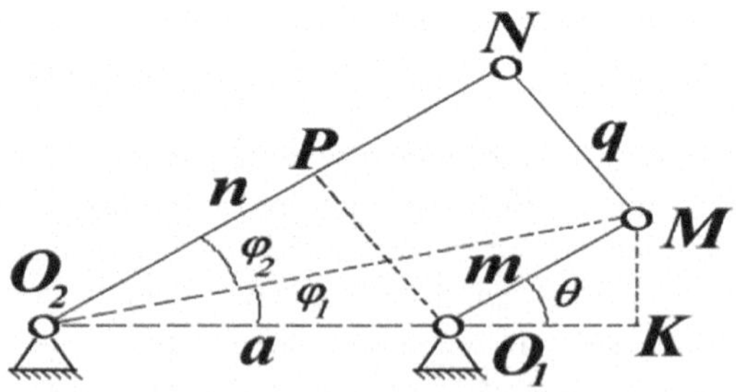

Fig. 7. Four bar linkages mechanism

It is also possible to determine the clamping angular positions of the rocker and crank. From the triangle O_1O_2P, it results:

$$\theta_c = \varphi_c = atan\frac{q}{n-m} \tag{14}$$

When the four-bar linkage mechanism leaves the particular position, in order to open the gripper, a relationship between the crank position, θ, and the rocker one. From Fig. 7, in triangle O_2MK:

$$tan\varphi_1 = \frac{msin\theta}{a+mcos\theta} \tag{15}$$

In the triangle O_1O_2M:

$$\overline{O_2M^2} = a^2 + m^2 - 2am \cdot cos(\pi - \theta) = a^2 + m^2 + 2am \cdot cos\theta \tag{16}$$

And also, from the triangle O_2MN:

$$cos\varphi_2 = \frac{\overline{O_2M^2} + n^2 - q^2}{2n \cdot \overline{O_2M}} = \frac{a^2 + m^2 + n^2 - q^2 + 2am \cdot cos\theta}{2n\sqrt{a^2 + m^2 + 2am \cdot cos\theta}} \tag{17}$$

By combining (15) and (17), it finally results:

$$\varphi = atan\frac{m\,sin\theta}{a+m\,cos\theta} + \text{acos}\frac{a^2 + m^2 + n^2 - q^2 + 2am \cdot cos\theta}{2n\sqrt{a^2 + m^2 + 2am \cdot cos\theta}} \tag{18}$$

If the dimensions of 2 arms (m, n) of the four bar linkage mechanism were already established, the 3rd mobile bar (q) was not yet defined, but could influence both the length of 4th bar (a) and the relationship between the crank angular position (input) and the rocker angular position (output). Being known that the mobile jaw opening should be larger than $e + y_{max}$, where e is e_{60} or e_{45}, the necessary angular displacement of the rocker is:

$$\Delta\varphi = asin\frac{e + y_{max}}{p + l + 2L} \tag{19}$$

The dimensions of 2 arms (m, n) of the four-bar linkage mechanism were already established, while the 3rd and 4th, could influence the relationship between the crank angular position (input) and the rocker angular position (output).

The length of the bar MN in Fig. 7 is denoted by q. A minimum value of this one, if the RC servo width and position of the O_2 joint are considered, is q = 14 mm. An increase of this parameter will be analyzed as concerns the influence upon the 4th bar length (a) and on the relationship between the angles θ and φ. With these data (13) becomes:

$$a = \sqrt{(n-m)^2 + q^2} = \sqrt{2025 + q^2} \tag{20}$$

Generally, in Fig. 7, $\varphi = \varphi_1 + \varphi_2$ and, if making use of notation, $u^2 = a^2 + m^2 + 2am \cdot cos\theta$, (18) becomes:

$$\varphi = atan\frac{msin\theta}{a + mcos\theta} + \mathrm{acos}\frac{n^2 + u^2 - q^2}{2n \cdot u} \quad (21)$$

The Eqs. (14), (20) and (21) were used for an investigation of the influence of the q bar length upon the output angle (movement of the mobile jaw), but it does not reveal a spectacular optimization, because the output angle is within [8°…9°] range, while the input angle is within [68.6°…72.7°].

The clamping mechanism proposes a novelty concerning the jaws which are grasping the pipe, by letting one of them to be both mobile and compliant. The jaw compliance is the solution for generating a grasping force, when the actuator can be controlled only in position (Fig. 8).

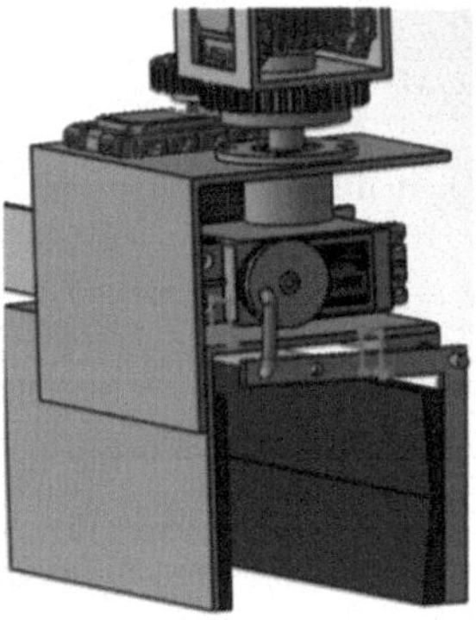

Fig. 8. Clamping mechanism with a rigid jaw and a mobile and compliant one

As previously told, the compliant jaw has to be also analyzed by Finite Element Method (FEM). The used FEM tool was LISA 8.0.0, "a user-friendly finite element analysis package for Windows with an integrated modeler, multi-threaded solver and graphical post-processor" [8]. It is to be mentioned that LISA software is "affordable", but also has a free version with 1300 nodes limit.

The geometry of the compliant jaw was shown in Fig. 4. In order to be analyzed with the methods of the Strength of Materials, the parallelepiped was considered fixed and the deflection of the V-shaped end was calculated, Even the real movement is vice versa, because the V-shaped end is blocked by the pipe, and the other end is displaced by the rocker, the FEM analysis was performed with the same constraints and load position as the analytical model. There is no influence of this change upon the deflection results, because the stationary reference frame considered, can have a virtual rotation with the rocker negative velocity, in order to fix the parallelepiped end and let the V-shaped end to deflect.

Because the analysis is performed on a 3D structure, a third dimension (b) of the jaw has to be chosen. One option is to use the values that were analytically calculated (b = 32.5 mm; h = 3 mm), but for a reasonably meshing and keeping the nodes number under 1300, as LISA free version allows, the chosen value was 10 mm. In these circumstances, the finite element analysis was performed with the load Q_{FEM} = 23 N, 3.25 times smaller than the analytical one, Q_{an} = 75 N. This ratio is the same one between the analytical width of 32.5 mm and FEM width of 10 mm.

The maximum tensile stress in the cross section of the rectangular bridge is σ_{FEM} = 36.96 MPa < 47 MPa, the allowable value of the material.

For comparison of the deflections, the displacement of the compliant jaw points is calculated with the results in Fig. 9.

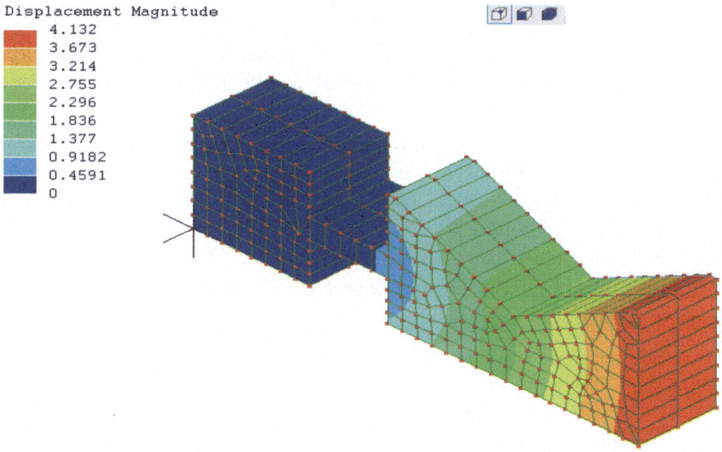

Fig. 9. Displacement magnitude

The displacement of the rectangular bridge end is in the range 0.4591….0.9182 mm, while its analytical value is 0.6 mm, which confirms matching of the results. As concerns the displacement of the V edge, where the load Q is applied, this one is in the range 1.836….2.296 mm. For comparison, this displacement is analytically calculated, by use of y_{max} and θ_{max}:

$$y_V = y_{max} + L \cdot \theta_{max} = 2.376 \text{ mm} \tag{22}$$

4 Conclusion and Future Work

The paper has presented the design and analyze of a clamping mechanism proposed for an external pipe inspection robot. The analyze prove the mechanism has the capability to achieve the task, and it is done for a small-scale demonstrator. The above presented structure is valid both for a small-scale demonstrator and for a large-scale robot, for which the adequate components (mechanical structure, power supply and actuators) should be chosen when it is built.

References

1. Shukla, A., Karki, H.: Application of robotics in onshore oil and gas industry-a review Part I. Robot. Auton. Syst. **75**, 490–507 (2016)
2. Sharma, S.L., Qavi, A., Kumari, K.: Oil pipelines/water pipeline crawling robot for leakage detection/cleaning of pipes. Glob. J. Res. Eng. H Robot. Nano-Tech. **14**(1), 31–37 (2014)
3. Schempf, H., Mutschler, E., Gavaert, A., Skoptsov, G., Crowley, W.: Visual and nondestructive evaluation inspection of live gas mains using the explorer^{TM} family of pipe robots. J. Field Robot. **27**(3), 217–249 (2010)
4. Kotawad, A., Lad, K., Jadhav, S., Mandlik, R.: Identify the deterioration in pipe by using wheel operated robot. Int. J. Res. Appl. Sci. Eng. Technol. **4**(II), 278–281 (2016)
5. Chatzakos, P., Markopoulos, Y.P., Hrissagis, K., Khalid, A.: On the development of a modular external-pipe crawling omni-directional mobile robot. Ind. Robot.: Int. J. **33**(4), 291–297 (2006)
6. Singh, P., Ananthasuresh, G.K.: A compact and compliant external pipe-crawling robot. IEEE Trans. Robot. **29**(1), 251–260 (2013)
7. Nițu, C., Grămescu, B., Hashim, A.S., Avram, M.: Inchworm locomotion of an external pipe inspection and monitoring robot. In: Machado, J., Soares, F., Veiga, G. (eds.) Innovation, Engineering and Entrepreneurship, HELIX 2018. Lecture Notes in Electrical Engineering, vol. 505. Springer, Cham (2019). https://doi.org/10.1007/978-3-319-91334-6_63. ISBN 978-3-319-91333-9, Online ISBN 978-3-319-91334-6
8. https://lisafea.com/

AI Based Voice Translator to Sign Language

Vlad Andrei Hanganu, Andra Daria Duță,
Constantin Daniel Comeagă, and Bogdan Grămescu(✉)

University Politehnica of Bucharest,
313, Splaiul Independenţei, 060042 Bucharest, Romania
bogdan.gramescu@upb.ro

Abstract. The paper presents a solution for an application that makes communication easier for people with hearing deficiencies, making a conversion from users' voice messages to the sign language.

Keywords: Mimico-Gestual language · Translator · Software

1 Introduction

The mimico-gestural language (see Fig. 1) is a language expressed by sign made by hand [1] combined with gesture, facial mimics, words spoken without sound and body posture, making natural configuration-space gestures and visual perception through which a deaf person can set up a communication channel in the social environment, whether they are deaf people who know the language or there is someone who knows the language of signs. While a communication channel for speech is voice-hearing, sign language is transmitted over a gesture-visual channel.

Interpreters using sign language work in meetings where the participants with hearing deficiencies participate and interpret from spoken language to sign language and vice versa. The interpreter is seated or standing in a place most visible to hearing impaired participants.

Although many gestures and mimic expressions, being similar or identical, can be recognized anywhere beyond the national or cultural boundaries, there is no unique or universal sign language to be understood by everyone. In fact, there are more than 100 dialects of signs in the world.

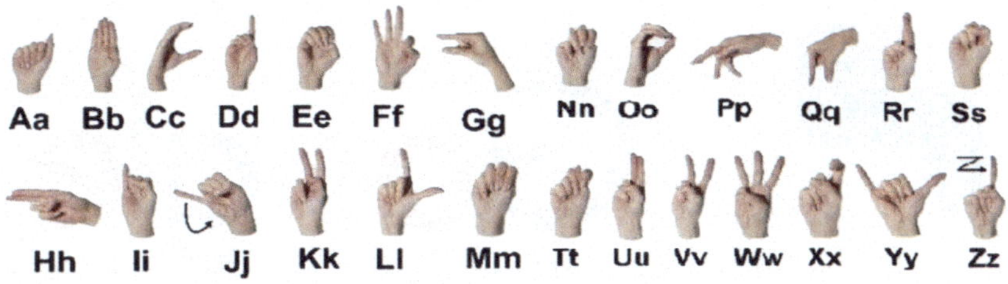

Fig. 1. Sign language

G. I. Gheorghe (Ed.): ICOMECYME 2019, LNNS 85, pp. 231–236, 2020.
https://doi.org/10.1007/978-3-030-26991-3_21

2 Fields of Use

From the very beginning, sign language was designed to facilitate the communication between people with hearing deficiencies and to make their lives easier.

Although sign language is currently used almost exclusively by deaf people, its origin is as old as oral language or even older in the history of humanity and has been and still is being used by hearing communities.

In fact, most indigenous people in the North American regions used the mimic-gestural language to understand each other, speaking different languages, with very different phonologies. The language was used even after the European conquests.

People have also used mimic gestures in various areas, such as road traffic policemen or scuba divers for underwater communication.

3 Analysis of Existing Solutions

There are various online applications that convert text into sign language and sign language into text.

To convert text into sign language, Artificial Intelligence modules are used to integrate all the nomenclature applied in this conversion into a database.

To convert from sign language into text, image processing is used through the TensorFlow or OpenCV library to real-time recognize gestures and assign the corresponding text structure.

There are currently few programs to help with this problem, the effectiveness being made by using a specific software.

The programs used in this project are Python [2] and the Google language modules using Artificial Intelligence. Both are Open Source platforms that any user can use (see Fig. 2).

Python [3] is a dynamic multi-paradigm programming language created in 1989 by the Dutch programmer Guido van Rossum. Van Rossum is also a leader of the software developer community that is working on Python language development and its basic implementation, CPython, written in C.

The **Google** Language Modules [4] are using neural networks for understanding and learning linguistic structures.

Fig. 2. Programe soft folosite

4 Own Solution

In principle, this program records the voice through the microphone and in the first stage performs a voice translation in the text.

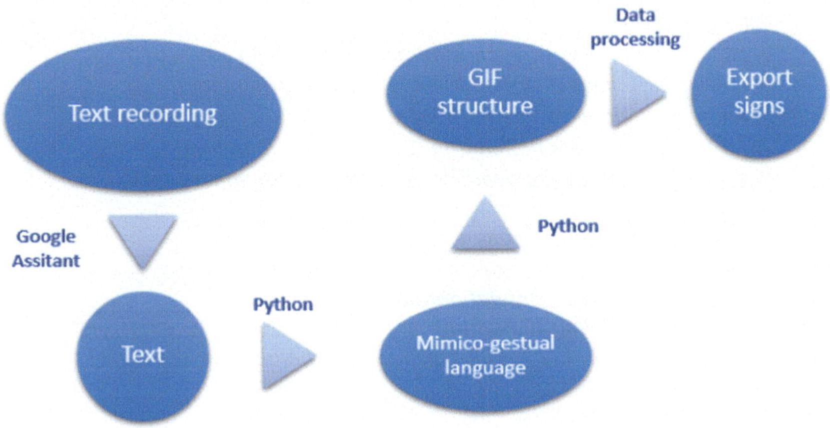

Fig. 3. Block diagram

Using the microphone as the source, the voice is recorded, followed by voice recognition software from Google to take over and display the actual text on the screen.

When the code picks up the text, it analyzes it letter by letter with the help of loops to assign to each structure the specific image in the sign language, followed by the storage of each attributed picture in memory.

Basically, the program indexes each image into a vector to subsequently transform it into a specific GIF (see Fig. 3).

The Graphics Interchange Format (GIF) [5], is a bitmap image format that was developed by a team at the online services provider CompuServe led by American computer scientist Steve Wilhite on June 15, 1987 (Fig. 4).

The format supports up to 8 bits per pixel for each image, allowing a single image to reference its own palette of up to 256 different colors chosen from the 24-bit RGB color space. It also supports animations and allows a separate palette of up to 256 colors for each frame.

After conversion, a new window opens in which the display is practically made (see Fig. 5).

CompuServe introduced GIF on June 15, 1987 to provide a color image format for their file downloading areas, replacing their earlier run-length encoding (RLE) format, which was black and white only. GIF became popular because it used LZW data compression, which was more efficient.

The program is easy to use because it is open from a bat file that actually opens the implemented code through the python console (see Fig. 6).

The code will interpret the message and transform it as such into the proposed view.

```
filenames = ['alfabet/A.png', 'alfabet/B.png', 'alfabet/C.png', 'alfabet/D.png',
'alfabet/E.png', ...]
images = []
...
for num in text:
    for let in num:
        let = let.lower()
        if let == 'a': images.append(imageio.imread(filenames[0]))
        if let == 'b': images.append(imageio.imread(filenames[1]))
        if let == 'c': images.append(imageio.imread(filenames[2]))
        if let == 'd': images.append(imageio.imread(filenames[3]))
        if let == 'e': images.append(imageio.imread(filenames[4]))
        ...
imageio.mimsave('movie.gif', images, fps=1)
ag_file = "movie.gif"
animation = pyglet.resource.animation(ag_file)
sprite = pyglet.sprite.Sprite(animation)
win = pyglet.window.Window(width=sprite.width, height=sprite.height)
green = 0, 1, 0, 1
pyglet.gl.glClearColor(*green)
@win.event
def on_draw():
win.clear()
sprite.draw()
pyglet.app.run()
```

Fig. 4. Source code

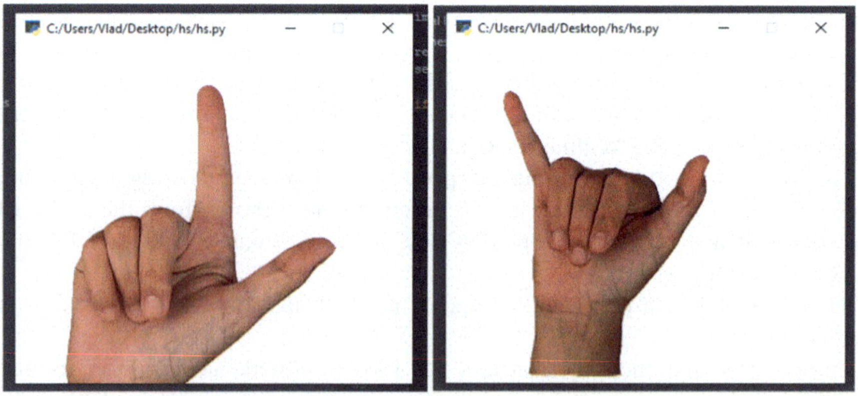

Fig. 5. Signs display

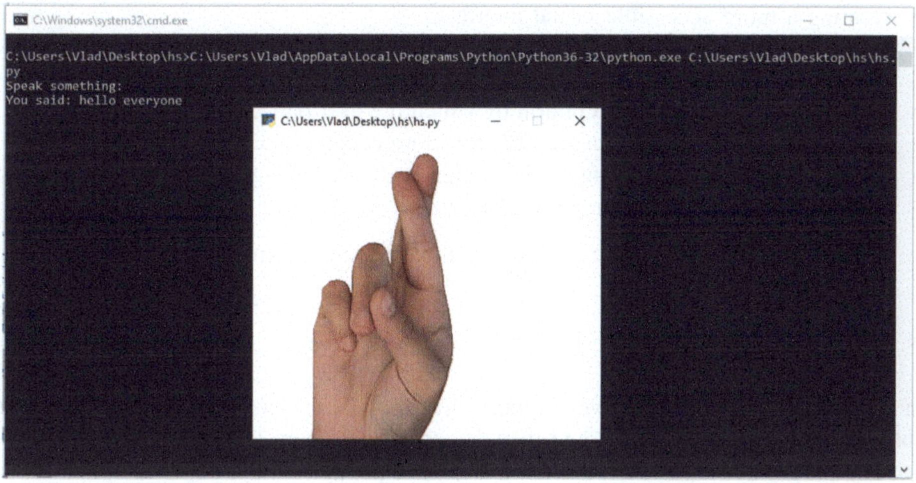

Fig. 6. Visual interface in command prompt

5 Subsequent Implementation

The solution that has been presented so far is a prototype. The program in its initial state proposes the possibility of effective communication for hearing impaired people in a simple form.

There are plenty ideas for a new visual interface in this program, like this: (see Fig. 7)

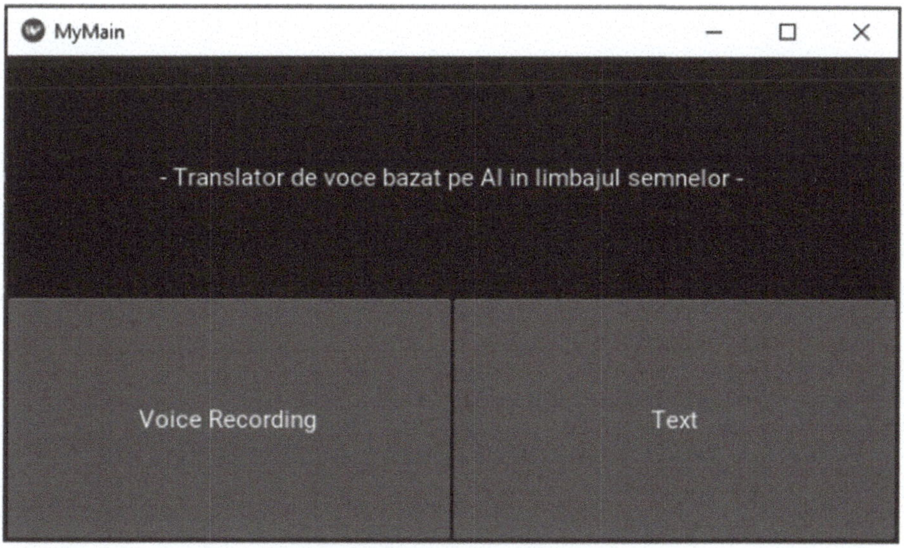

Fig. 7. Visual interface model

- creating a more accessible visual interface.
- reshaping the images displayed in a single window so that the application is compact.
- the possibility of inserting a button by which the user pauses between image views.
- adding more languages than English for more accessible communication.
- the possibility of introducing complex sign structures using a database of all words and a network by which they are determined.

These will appear in future versions of the program for more user-friendly use.

6 Conclusions

In addition to other constructive solutions, the application has the advantage of being able to record voice structures and then to convert into sign language.

The program is optimal for hearing impaired people, only able to speak and see, facilitating much easier communication without the user actually knowing this language.

References

1. Sign language. https://ro.wikipedia.org/wiki/Limba_semnelor
2. Python IDE. https://www.python.org/
3. https://ro.wikipedia.org/wiki/Python
4. Google language. https://ai.google/research/teams/language/
5. GIF. https://ro.wikipedia.org/wiki/GIF

Pneumatic Incremental Proportional Valve

Mihai Avram, Constantin Bucsan, Lucian Bogatu(✉),
and Daniel Besnea

Politehnica University of Bucharest, Bucharest, Romania
lucian.bogatu@upb.ro

Abstract. Performant actuating systems use both classic and proportional valves. Proportional valves are based on an electro-mechanical actuator. The characteristic of the electro-mechanic actuator $x_{ia} = f(x_i)$ may be continuous or discontinuous. This influences the characteristic of the whole device. The use of a discontinuous actuator has some advantages, as: it can be used in open loop, it allows starting, stopping and inverting the rotating direction without losing steps, it memorizes the position and it is compatible with the numerical technique. The paper proposes a proportional valve containing a classic pneumatic control valve with preferential position and electric control, and a proportional device with a special construction. The proposed construction and the functioning of the device are described, and the flowing area variation is calculated. The static characteristic of the proposed valve is determined, and the real profile of the control cam is calculated.

Keywords: Pneumatic · Incremental · Proportional valve

1 Introduction

Performant actuating systems use both classic and proportional valves [1]. Proportional valves are based on an electro-mechanical actuator that transforms the electrical control energy into a mechanical quantity to control the movement of the mobile subassembly. This can be displaced in any position within the working stroke, determining the flowing paths through the device, the values of the flowing areas and so the values of the flow rates to the consumer orifices. This way the output flow rates can reach any value between zero and the nominal flow and the characteristic $\dot{m} = f(x_i)$ is continuous, x_i being the electrical control quantity (voltage or current).

Figure 1 shows the specific characteristic of the proportional valve type MPYE-5…-010-B produced by FESTO [2]. It is a valve with 3 positions and 5 orifices with the functioning scheme shown in Fig. 2.

The x_i signal (see Fig. 2) is the input signal to the electronic amplifier of the proportional device, unlike the input signal of the actuator x_{ai}.

The characteristic of the electro-mechanic actuator $x_{ia} = f(x_i)$ may be continuous or discontinuous. This influences the characteristic of the whole device.

In practice there are many applications where a discreet control of the flow is not inconvenient. In such a case one can choose an actuator with a discontinuous actuator, such as a stepper motor.

G. I. Gheorghe (Ed.): ICOMECYME 2019, LNNS 85, pp. 237–246, 2020.
https://doi.org/10.1007/978-3-030-26991-3_22

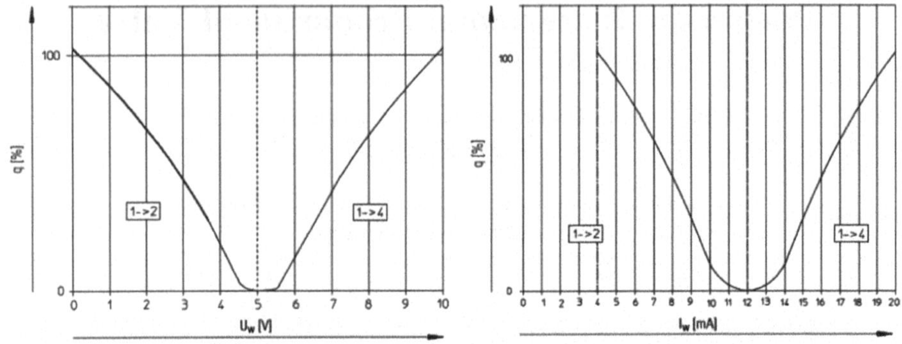

Fig. 1. The specific characteristic of the proportional valve type MPYE-5…-010-B [2].

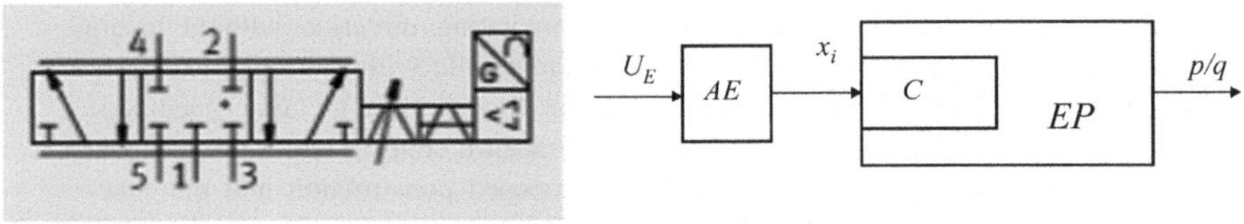

Fig. 2. The functioning scheme of the valve.

This one converts the control signals applied to its phases into a rotation movement consisting of discreet angular displacements representing the steps of the motor. The number of steps must be in concordance to the number of control impulses applied to the phases of the motor; this implies controlling the motor with a frequency lower than the limit frequency of the working regime. This type of actuator has some advantages [3], as: it can be used in open loop, it allows starting, stopping and inverting the rotating direction without losing steps, it memorizes the position and it is compatible with the numerical technique. There are also some disadvantages: the angular step is fixed, and the control scheme must be adapted to the type of the motor and it is relative complex when high speeds must be obtained.

2 The Working Scheme of the Proposed Valve

Figure 3 shows the principle scheme of the proposed valve. The starting point for the proposed solution is the work [4].

The proposed proportional valve contains a classic pneumatic 5/3 control valve with preferential position and electric control *DC* and a proportional device with a special construction, consisting of:

- the conical valves S_1 and S_2, the helicoidal springs a_1 and a_2, the body c_1, the cap supporting the bearing cr_1 and the contact bearing r_1;
- the actuating subassembly containing the electric stepper motor *MPP* and the control cam *c*.

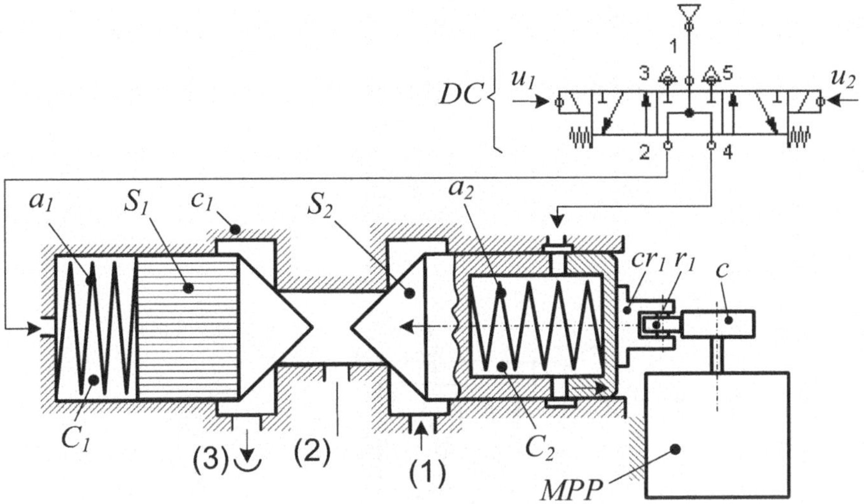

Fig. 3. The principle scheme of the proposed valve.

The working of this device is presented using Table 1:

- the central position *(0)* is obtained when the control valve *DC* is not controlled, in the absence of the voltages u_1 and u_2; in these conditions the two chambers C_1 and C_2 are supplied with pressure and the two valves S_1 and S_2 are pressed to their seats so the orifices *1*, *2* and *3* are blocked;
- the controlled position *(1)* is obtained when the control valve *DC* is commanded with the voltage u_1 and the stepper motor *MPP* is programmed to execute the desired number of steps n_{p0}, in the direction ω_1 (see Fig. 4); first the control valve *DC* is commuted, then the motor *MPP* is controlled; in these conditions the chamber C_1 remains supplied with pressure and the chamber C_2 is connected to the atmosphere; so the valve S_1 remains in contact with its seat and blocks the connection $2 \rightarrow 3$; due to the supply pressure the valve S_2 will be displaced from its seat with a stroke equal with the offset *h* of the command cam *cc* corresponding to the angle $\alpha = n_{p0} \cdot \alpha_p$, where α_p represents the value of the angular step of the motor; in consequence, a flowing area is generated between the conical surface of the valve S_2 and its cylindrical seat, connecting orifices *1* and *2*, the value of the flowing area depending on the number of programmed steps;
- the controlled position *(2)* is obtained when the control valve *DC* is controlled with the voltage u_2; in these conditions the chamber C_2 remains supplied with pressure and the chamber C_1 is connected to the atmosphere; so the valve S_2 remains in contact with its seat and blocks the connection $1 \rightarrow 2$; due to the pressure from the consumer *(2)* the valve S_1 will be displaced from its seat; in consequence, a flowing area is generated between the conical surface of the valve S_1 and its cylindrical seat, having a value equal with the nominal flow area, establishing the connection $2 \rightarrow 3$.

Table 1. The working steps of the valve.

Position	u_1	u_2	Stepper motor MPP
(0)	0	0	STOP
(1)	24 V	0	RUN $S_{c,1\rightarrow 2} \sim \varphi_M$
(2)	0	24 V	STOP

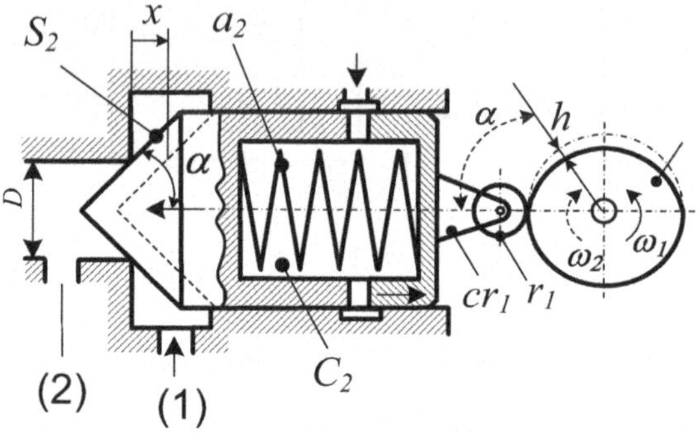

Fig. 4. The displacement of a conical valve depending on the rotation angle of the cam.

For a displacement *h* of the valve from its seat a flowing area is generated, with a value given by:

$$S_c(h) = k_1 \cdot h - k_2 \cdot h^2 \tag{1}$$

where k_1 and k_2 are constants:

$$k_1 = \pi \cdot D \cdot sin\ \alpha \tag{2}$$

$$k_2 = \frac{\pi}{2} \cdot sin\ \alpha \cdot sin\ 2\ \alpha \tag{3}$$

where *D* represents the diameter of the seat and α is the angle of the valve cone.

Instead of the conical valve a cylindrical control part *ER* may be used (see Fig. 5).

The cylindrical portion of the control part *ER* features *n* longitudinal channels with the inclination β and the width *b* (see Fig. 6).

In this case the control area approximative value is given by:

$$S_c(h) = k_3 \cdot h \tag{4}$$

where:

$$k_3 = n \cdot b \cdot sin\ \beta \tag{5}$$

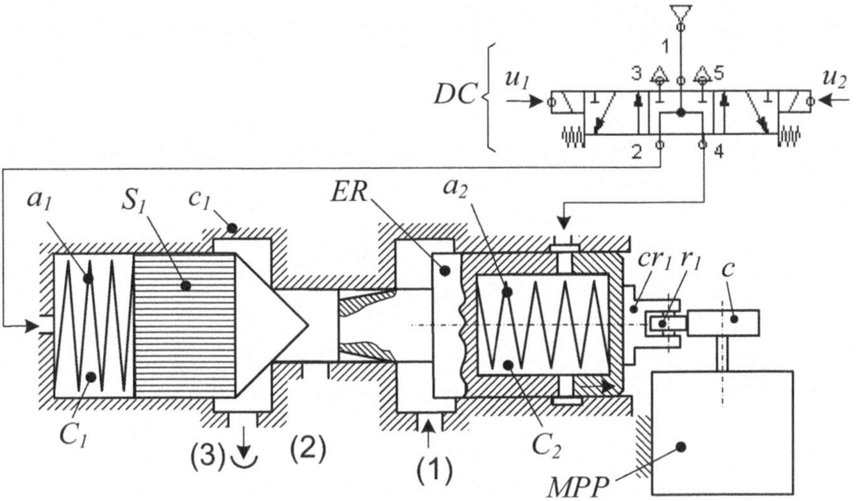

Fig. 5. The valve using a cylindrical control part *ER*.

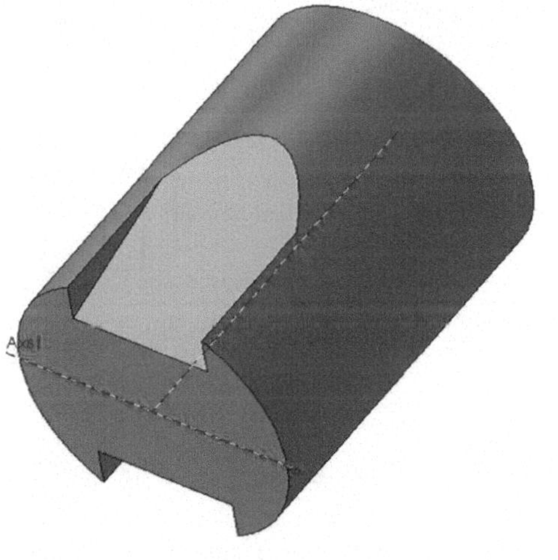

Fig. 6. Cylindrical control part ER featuring 2 longitudunal channels.

The position *(2)* is obtained when the control valve *DC* is controlled with the voltage u_2 and the motor *MPP* is not programmed.

Due to the counterpressure from the orifice *2* the valve S_1 will be displaced from its seat, generating a flowing area and connecting orifice *2* to the atmosphere orifice *3*.

Figure 7 presents the variation of the flowing area S_c depending on the displacement h for three cases, as following:

- *(I)* – for the values: $d_n = 3$ mm, $D = 6$ mm, $\propto = 45°$, and a maximum displacement of the valve $h_{max} = 0.55$ mm;

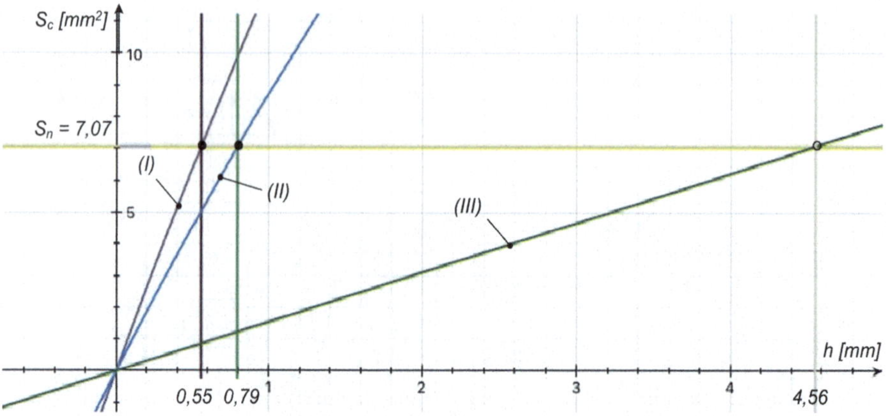

Fig. 7. The variation of the flowing area S_c depending on the displacement h.

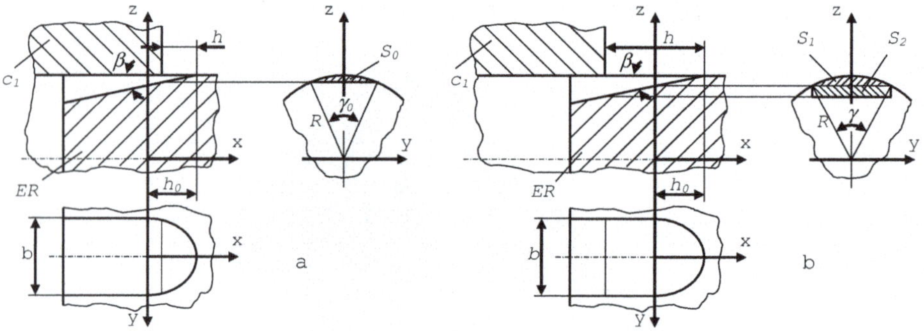

Fig. 8. The positions of the control part *ER* against the valve seat: (a) $0 \leq h \leq h_O$, (b) $h_O < h \leq h_{max}$.

- *(II)* – for the values: $d_n = 3$ mm, $D = 6$ mm, $\propto = 30°$, and a maximum displacement of the valve $h_{max} = 0.79$ mm;
- *(III)* – for the values: $d_n = 3$ mm, $b = 3$ mm, $\beta = 15°$, and a maximum displacement of the valve $h_{max} = 4.56$ mm.

The exact calculus of the flowing area is given by:

- for $0 \leq h \leq h_O$ (see Fig. 8a):

$$S_c(h) = n \cdot S_0 \cdot cos\,\beta = n \cdot \frac{R^2}{2}(\gamma_1 - sin\,\gamma_1) \cdot cos\,\beta \tag{6}$$

– for $h_0 < h \leq h_{max}$ (see Fig. 8b):

$$S_c(h) = n(S_1 + S_2)\cos\beta = n\left[\frac{R^2}{2}(\gamma - \sin\gamma) + \left(h \cdot tg\beta - R + R\cos\frac{\gamma}{2}\right)\right]\cos\beta \quad (7)$$

where:

$$h_0 = R\frac{1 - \cos\frac{\gamma}{2}}{tg\beta}; \quad (8)$$

$$\gamma_1 = 2 \cdot \arcsin\frac{\sqrt{2 \cdot R \cdot h \cdot tg\beta}}{R}; \quad (9)$$

$$\gamma = 2 \cdot \arcsin\frac{b}{2 \cdot R}; \quad (10)$$

and R is the *ER* part radius.

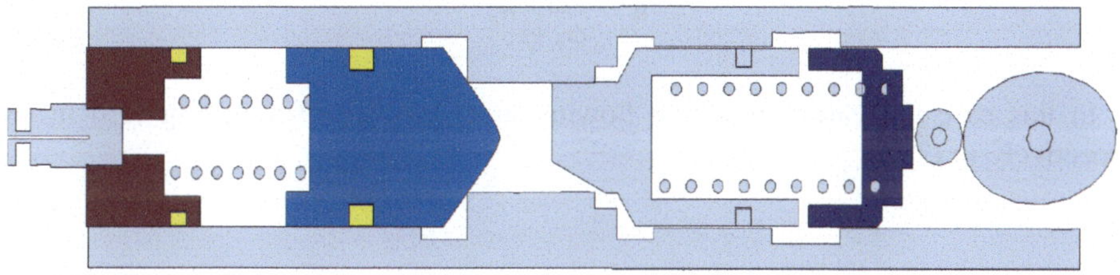

Fig. 9. A longitudinal section of the control valve.

In Fig. 9 is presented a longitudinal section of the 3D model of the control valve.

3 The Static Characteristic of the Proposed Valve

The static characteristic represents the dependence of the mass flow $\dot{m}$ on the number of programmed steps n_p, given by:

$$\dot{m} = \frac{K \cdot P_a \cdot S_n(n_p)}{\sqrt{T_a}} \cdot N\left(\frac{P}{P_a}\right) \quad (11)$$

where:

$$N\left(\frac{P}{P_a}\right) = \begin{cases} 1 & \text{if } 0 \leq P/P_a \leq 0.528 \\ a \cdot \left[\left(\frac{P}{P_a}\right)^{2/\chi} - \left(\frac{P}{P_a}\right)^{(\chi+1)/\chi}\right]^{1/2} & \text{if } \frac{P}{P_a} > 0.528 \end{cases} \quad (12)$$

and the values of the coefficients are [1]: $\chi = 1.4$, $R = 287.04\frac{m^2}{s^2 \cdot K}$, $K = 0.04042\frac{\sqrt{K} \cdot s}{m}$, $a = 2.6143$.

If the variation of the displacement of the valve depending on the rotation angle of the cam is imposed to be liniar, the movement of the valve is given by:

$$h = Ct_1 \cdot \varphi + Ct_2 \tag{13}$$

where Ct_1 and Ct_2 are constants determined using the limit conditions:

- $\varphi = 0$, $h = 0$ rezulta $Ct_2 = 0$;
- $\varphi = \varphi_1$, $h = h_{max}$ and $Ct_1 = \frac{h_{max}}{\varphi_1}$.

In this case the valve movement is given by:

$$h = \frac{h_{max}}{\varphi_1} \tag{14}$$

If the cam is actuated by a stepper motor with N steps per rotation, then:

$$h = \frac{h_{max}}{\varphi_1 \frac{2\pi}{N}} n_p \tag{15}$$

In this case the variation of the flowing area is obtained replacing (15) in (6), respectively in (7).

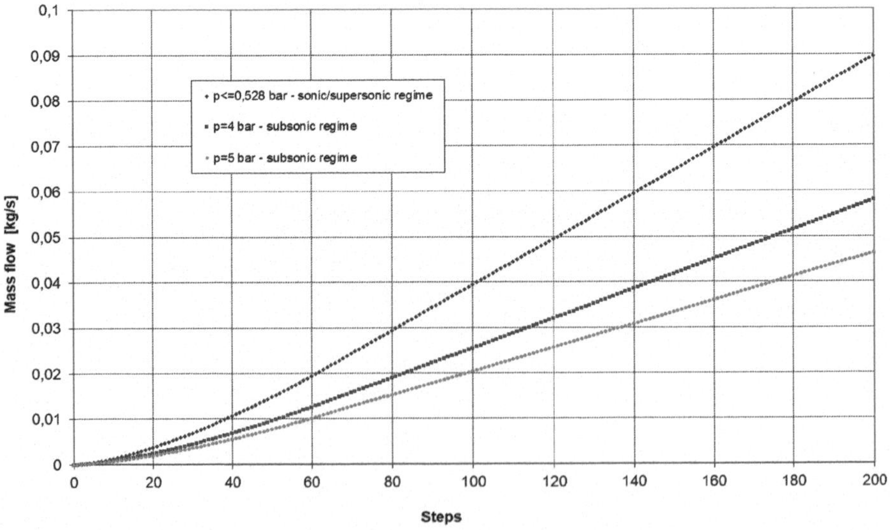

Fig. 10. The flow characteristic.

The flow characteristics obtained for $N = 200$ steps, $p_a = 6$ bar, $\varphi_1 = 2\pi$ rad, $h_{max} = 4.56$ mm, $b = 3$ mm, $\beta = 15°$, $p \leq 0{,}528$ bar, $p = 4$ bar and $p = 5$ bar are shown in Fig. 10.

For a linear movement of the valve (11) and if the follower is centric, the real profile of the cam is given by:

$$\begin{cases} x_r = \mathrm{A} \cdot \cos\varphi - r_r \dfrac{\mathrm{A} \cdot \cos\varphi + \mathrm{B} \cdot \sin\varphi}{\sqrt{[\mathrm{B} \cdot \cos\varphi - \mathrm{A} \cdot \sin\varphi]^2 + [\mathrm{B} \cdot \sin\varphi + \mathrm{A} \cdot \cos\varphi]^2}} \\ y_r = \mathrm{A} \cdot \sin\varphi - r_r \dfrac{\mathrm{A} \cdot \sin\varphi - \mathrm{B} \cdot \cos\varphi}{\sqrt{[\mathrm{B} \cdot \cos\varphi - \mathrm{A} \cdot \sin\varphi]^2 + [\mathrm{B} \cdot \sin\varphi + \mathrm{A} \cdot \cos\varphi]^2}} \end{cases} \tag{16}$$

where:

$$A = \left(r_0 + \frac{h_{max}}{\varphi_1} \varphi \right) \tag{17}$$

$$B = \frac{h_{max}}{\varphi_1} \tag{18}$$

r_O is the minimum radius of the theoretical profile of the cam;
r_r is the radius of the follower.

Figure 11 shows the real profile of the cam for $r_O = 10$ mm and $r_r = 2$ mm.

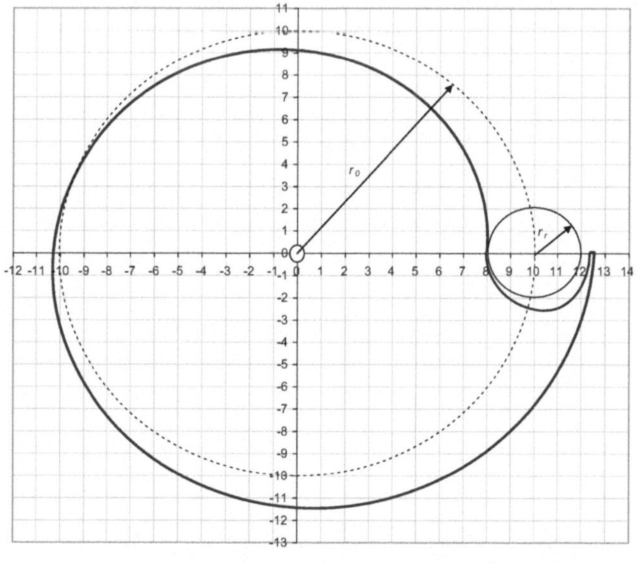

Fig. 11. The cam profile.

4 Conclusions

The use of a discontinuous actuator has some advantages, as: it can be used in open loop, it allows starting, stopping and inverting the rotating direction without losing steps, it memorizes the position and it is compatible with the numerical technique, and also some disadvantages: the angular step is fixed, and the control scheme must be adapted to the type of the motor. The proposed valve using a cylindrical control part with longitudinal channels instead of a conical control part, can have a linear characteristic if the real profile of the control cam is properly calculated.

References

1. Avram, M., Bucşan, C.: Smart Pneumatic Drive Systems. Editura Politehnica PRESS Publishing House, Bucharest (2014). (in Romanian)
2. https://www.festo.com/cat/ro_ro/data/doc_engb/PDF/EN/MPYE_EN.PDF. Accessed 29 May 2019
3. Kuo, B.C., Kelemen, A., Crivi, M., Trifa, V.: Command and Incremental Position Control Systems. Editura Tehnica Publishing House, Bucharest (1981). (in Romanian)
4. Avram, M., Bucşan, C., Bogatu, L., Besnea, D., Prisăcaru, G.: Proportional pneumatic distributor with electromechanical actuator, RO 129892 B1 Patent (2014)

Energy Harvesting from Renewable Energy Sources

Marian-Alin Bănică[1,2(✉)]

[1] National Institute of Research and Development in Mechatronics and Measurement Technique, Şos. Pantelimon 6-8, Bucharest, Romania
alin.banica01@gmail.com
[2] University "POLITEHNICA" of Bucharest, Splaiul Independeței 313, 060042 Bucharest, Romania

Abstract. Energy harvesting is systematic process and the main activity (solar energy harvesting, wind energy harvesting, ocean energy harvesting etc.) where a lot of energy is harvested and supplied to the main energy grid. Harvesting on the other hand means a bountiful process and systematic process and there is a major energy gain (100% activity). Energy harvesting technologies supply unlimited operating life of low-power equipment and even remove the need to replace batteries where it is costly, unfeasible, or unsafe. The performance of energy harvesting is fundamentally linked to the amount and nature of the source energy present in the environment. When designing an energy harvesting solution, knowledge of the application constraints and the details of the energy source have to be known in advance.

Keywords: Energy harvesting · Solar energy harvesting · Wind energy harvesting · Wave energy harvesting

1 Energy Harvesting

1.1 Models and Application for Energy Harvesting

At present there are considerable and continuing research efforts worldwide to support the energy harvesting paradigm and self-powered electronics. The majority of the reported research in energy harvesting has been on improving the efficiency of the energy harvesters through the design and fabrication of novel micro-generators, materials and devices. The amount of power that can be harvested in a particular application is highly dependent upon the energy source being harvested [1].

An energy harvester has normally three main components:

- the micro-generator which converts ambient environment energy into electrical energy
- the voltage booster which pumps up and regulates the generated voltage
- the storage element

Therefore, the performance optimisation should only be based on a model that describes the energy harvester as an integrated system.

G. I. Gheorghe (Ed.): ICOMECYME 2019, LNNS 85, pp. 247–254, 2020.
https://doi.org/10.1007/978-3-030-26991-3_23

The most existing modelling and optimisation methods are concentrating on either the micro-generator or the external circuits separately while the design tools for an integrated system are missing. MATLAB and finite element analysis (FEA) packages are being used to simulate and optimise the performance of the microgenerator part of the self-powered system [1].

2 Commercial Energy Sources

Renewable energy sources such as solar, wind, wave etc. are forecast to play an increasing role in the overall energy mix [2].

Commercial energy sources are offset by:

- solar energy
- wind energy
- wave energy

2.1 Solar Energy Harvesting

Solar energy is one of the most important renewable energy sources. Solar energy is plentiful; that is the greatest availability compared to other energy sources. The amount of energy supplied to the earth in one day by the sun is sufficient to power the total energy needs of the earth for one year. Solar energy is clean and free of emissions, since it does not produce pollutants or by-products harmful to nature. The conversion of solar energy in to electrical energy has many application fields. Residential, vehicular, space and aircraft, and naval applications are the main fields of solar energy [3].

Sunlight has been used as an energy source by ancient civilizations to ignite fires and burn enemy warships using "burning mirrors." Till the eighteenth century, solar power was used for heating and lighting purposes.

During the 1800s, Europeans started to build solar-heated greenhouses and conservatories. In the late 1800s, French scientists powered a steam engine using the heat from a solar collector. This solar-powered steam engine was used for a printing press in Paris in 1882. A highly efficient solar-powered hot air engine was developed by John Ericsson, a Swedish-American inventor. These solar driven engines were used for ships.

The first solar boiler was invented by Dr. Charles Greely, who is considered the father of modern solar energy.

The first working solar cells were invented in 1883 by Charles Fritts. Selenium was used to build these prototypes, achieving efficiencies of about 1%. Silicon solar cells were developed in 1954 by researchers Calvin Fuller, Daryl Chapin, and Gerald Pearson. This accomplishment was achieved by following the fundamental work of Russel Ohlin the 1940s. This breakthrough marked a fundamental change in the generation of power. The efficiency of solar cells increased from 6% up to 10% after the subsequent development of solar cells during the 1950s; however, due to the high costs of solar cells ($300 per watt) commercial applications were limited to novelty items [3].

Solar energy harvesting is the process of capturing and storing solar energy which is radiated from the sun. Then it is converted form light of heat energy to electrical energy by suitable method. Solar energy is free and rather available in many parts around the world.

In just one year, the sun can provide the earth with 15.000 times more energy than the atomic and fuel energy actually needed during the year.

If the fossil fuel resources are consumed at the present rate, all the fossil fuels will exhaust. This will create an unprecedented energy crisis on the earth.

Methods of Solar Energy Harvesting

- solar thermal collectors which is a heat-absorbing panels and a series of attached circulation tubes to generate electricity
- concentrating solar power that uses mirrors or lenses to concentrate the sun's rays to heat the fluid and produce steam which drives a turbine and generate power
- photovoltaic technology [4]

Applications of Solar Energy Harvesting

- Scott Brusaw, a former electrical engineer, has figured an innovative solar panels in hexagons that can connect to create a road way
- A project called "Solar Roadways" which can generate power and can use its power to light up to the road at night time or to send driving alerts to drivers who are on the road. It can be used to keep roadways cleaned and melt the snow that falls on them.
- He launched a highly successful campaign which raised over 2 million dollars. These roadways will be seen soon in U.S.A. [4].

Solar Energy Harvesting System for Wireless Sensor Network Nodes. A basic solar energy harvesting system consists of a Solar Panel, DC-DC converter, rechargeable battery, a battery charge protection circuit called battery management system (BMS) and DC-DC converter control unit. Generally, there are two types of DC-DC converter control methods:

- pulse width modulation (PWM) control
- maximum power point tracking (MPPT) control

The Fig. 1a shows a block diagram of a pulse width modulation (PWM) controlled DC-DC buck converter. Similarly, the Fig. 1b shows the block diagram of Perturb & Observation (P&O) maximum power point tracking (MPPT) controlled solar energy harvester (SEH) system [5]. In Fig. 1b, the SHE system consists of a solar panel, a DC-DC buck converter, a rechargeable battery, a maximum power point (MPPT) controller, and a WSN sensor node connected as a DC load. The ambient solar light energy is harvested using the solar panel and converted into the electrical energy. The DC-DC Buck converter steps down and regulates the magnitude of this harvested voltage, and supplied to the rechargeable battery. The MPPT controller tracks the voltage and current from the solar panel and adjusts the duty cycle accordingly for the MOSFET of

DC-DC Buck converter. Finally, the battery voltage is utilized to operate the wireless sensor node.

The WSN performs the function of sensing, computation, and communication with other similar characteristics nodes. Thus, autonomous operation of monitoring and control of any physical phenomenon such as temperature, humidity, pressure or acceleration can be achieved using the SEH-WSN nodes.

In this whole scenario, the efficiency of the solar energy harvester circuit plays a very important role. If the efficiency of the solar energy harvester system is poor, then the battery will not get recharged properly and hence the wireless sensor network lifetime will reduce [5].

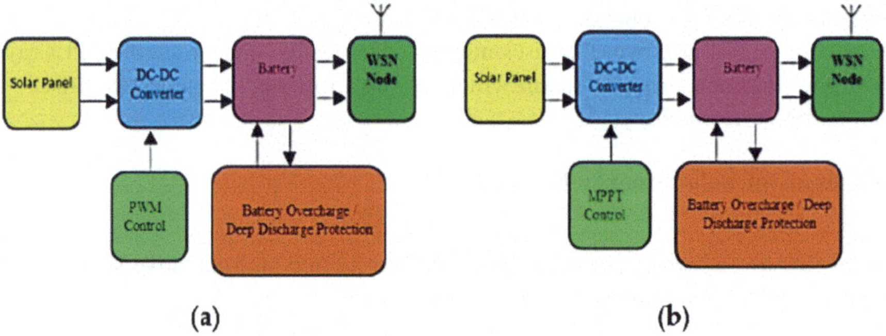

Fig. 1. Block diagram of solar energy harvesting system using PWM and MPPT control. (a) Using PWM control; (b) using MPPT control

2.2 Wind Energy Harvesting

Wind turbines capture the kinetic energy of winds and convert it into a usable form of energy. The kinetic energy of winds rotates the blades of a wind turbine. The blades are connected to a shaft. The shaft is coupled to an electric generator. The generator converts the mechanical power in to electrical power.

Wind energy conversion systems (WECS) involve many fields of various disciplines such as kinematics, mechanics, aerodynamics, meteorology, power electronics, power systems, as well as topics covered by structural and civil engineering [3].

In order to efficiently capture wind energy, several key parameters need to be considered:

- air density
- area of the blades
- wind speed
- rotor area

The force of the wind is stronger at higher air densities. Wind force generates torque, which causes the blades of the wind turbine to rotate. Therefore, the kinetic energy of the wind depends on air density; therefore heavier (denser) winds carry more kinetic energy. At normal atmospheric pressure and at 15°C (59°F), the weight of the air is 1.225 kg/m^3, but if the humidity increases, the density decreases slightly. Air density is also influenced by temperature; therefore warmer winds are less dense than cold ones, so at high altitudes the air is less dense [3].

The area of the blades (or air-swept area) plays an important role in the captured wind energy. The wind speed is the other parameter. It is expected that wind kinetic energy rises as wind speed increases.

The kinetic energy of the wind can be expressed as:

$$Ek = \frac{1}{2}mv^2 = \frac{1}{2}\rho V v^2 = \frac{1}{2}\rho A d v^2 = \frac{1}{2}\rho R^2 \pi d v^2 \tag{1}$$

where: E_k is the wind kinetic energy
m is the wind mass
v is the wind speed
ρ is the air density
A is the rotor area
R is the blade length
d is the thickness of the "air disc" (see Fig. 2)

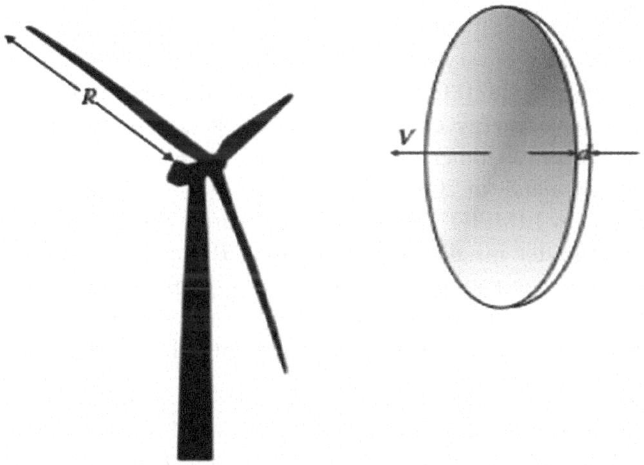

Fig. 2. Kinetic energy of wind

The overall power of (P) is:

$$P = \frac{E_k}{t} = \frac{1}{2}\rho R^2 \pi \frac{d}{t} v^2 = \frac{1}{2}\rho R^2 \pi v^3 \tag{2}$$

$$P = \frac{1}{2}\rho R^2 \pi v^3 \tag{3}$$

The power content of the wind varies with the cube (or the third power) of the average wind speed (see Fig. 3) [3].

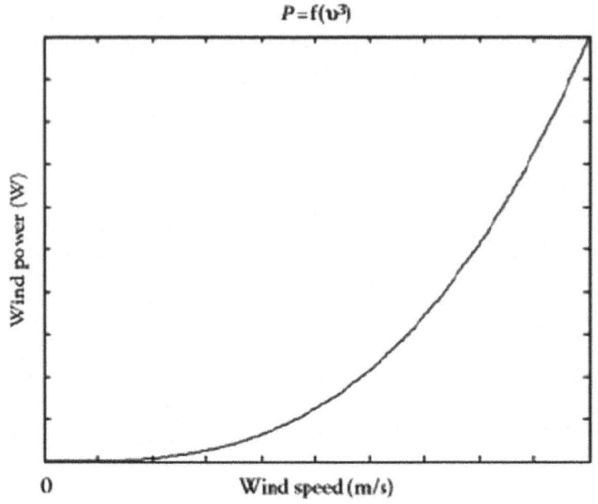

Fig. 3. Specific wind power due to wind speed variation

2.3 Wave Energy

The concept of "kinetic recovery of wave energy" began to be discussed to explore this new field at the end of the 18th century. In 1799, the scholar Pierre-Simon Girard patented the first wave energy concept, but the first large-scale prototype was conceived 200 years later. In other words, the first wave energy trading system was installed in Islay, Scotland, in 2000.

It was called "Islay LIMPET" (Land Installed Marine Power Energy Transmitter), a marine power transmitter installed on the ground. Basically, it was a 500 kW wave collector connected to the national grid. It was totally disbanded in 2018.

Ocean wave energy systems convert the kinetic and potential energy contained in the natural oscillations of ocean waves into electricity. There are a variety of mechanisms for the utilization of this energy source.

Wave Energy Harvesting. The first wave waves patent took place in Paris in 1799. Monsieur Girard and his son proposed the use of direct mechanical action to drive heavy machinery, including mills, saws and pumps.

After the first patent, thousands of inventors followed in this area. There were 340 from 1855 to 1973 only in the UK.

The pioneer of modern wave energy was Yoshio Masuda, Japanese naval commander. Masuda tested several different wave energy devices at sea and hundreds of these units have been used to activate the navigation lights. Masuda is also credited for various wave energy inventions, such as KAIMEI, a large barge used as a test platform and oscillating water column, which was initially used for small-scale navigation.

Wave energy had a new interest in the 1970 s during the oil crisis of 1973. In 1973, members of the Organization of Petroleum Exporting Arab Countries (OAPEC) decided to ban oil exports, which in turn led people to seek energy alternative [6].

One of these pioneers was Stephen Salter, famed for the "Salter Duck" invention, which is the duck prototype that trained the wave energy conversion into kinetic energy.

The device incorporates an opposite wave power generation system based on a pendulum connected to a generator. As the device moves up and down the waves, the pendulum binds back and forth to generate electricity (see Fig. 4).

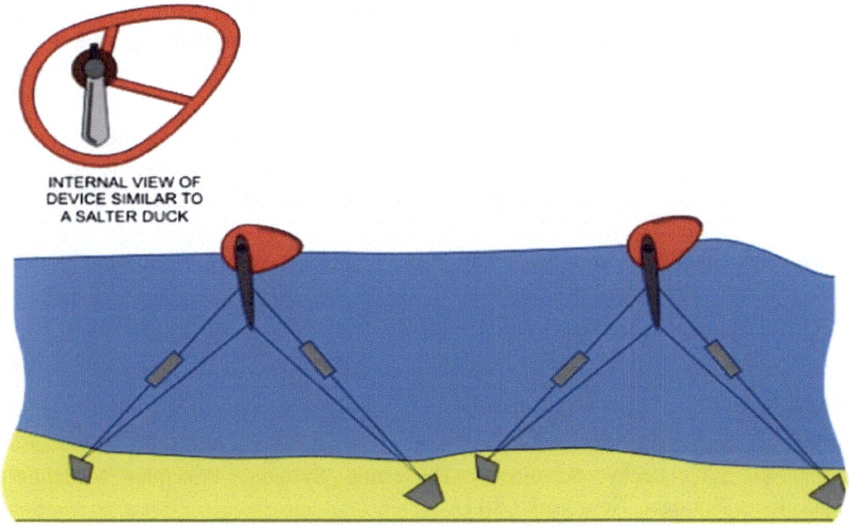

Fig. 4. Prototype "Salter Duck" [7]

Applications of Harvesting Kinetic Energy. The proposed various methods have been for converting wave energy into usable electrical energy practically.

Current experimental and theoretical researches that indicated up to 90% of the wave's power can be extracted given certain conditions. Thus, the ocean wave power can be efficiently be converted into electrical energy. The total conversion of wave energy efficiency is as round 35% of the ancillary when considering all the year throughout conversion processes [3].

There are two types of wave energy generation sites with respect to their distance from the shore, which are discussed in detail in the following subsections:

- offshore energy harvesting topologies
- nearshore energy harvesting topologies

Offshore applications are located away from the shore and they generally use a floating body as wave power absorber and another body that is fixed to the ocean bottom. Salter Cam and buoys with air-driven turbines are the only applications involving rotational generators in offshore applications.

Nearshore topologies are applied to the shore or within the surfing zone of the ocean. Nearshore applications have some advantages and disadvantages in comparison to the off-shore applications with respect to the selected method [3].

3 Conclusion

Energy harvesting has been well studied in the past decade. It has been regarded as one of the best alternatives to energy source for wireless sensor networks.

With the recent development of adaptive kinetic energy harvesting, this drawback will eventually overcome, which will bring kinetic energy harvesting to much broader applications.

Energy harvesting is the conversion of ambient energy present in the environment into electrical energy. It is identical in principle to large-scale renewable energy generation, for example, solar or wind power, but very different in scale.

Acknowledgements. This work was supported by a grant of the Romanian Ministry of Research and Innovation, Research program NUCLEU, contract no. 17 N/2019, project number PN 19.24.01.01.

References

1. Kaźmierski, J.T., Beeby, S.: Energy Harvesting Systems: Principles Modeling and Applications. Springer, New York (2011)
2. Sah, S.L.: Renewable and Novel Energy Sources. M D Publications PVT LTD, New Delhi (1995)
3. Khaligh, A., Onar, C.O.: Energy Harvesting: Solar, Wind, and Ocean Energy Conversion Systems. USA (2010)
4. https://www.slideshare.net/AfrinNirfa1/solar-energy-harvesting-and-its-applications. Accessed 06 May 2019
5. Sharma, H., Haque, A., Jaffery, Z.A.: Modeling and optimisation of a solar energy harvesting system for wireless sensor network nodes. J. Sens. Actuator Netw. **7**, 40 (2018)
6. https://academic.oup.com/jah/article/99/1/236/854867. Accessed 06 May 2019
7. https://permaculturenews.org/2010/02/23/wave-power. Accessed 06 May 2019

Electromechanical Structure of the Experimental Model of a Robotic Head

Tudor Catalin Apostolescu[1], Georgeta Ionascu[2], Silviu Petrache[2(✉)], Lucian Bogatu[2], and Laurentiu Adrian Cartal[2]

[1] Faculty of Informatics, Titu Maiorescu University, Bucharest, Romania
[2] Faculty of Mechanical Engineering and Mechatronics, POLITEHNICA University of Bucharest, Bucharest, Romania
silviupsf@yahoo.com

Abstract. The electromechanical structure of the experimental model of a robotic head, as well as the computer-assisted control of drive elements are described in this paper. The number of degrees of freedom for the anthropomorphic head was limited to seven, as follows: the oscillation movement of the eyeballs in the vertical plane (up-down); the oscillation movement of the eyeballs in the horizontal plane (left-right); the up-down movement of the eyelids; the up-down movement of the jaw; lateral extension of the lips; turning the head horizontally; the tilt movement of the head. The motors used in the construction of the experimental model of the humanoid head are of two types: DC motors with built-in reducer and with encoder, and stepper motors. All commands must be correlated with the voice signal from an auxiliary sound source, more precisely with the vocal sequence of the emitted sound signal, which the robot must simulate it by mimics of the eyes, mouth and neck. For this purpose, all motors are controlled through a data acquisition board, which receives the orders provided by LabVIEW. In this program, specific files were developed for the set of responses that robot must simulate them.

Keywords: Electromechanical structure · Robotic head · Motion control

1 Introduction

The humanoid robots are complex mechatronic systems made of basic components as structural parts and specific mechanisms that simulate different features of the human being, and, also, sensors and actuators with a control architecture, respectively hardware and software systems, which ensure planning and control [1–3]. Specialists from different fields as precision mechanics, electronics, computer science, artificial intelligence, medicine and, also, humanities combine their efforts to design and to fabricate such robots [4–6].

The robotic head, which is presented in this paper, has seven degrees of freedom (DOF) that allow the up and down movement of the eyes, eyelids, jaw and head, the left to right movement of the eyes and head, and the horizontal extension of the lips also. Stepper motors and servomotors are used as individually controlled actuators.

G. I. Gheorghe (Ed.): ICOMECYME 2019, LNNS 85, pp. 255–272, 2020.
https://doi.org/10.1007/978-3-030-26991-3_24

The robot head assembly consists of four main mechanisms [7] to generate the desired movements, one for the eyes (11), one for the jaw (15) and lips (14), one for the eyelids Fig. 1, representing 3D model designed with CATIA software.

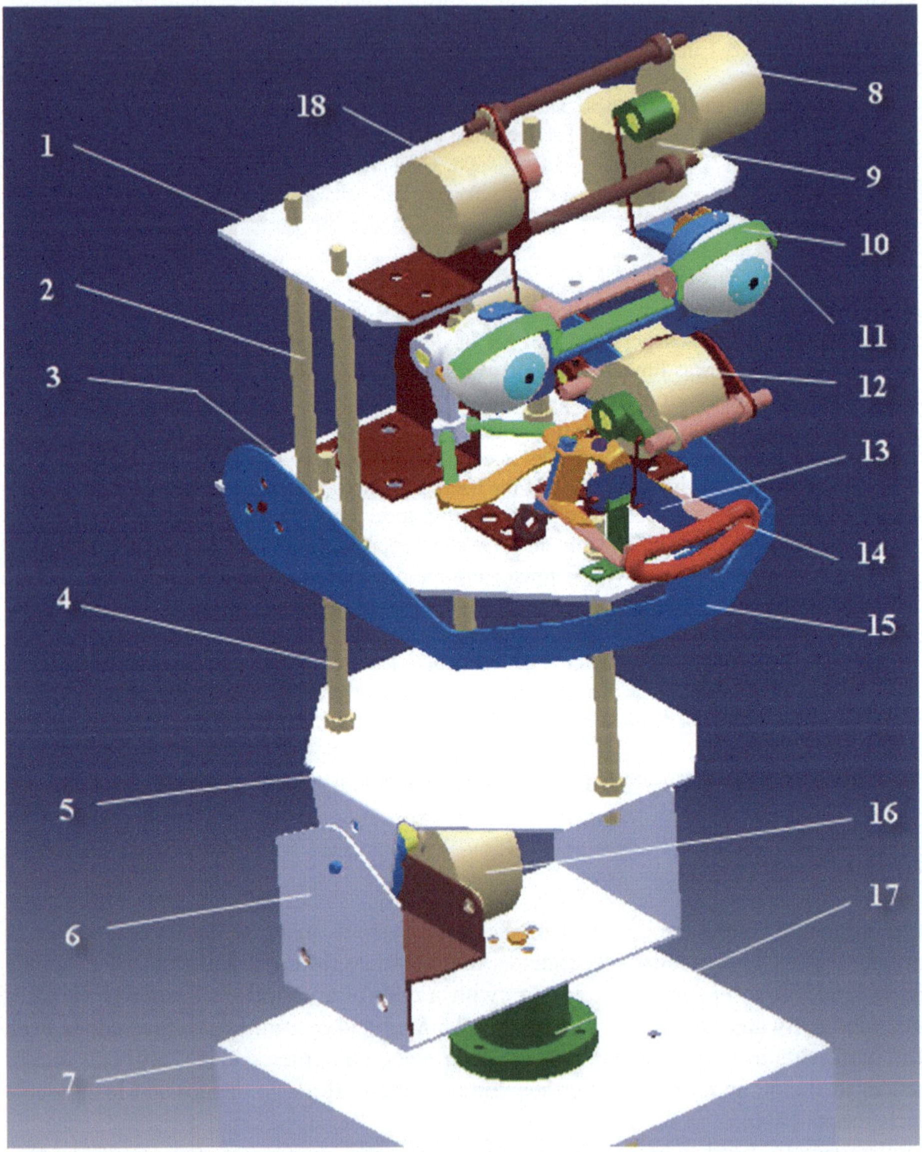

Fig. 1. Overview of the front-side of 3D model of the robot head assembly: 1, 3, 5, 6 – plates; 2, 4 – thrusts; 7 – fixed part of the device; 8, 9, 12, 16, 18 – gearmotors; 13 – lever.

2D model and the overall sizes are given in Fig. 2. The structure of the experimental model of robotic head, as well as the computer-assisted control of drive elements are described further.

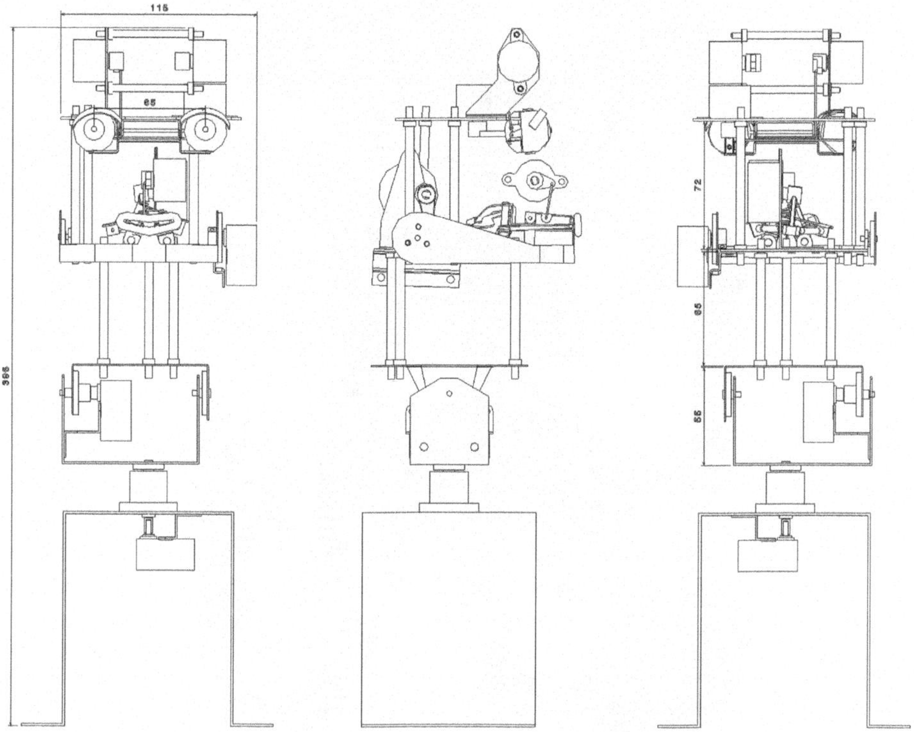

Fig. 2. 2D model of the robot head assembly

2 The Electromechanical Structure of the Robotic Head's Experimental Model

The device was designed and created to simulate the human face functions (Fig. 3). Regarding from the engineering view, the human face can be considered as a group of information receivers (optical, acoustical and chemical sensors) but also as a data transmitter using mimics and speech.

Fig. 3. The anthropomorphic head assembly (front view).

During the research, the number of degrees of freedom for the anthropomorphic head was limited to seven, namely:

- the oscillation movement of the eyeballs in the vertical plane (up-down);
- the oscillation movement of the eyeballs in the horizontal plane (left-right);
- the up-down movement of the eyelids;
- the up-down movement of the jaw;
- lateral extension of the lips;
- turning the head horizontally;
- the tilt movement of the head.

All these movements are achieved by using stepping motors and DC servomotors. They use gearboxes to obtain a high torque at the output shaft for the same power.

The assembly includes a four subassemblies, organized according to the function to be achieved:

- the eyeballs mechanism subassembly;
- the eyelid mechanism subassembly;
- the subassembly of the mouth actuator;
- the neck mechanism subassembly.

From the point of view of the electronic control, each motor is individually controlled.

Step-by-step motors are of two types, which differ by the intensity of the current. At the eyeball and eyelid mechanisms, where the torque at the motor shaft is low, the current is low. At the neck mechanism, the motors are of higher current, therefore generating a major torque, able to overcome the moment of high inertia of the mobile assembly.

The DC servomotors are controlled via PC, using the LabView programming environment and two data acquisition boards.

2.1 The Mechanism of Eyeballs and Eyelids

The eyeballs perform two main moves:

- the oscillation movement in vertical plane (top-down);
- the oscillation movement in horizontal plane (left-right).

The eyelids perform a third movement of oscillation in the vertical plane.

The two eyeballs are constructively different, because each is an element driven by a distinct mechanism. The left eyeball is an element of the mechanism that generates the oscillation movement in the horizontal plane, while the right eyeball is the element of the mechanism that generates the oscillation in the vertical plane.

To provide support for the eyeball in order to execute the oscillation movement in the horizontal plane, there are two pins provided, situated at the extremities of the eyeball vertical plane. The left-right movement of the left eyeball will be transmitted to the right eyeball with a horizontal lever which is jointed with the left eyeball using a circular hole.

Similarly, the left eyeball has a spherical surface oriented towards the exterior of the humanoid head, but also a flat surface located behind it. The mechanisms that provide eyeballs movement and eyelids movement along with their stepping motors used in the structure of the human head simulator are shown in Fig. 4.

The Oscillation Mechanism in Vertical Plane of the Eyeballs

Making a simultaneous up and down movement by the eye was possible by actuating a horizontal lever (pos. 4, Fig. 4) to which the left and right eyeballs are attached, using joints.

Relative to common lever, the two eyeballs are able to rotate around their own vertical axis, which will allow them to move horizontally using a mechanism which is described below.

Fig. 4. The assembly of the mechanisms that actuate the eyeballs and the eyelids: 1 - the motor that actuates the eyes in vertical plane; 2 - the eyelid drive motor; 3 - the lever of the eyes oscillation mechanism in horizontal plane; 4 - the eyeballs coupling lever; 5 - the eyelids lever; 6 - left eyeball; 7 - right eyeball.

In the up and down movement, the eyes can only move if the support lever performs that stroke. The lever movement is the result of actuating a quadrilateral mechanism.

The rotation of the driven element is in the opposite direction to the rotation of the motor shaft. This is because the eyeballs have to do only a downward movement.

The rotation angle generated by the motor shaft is limited by a microswitch.

The connecting rod of the mechanism is provided with a protrusion which, after performing the required rotation angle, will actuate the lever of the corresponding microswitch. When the circuit opens, the motor power supply stops, so the eyeballs will remain locked to the lowered position. The up movement of the eyeballs occurs when powering the stepper motor 1 in opposite direction until the lever 4 (Fig. 4) regains its initial position.

The motor used to actuate the eyeballs rotation mechanism is a stepper motor that has a built-in gearbox in order to provide a high torque at the motor shaft.

The Horizontal Oscillation Mechanism of the Eyeballs. The mechanism chosen to achieve this movement is the oscillating slide type and acts only on the left eyeball. It is shown in Fig. 5.

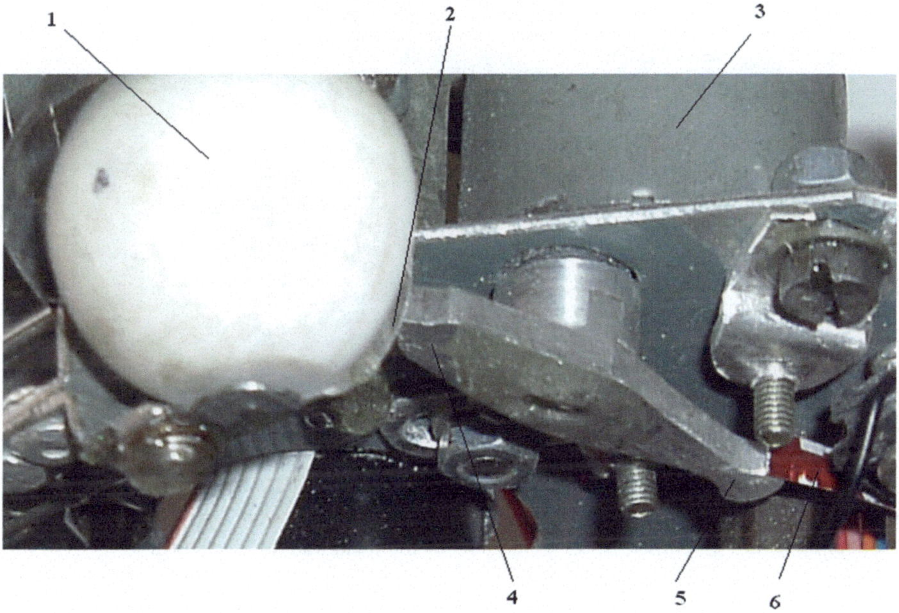

Fig. 5. Horizontal oscillation mechanism of the eyeballs; 1-eyeball, 2-back element of the eyeball 3-motor, 4-driving rod, 5-the protrusion end of the rod, 6-microswitch.

The lever 4 receives the rotation motion of the stepper motor (pos. 3, Fig. 5), whose shaft performs a number of steps until the power supply is interrupted. This finite angle rotation is transmitted to the left eyeball using the lever 4 which is in contact with the slider 2, which is fixed to the left eyeball. It will rotate according to a rule that has been analyzed in the design phase of the mechanisms.

The lever attached to the motor is provided with an extension (pos. 5, Fig. 5) which, at the end of the eyeballs oscillation movement, comes into contact with the microswitch lever (pos. 6, Fig. 5), which opens power supply circuit of the stepper motor that actuates the mechanism. As a result, the horizontally movement of the eyeballs stops.

The right eyeball turns horizontally simultaneously with the left one, because a parallelogram mechanism was provided for this purpose. Its image on the experimental stand is shown in Fig. 6.

Fig. 6. The mechanism for correlation the eyeballs movement: 1 - left eyeball, 2 - right eyeball, 3 - coupling lever; 4 - corner, 5, 6 - joints pins of the quadrilateral mechanism.

The eyeballs 1 and 2 can move simultaneously in the horizontal plane because they are mounted using the joints 5 and 6 on a common support lever. The lever 3 is a constituent element of a parallelogram mechanism that was described in the previous step. It will allow the transmission of the rotation movement from the input shaft of the mechanism (attached to the left eyeball) to the output shaft of the mechanism (attached to the right eyeball).

The dimensions of the constituent elements of this mechanism were established in the previous design phase and remained unchanged after experimental testing.

The mechanism of the eyelids up and down movement is a quadrilateral mechanism with levers of unequal lengths. The main element is the lever that materializes both eyelids jointed with the chassis of the robot head assembly.

The construction of the eyelid mechanism according to the experimental model can be seen in Fig. 7.

The stepper motor 1 rotates the lever attached to its output shaft. Using the element 3, jointed to the above mentioned lever, the rotation of the shaft is transmitted to the eyelid lever 3.

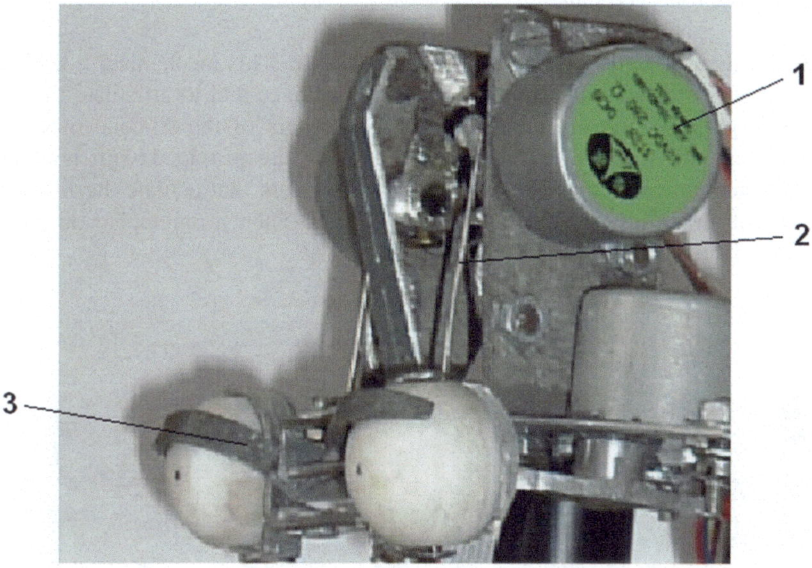

Fig. 7. The quadrilateral mechanism of eyelid oscillation: 1 - stepper motor; 2 - crank; 3 - eyelid lever.

Stopping the simultaneous movement of the eyelids is done exactly as with the oscillatory mechanism of the eyeball, using a microswitch. The switch lever is driven by a protrusion of the quadrant crank, which is fixed to the drive motor shaft.

The minimum motor step is 5,625°. On the output shaft of the gearmotor, the minimum step decreases 64 times, due to the reduction ratio of the gears mechanism. Also from the constructive data, it can be observed that in load, the maximum starting frequency is 550 steps/s. It follows that the output shaft rotation angle of 35° will be performed during the following time:

$$t = \frac{35^{\circ} \cdot 64}{5,625^{\circ} \cdot 550} = 0,724\,\mathrm{s} \tag{1}$$

This time is approximately the same for the eyeballs movements mentioned above. If for the eye this value is very close to reality, for the eyelid the time of 0.7 s gives the feeling of some slowing in the robot's reaction.

The tests performed showed that the eyelid movement is fast enough and is properly controlled by the electronic circuit of the stepper motor.

2.2 The Mechanism for Actuating the Mouth

The movements performed by the mouth are the following:

- lateral extension of the hole;
- opening the hole vertically.

The Mechanism of Lateral Extension of the Mouth. This mechanism is designed to pull the corners of the lips to simulate the smile grimace and to articulate the vocal i.

For this purpose, the inner parts of the lip corners of the silicone mask, which constitute the "skin" of the anthropomorphic robot, are provided with hook-shaped textile fasteners attached to this elastic mask. The levers that actuate the lips are also provided with this kind of pieces of textile material. Their arrangement on the levers can be seen in Figs. 8 and 9.

Fig. 8. The lateral extension of the mouth; 1 - motor; 2 - lever for right extension; 3 - lever for left extension; 4 - fixed support plate for the mechanism; 5 - gear; 6 - the ends of the extension levers provided with textile pieces adherents to the device mask.

The construction of the lips extension mechanism is shown in Fig. 8. It should be noted that all geometric elements have been dimensioned in the previous design stage and have not been modified in the experimental test phase.

The lever 2 and the driving wheel of the gear 5 are fixed on the motor shaft 1. The rotation of the motor is transmitted to the driving wheel and by means of the gear 5, the lever 3 will rotate in the opposite direction, because it is fixed to the driven wheel of the gear 5. The transmission ratio of the gear unit equals 1.

Therefore, the two levers 2 and 3 will have equal angular displacements, but of opposite directions. The driven motor and toothed wheel work together with the chassis of the mechanism (item 4, Fig. 8).

The turn-off command of the motor 1 is provided by a microswitch which is not visible in Fig. 8.

The drive motor is a DC servomotor incorporating a gearbox and an encoder. The engine is MAXON type.

This type of mechanism actuation has been chosen because the servo gearbox provides a high torque at the output shaft as well as a rigorous control of the different positions that the corners of the lips have to reach when simulating the various sounds or the grimace of the smile.

The Opening Mechanism of the Mouth. The opening movement of the mouth to a human subject occurs by moving the mandible downwards and closing by moving the mandible upwards. The projected anthropoid head was obtained from this type of solution by building a lever that materializes the lower jaw (Fig. 9).

As with the human skull, the lever of the mandible, having its specific shape, will give a certain aspect to the lower contour of the face. Therefore, in order not to suggest aggressiveness, all its edges were rounded.

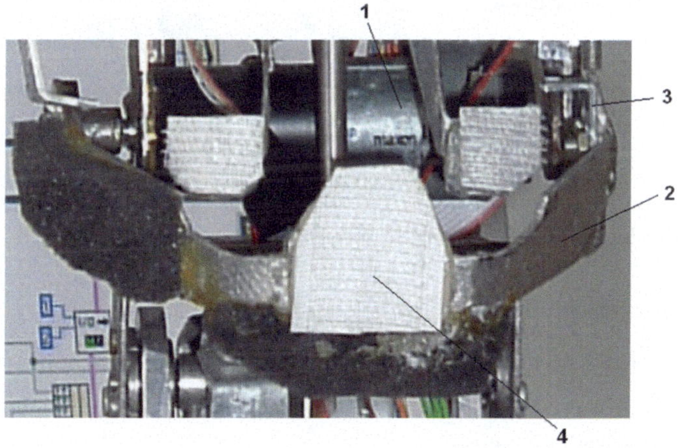

Fig. 9. Opening-closing mechanism of the mouth: 1 - motor, 2 - maxillary lever; 3 - element that controls the microswitch; 4 - a textile piece adhering to the silicone mask of the device.

The mandible lever performs a vertical oscillation, being driven by a DC motor (item 1, Fig. 9). These movements are necessary in order to simulate the vocal articulation movements. The lever is fastened by the silicone mask that covers the “crown cap” of the robot’s head, and elastically deforms it when it moves downwards, as the mask is molded on the mandible lever.

The horizontally arranged motor 1 transmits the rotation movement of its shaft to the mandible lever 2 which is attached to with the latter. The mandible lever rotates vertically with an angle the value of which is limited by the position of a microswitch located in the left side (pos. 2, Fig. 10), which is actuated by the mandible lever by a protrusion thereof, noted 1.

The engine, 0 lever, is secured to a platinum, which is fixed on the robot chassis.

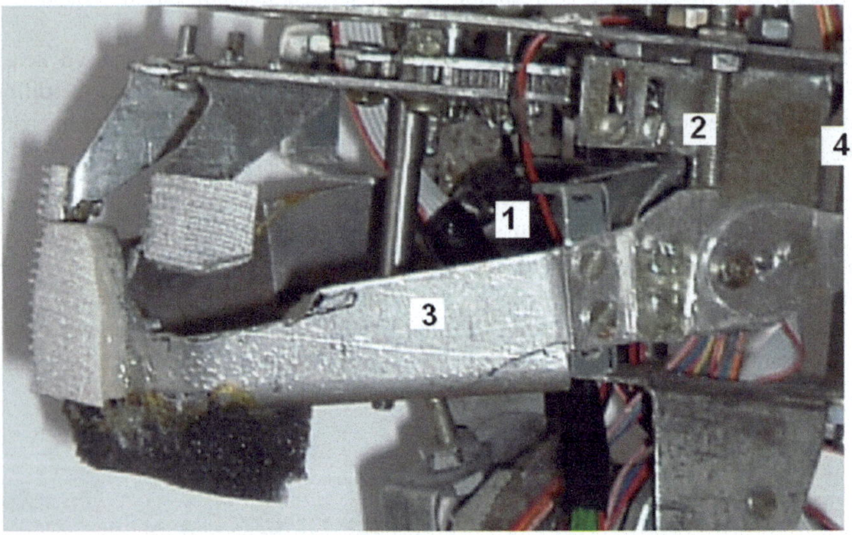

Fig. 10. Opening-closing mechanism of the mouth (lateral view): 1 - lever of the microswitch, 2 - the microswitch; 3 - maxillary lever; 4 - fixed lever, attached to the chassis.

The actuation is made by using a DC servomotor of the type used to actuate the elements of the lateral extension of the mouth.

2.3 The Neck Mechanism

The anthropoid head was designed to perform two main neck movements:

- vertical (top-down) oscillation;
- horizontal (left-right) rotation.

Every degree of freedom is achieved driving by separate motors.

The Vertically Oscillation Mechanism of the Head. Head movement up and down is a small amplitude movement, but requires a high speed of execution. This has made it necessary to use a high torque actuator. For reasons of control, the angle of rotation of the head on both sides of the axis of symmetry of the head is the same and has a fixed value. The construction of the mechanism can be seen in Fig. 11.

Fig. 11. The mechanism of rotation of the head in a horizontal plane: 1 - motoreductor; 2 - the lever for actuating the microswitch; 3 - the microswitch; 4 - the bearing supporting the vertical shaft of the device; 5 - the support of the mechanism for the head's horizontal rotation.

The mechanism functions as follows: the gearmotor 1, which has a large torque, engages the fixed parts of the shaft, a coupling and a connecting element. This one is fastened by bolts to the lever body 5 which supports the entire mechanical structure of the head. This makes the rotation move from the gearhead to the upper head assembly. The limitation of this angular displacement of the head is achieved by disconnecting the drive of the gearmotor when the contact of the microswitch 3 is released.

The entire mechanism is rigidly positioned on the lever 5, which connects to the head rotation mechanism in the horizontal plane.

The kinematic scheme of the mechanism includes, in addition to the main elements, the motor and the support lever (pos. 5, Fig. 11) of the robot face assembly, a spring whose role is to take over and balance the weight of the upper assembly. In the absence of this helical spring, the entire head of the robot would rotate around the motor axis under its own weight. This static balancing device is presented with all the constructive details in Fig. 12.

The spring in the construction of the balancing device was dimensioned as follows: the spring wire diameter 1 mm, the winding index equal to 8, the number of turns equal to 27.

Fig. 12. The static balancing mechanism of the robot head weight: 1 - articulated lever; 2 - spring; 3 - maxillary lever; 4 - the support of the mechanism for the head's horizontal rotation.

When performing the tests, it was found that the arc manages to statically and dynamically balance the upper part of the device, but allows vertical oscillations, the damping of which is possible only by the action of the actuators.

The engine with which the mechanism is actuated is a stepper motor with a built-in gearbox, capable of generating a sufficient torque to the output shaft for the rotation of the mechanical assembly.

The Horizontal Oscillation Mechanism. This movement consists of a small angle rotation of about ±15° of the robot's head. In order to achieve this rotation, the entire assembly, including the rotation mechanism of the head in the vertical plane, is fixed to a platinum, and this is trained in a left-right oscillation movement using a stepper motor (Fig. 13).

The motor that actuates the mechanical structure is a stepper one. The gear unit incorporated in the stepper motor assembly gives the actuator a large output torque capable of overcoming the inertia of the robot head.

The motor 1 transmits the rotation movement to the upper part of the device which has as the main bearing element lever 4. The transmission of the movement is achieved by means of a coupling 5. The motor shaft is pulled into the bush 3.

The motor is mounted on a rigid chassis with a large support base, marked 2.

Fig. 13. Horizontal rotation drive system of the head: 1 - stepper motor; 2 - chassis; 3 - the bearing of the humanoid head; 4 - the support of the mechanism for the head's horizontal rotation; 5 - coupling; 6 - microswitch.

The rotation movement stops when the microswitch 6 is actuated by a lever fixed to the support element 4 of the upper part of the device. The dimensioning of the constituent parts of the mechanical part was correct and no significant changes were necessary following the experimental model.

3 Computer-Assisted Control of Drive Elements

The simulation of speech and mimics of a humanoid head is possible by simultaneous and correlated action of all the actuators of the constitutive mechanisms of the device.

The motors used in the construction of the experimental model of the humanoid head are of two types: DC motors with built-in reducer and with encoder and stepper motors. At the latter, controlling the position, as well as bringing it to the initial position, considered as a zero position, can only be performed by commanding a certain number of steps. All commands must be correlated with the voice signal from an auxiliary sound source, more precisely with the vocal sequence of the emitted sound sequence, which the robot must simulate by mimics the eyes, mouth and neck.

That's why all engines are controlled through a data acquisition board, which receives the orders provided by LabVIEW. In this program, specific files were developed for the set of responses the robot must simulate.

Data Acquisition Board. Acquisition board 7344 (Fig. 14) is a combination of servo and step-by-step controller. It provides fully programmable motion control for up to four independent motion axes. It includes motion limitation and insertion of switches as well as I/O terminals for general purpose functions. A particular feature of the 7344 acquisition plate is that it has the ability to command the execution of some movements through complex trajectories by step-by-step or servo drives.

The servo axes can control servomotors and other servo devices using exclusive closed-loop commands. These axes use quadrature encoders or analog pulses for negative position or speed feedback and provide analog outputs within ±10 V ranges. Step by step axes control the stepper motors. These axes can work in open or closed loops. They use quadrature encoders for position or speed feedback and offer digital outputs of step/directional control or clockwise/anti-clockwise. All step-by-step axes are designed to achieve full steps, half steps or microsteps.

All of the features of the 7344 purchase card are based on the use of a 32-bit CPU-based dual-processor Motorola MC68331 combined with an ADSP-2185 digital signal processor and a set of FPGA programmable gate arrays. The FIFO first-in-first-out interface and the implemented features provide high-speed communication.

The board allows for the execution of up to 10 simultaneous real-time multi-purpose movements.

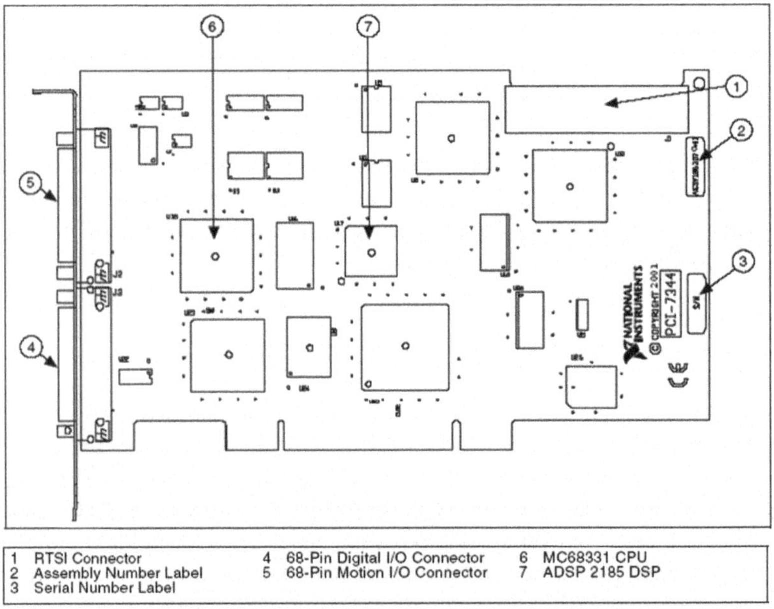

Fig. 14. Diagram of the components of the acquisition board.

Programming the acquisition board is done by using a set of functions implemented in the programming interface. All setting and motion control functions are executed by calling static or dynamic type libraries. These libraries are called by common languages. The implementation of the complete set of functions can be done through LabVIEW.

In LabVIEW, can be created 32-bit programs and executable programs for common type automation, data acquisition, experimentation, measurement, and automated control.

The 68-pin Motion I/O connector distributes all the signals for the four axes of closed-loop motion control, including encoder feedback, the limit inputs and initial inputs, the trigger inputs, and the analogue-to-digital converters of the signals. The 68-pin Digital I/O connector distributes 32 bits to be configured by the user.

An axis of the acquisition board consists of a trajectory generator and a PID control block for the servo axes or a step control block as well as at least one output. It is either a DAC output (for servo axes) or a pulse generator output. The servo axes (Fig. 15) must have either an encoder or an ADC channel for the feedback. Axes for closed-loop steps (Fig. 16) also require components to make the feedback.

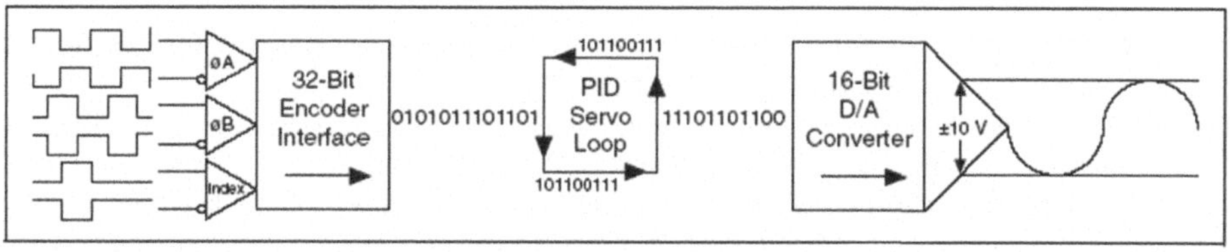

Fig. 15. Components of the servo axis.

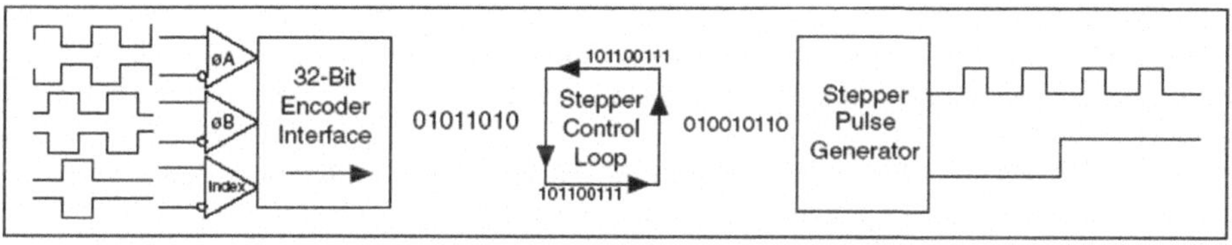

Fig. 16. Axis components for step-by-step control.

The acquisition board has three connectors for handling all signals from and to the outside system ordered:

- 68-pin Motion I/O Motion Connector
- The 68-pin Digital I/O Connector
- The RTSI connector

The connection of the board with the mobile system is done by cables and accessories.

4 Conclusions

The performed robotic humanoid head has seven degrees of freedom that allow simulation of human specific actions in order to transmit information and the expression of positive emotions, while interacting with a human subject. For this purpose, mechanisms and actuators, which are programmed according to voice message to be reproduced, are used. The experiments performed have demonstrated that the degree of similarity of the human robotic head with of the human facies is appropriate, especially in terms of speed of response of the machine. It depends mainly on the quality of the used motors and the electronic control scheme, determinant elements in the ability of the robot simulation.

References

1. Yan, J., Wang, Z., Yan, Y.: Humanoid robot head design based on uncanny valley and FACS. J. Robot. (2014). http://dx.doi.org/10.1155/2014/208924
2. Beira, R., Lopes, M., Praça, M., Bernardino, A., Santos-Victor, J.: Design and Development of a Robot Head, Project no. 004370, Internal Report, Instituto Superior Técnico – IST (2005)
3. Cid, F., Moreno, J., Bustos, P., Nunez, P.: Muecas: a multi-sensor robotic head for affective human robot interaction and imitation. Sensors **14**, 7711 (2014). https://doi.org/10.3390/s140507711
4. Cavallo, F., Semeraro, F., Fiorini, L., Magyar, G., Sinčák, P., Darioet, P.: Emotion modelling for social robotics applications: a review. J. Bionic Eng. **15**, 185 (2018). https://doi.org/10.1007/s42235-018-0015-y
5. Hoffman, G., Forlizzi, J., Ayal, S., Steinfeld, A., John Antanitis, J., Hochman, G., Hochendoner, E., Finkenaur J.: Robot Presence and Human Honesty: Experimental Evidence. In: Human-Robot Interaction HRI, Portland, OR, USA (2015). http://dx.doi.org/10.1145/2696454.2696487
6. Omrcen, D., Ude, A.: Redundant control of a humanoid robot head with foveated vision for object tracking. In: Proceedings of IEEE International Conference on Robotics and Automation, pp. 4151–4156, Anchorage, Alaska (2010)
7. Udrea, C., Alexandrescu, N., Panaitopol, H., Avram, M., Apostolescu, T.C.: The Constructive Bases of Industrial Robots (in Romanian). Editura Universitara Publishing House, Bucharest (2006)

Analysis and Modal Testing

Marian-Alin Bănică[1,2(✉)]

[1] National Institute of Research and Development in Mechatronics and Measurement Technique, Şos. Pantelimon 6-8, Bucharest, Romania
alin.banica01@gmail.com
[2] University "POLITEHNICA" of Bucharest, Splaiul Independeţei 313, 060042 Bucharest, Romania

Abstract. Modal analysis is recognized as one of the most powerful tools, available to the engineer, for the dynamic analysis of structures. In the past two decades, modal analysis has become a major factor in technology, especially in research for the determination, improvisation, optimization of dynamic features of engineering structures. Not only has it been recognized in mechanical and aeronautical engineering, but it has been discovered that modal analysis has numerous applications for civil and building structures, biomechanical problems, space structures, acoustic instruments, transport and nuclear power plants. In order to appreciate the importance of modern engineering and the potential for the future of science and technology, it is opportune to capture background facts that will contribute to high lighting this unique technology.

Keywords: Analysis testing · Modal testing · Frenquency response functions (FRF)

1 Analysis and Modal Testing

1.1 Generalities About Modal Analysis

The modal analysis is the process of determining the inherent dynamic characteristics of a system in forms of natural frequencies, damping factors and modelling forms, and their use to formulate a mathematical model for its dynamic behaviour.

The modal analysis is based on the fact that the vibration response of a temporal-invariant linear dynamic system can be expressed as a linear combination of a set of simple harmonic motions called natural vibration modes. This concept is similar to the use of a sinusoidal and cosine waveform Fourier combination to represent a complicated waveform.

Also, the modal analysis comprises both theoretical and experimental techniques. The theoretical modal analysis is anchored on a physical model of a dynamic system that includes its mass, stiffness and damping properties. These properties can be given in forms of partial differential equations. An example is the wave equation of a uniform vibration string set by mass distribution properties and elasticity.

The equation solution ensures the natural frequencies and shapes of the string mode and its responses to forced vibrations. However, a more realistic physical model will usually include mass, rigidity and damping properties in terms of spatial distributions,

G. I. Gheorghe (Ed.): ICOMECYME 2019, LNNS 85, pp. 273–280, 2020.
https://doi.org/10.1007/978-3-030-26991-3_25

namely mass matrices, stiffness and damping. These matrices are embedded in a set of normal motion differential equations [1].

1.2 Modal Testing

Modal testing is an experimental technique used to derive a modal model of an invariably linear vibration system.

The theoretical basis of the technique is ensured by establishing the relationship between the vibration response at a location and the excitation at the same location or at another location as a function of the excitation frequency.

This relationship, which is actually a complex mathematical function, is known as the frequency response function, or FRF in brief.

Excitation combinations and response at different locations lead to a complete set of Frequency Response Functions (FRFs) that can be collectively represented by a FRF array of the system. This matrix is usually symmetrical, reflecting the structural reciprocity of the system.

The practice of modal testing involves the measurement of FRFs or the impulses of a structure. Measurement of FRF can be done simply by affirming a measured excitation at a location of the structure in the absence of other excitations and measuring vibration responses at one or more locations [1].

Applications of Modal Analysis. Both the theoretical and experimental modal analysis finally arrive at the modal model of a dynamic system. Compared to FRF or vibration response, the modal model explicitly describes the dynamic characteristics of a system.

Consequently, modal analysis applications are closely related to the use of modal derivation in design, problem solving and analysis. Before you start discussing applications, it is important to refresh the two different paths from which a modal model derives.

The theoretical modal analysis is based on the description of the physical properties of a modal model derivation system. Such a description typically contains the mass, rigidity and damping matrices of the system. Thus, it is a path from the spatial data to the modal model. Experimental modal analysis obtains the modal model from the measured FRF data or the measured response data of free vibrations [1].

Some modal analysis applications involve the direct use of modal data from the measurement, while others use this data for further analysis. In the following, some of the modal analysis applications are examined.

Troubleshooting. Troubleshooting using experimental modal analysis is to gain a perspective on a dynamic structure that is problematic. This is the most popular application of experimental modal analysis since its occurrence. Also, many additional modal analysis applications are often discussed. Troubleshooting is based on experimentally derived natural frequencies, damping factors and modelling patterns of the structure. These data provide a fundamental understanding of the structural features and often reveal the root of the dynamic problems encountered in real life.

Sensitivity Analysis. A modal model of a dynamic system can be used to predict the sensitivity of its modal parameters due to changes in the physical parameters of the system. This sensitivity analysis is intrinsically linked to structural changes. However, the focus here is to identify which physical change is most effective at a proposed change in the modal parameter, such as changing a natural frequency. Instead, the structural change consists in studying variations in modal parameters due to a selected physical change. This analysis is particularly useful in redesigning a dynamic structure when a target is set on dynamic features and is sought the most effective way to achieve it [1].

2 Modalities for Modal Analysis

The modal analysis is based on mathematics to establish theoretical models useful in a dynamic system for example and to analyze data in various forms.

Mathematics involved is widespread, partly because modal analysis involves both time domain and frequency domain analysis. Mathematics deals with dynamic mesh systems and continuous structures.

Using mathematics is used to:

- the analytical and numerical study performed extends to the assembly of the curve,
- Matrix manipulation,
- statistical analysis,
- identifying parameters, etc. [2]

The theoretical matrix plays an essential role in modal analysis. This is because the theory of modal analysis is largely based on the analysis of a multi-degree dynamic freedom system (MDoF) that essentially relies on matrix theory. The main analyzes of the matrix involved are the solution of the linear equations, of the matrix and equation reversal problems. More specific topics, such as matrix derivatives, are also used in modal analysis and in its applications.

Fourier transform is fundamental for signal processing in modal analysis. It is customary to believe that Fourier rapid transformation is the benchmark for the development of modal analysis technology. Without this, modal analysis would still remain an academic and analytical question.

During its evolution, modal analysis was fueled by modern control theory. This is not surprising, since control theory subjects are systems while modal analysis deals with dynamic systems. The common theme of system identification allows modal analysis to adopt a number of useful techniques from modern control theory. These include state space theory, Laplace's transformation, time series analysis, and Hilbert's transformation [2].

3 A General Presentation of the Foundations of Modal Analysis

The modal analysis is an increasingly important engineering tool that was applied for the first time around 1940 in search of a better understanding of the aircraft's dynamic behavior. By the late 1960s, evolutions were slow, and experimental techniques were based on the use of expensive and cumbersome analogue spectrum analyzers. The modern model of modal analysis can be taken as of the early 1970s, relying on the commercial availability of Fast Fourier Transform (FFT) spectrum analyzers, transformation function analyzers (TFA) and discrete of increasingly smaller, less expensive and more powerful digital computers to process the data.

The modal analysis is primarily a tool for obtaining reliable models to represent the dynamic behavior of the structures. In general, we can say that modal analysis applications cover a wide range of objectives such as identifying and evaluating vibration phenomena, validating, correcting and updating dynamic analytical models, developing dynamic experimental models, assessing structural integrity, modifying and detecting damage, integrating the model with other areas of dynamics such as acoustics and fatigue, etc. [3].

3.1 FRFs Representation and Properties (SDOF)

The previously defined $H(\omega)$ response function is only one of the possible forms of FRF. It is called Reception and is usually represented by $\alpha(\omega)$ or $\alpha(i\omega)$. This complex quantity fully describes the relationship between the motion response and the excitation force applied to a system and thus fully characterizes its dynamic properties [4].

Since the function $\alpha(\omega)$ is a complex function of the frequency, there are three quantities (the real part, the imaginary part and the frequency) to be taken into account whenever the FRF data is drawn. Thus, a complete representation of a FRF in a single graph should be made using a three-dimensional display. It is obvious that this is not a convenient way to graphically represent the FRF. Alternatively, we can display FRF data in two separate plots - real and imaginary parts to frequency - as shown in Fig. 1.

It is interesting to note that the real part of the reception of $\alpha(\omega)$ crosses the frequency axis at resonance in time in the same frequency region, the imaginary part reaches a minimum (this is true only for hysteresis damping. In the case of viscous damping, the assertion can be considered as true only for low damping values) [4].

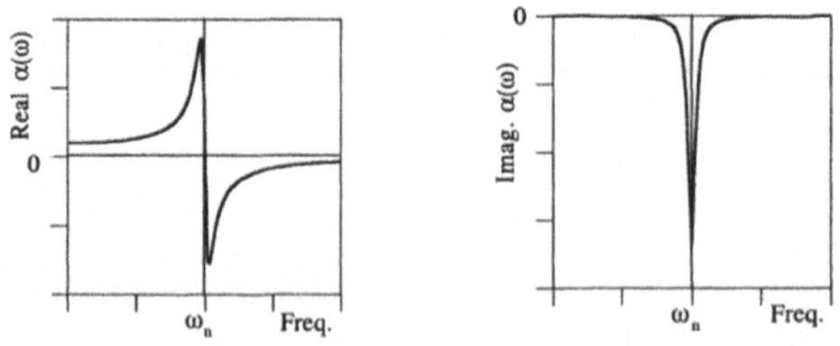

Fig. 1. Real and imaginary parts of the receptance plotted against frequency

The most common representation of a frequency response function is the Bode diagram, where FRF size and phase (instead of real and imaginary pieces) are represented as frequency functions (Fig. 1). This representation allows a light visual interpretation of the information contained in $\alpha(\omega)$.

Finally, if a graphical representation $\alpha(\omega)$ in real/planar imaginary (complex plan) the result is a circular loop containing all the information. However, in such a chart, it is normally not possible to identify the value of the frequency corresponding to any point on the curve, unless each data point (as shown in Fig. 2) is accompanied by a legend indicates the value of the appropriate frequency.

This representation is known as a Nyquist graph and has the special advantage of improving the resonant region since the circular loop only occurs near the resonance (180° phase shift of the FRF).

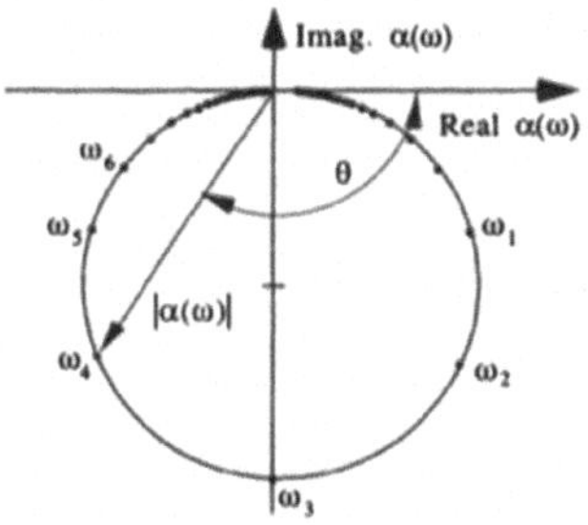

Fig. 2. Nyquist plot of the receptance

Bode diagrams are most commonly used in modal analysis, although the magnitude frequency graph is not shown as in Fig. 3, but rather using logarithmic scales or at least a magnitude (vertical) logarithmic scale. This is because the dynamic range of responses can very easily cover a relatively wide range of values and, as a consequence, linear scales tend to completely lose details at lower response levels.

If logarithmic scales are used for both magnitude and frequency axes, data that is displayed as curves on linear scales becomes asymptomatic to straight lines.

This feature provides a simple means of checking the validity of a graph and also allows for easy identification of the mass and rigidity characteristics of the studied system. For example, one can easily see that if we take into consideration a rigid mass m, free in space, upon which a harmonic force f is applied, the reception of this simple system is given by:

$$\alpha(\omega) = -\frac{1}{\omega^2 m} \tag{1}$$

A log-log plot of the magnitude of $\alpha(\omega)$ against frequency is therefore a straight line with a slope of −2. Applying the same reasoning to a simple isolated massless spring element, we obtain:

$$\alpha(\omega) = \frac{1}{k} \tag{2}$$

and therefore, a log-log plot of the magnitude of $\alpha(\omega)$ against frequency is now a straight line with zero slope.

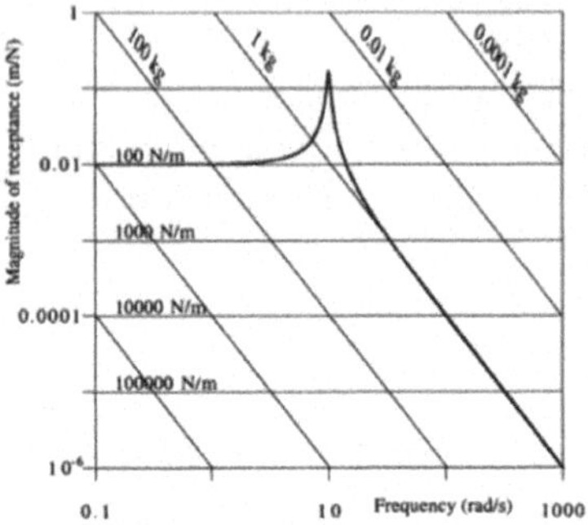

Fig. 3. Example of an annotated log-log plot of the magnitude of the receptance versus frequency

In order to illustrate these properties of FRF log-log plots, Fig. 3 displays the magnitude of the receptance of a SDOF system for which m = 1 kg, k = 100 N/m and c = O.6 Ns/m. Constant mass and constant stiffness lines have been added to the plot. It is obvious that, outside the resonance region, the receptance is asymptotic to straight mass and stiffness lines. It is very easy to understand the physical meaning of this: while at very low frequencies (below resonance), the system approaches a static load/deflection behaviour which is dominated by the spring stiffness, at frequencies above resonance it is the inertia of the mass that dominates the system response [3].

It is therefore obvious that one can extract the mass and stiffness characteristics of a SDOF system from a log-log plot of experimental data. Damping characteristics can be obtained as well, as will be explained further on. Though at a very simple level, this reasoning can be viewed as a first step towards the so-called system identification techniques which aim at deriving the dynamic characteristics from experimental data.

Usually vibration is measured in terms of motion and therefore the corresponding FRF may also be presented in terms of velocity or acceleration and not necessarily in terms of displacement as we have done so far. The terminology used in this text is:

$$\alpha(\omega) = \frac{\textit{displacement response}}{\textit{force excitation}} = \text{Receptance}$$

$$Y(\omega) = \frac{\textit{velocity response}}{\textit{force excitation}} = \text{Mobility}$$

$$A(\omega) = \frac{\textit{acceleration response}}{\textit{force excitation}} = \text{Accelerance [5].}$$

The designation mobility is also widely accepted as a general designation for any of the motion/force FRF forms.

Displacement, velocity and acceleration are mathematically interrelated response quantities. Therefore, knowledge of an FRF in terms of anyone of the motion parameters will allow immediate derivation of any of the other FRF forms. Considering harmonic vibration:

$$Y(\omega) = \frac{\dot{x}(t)}{f(t)} = \frac{i\omega\bar{X}e^{\omega_n t}}{Fe^{i\omega t}} = i\omega\frac{\bar{X}}{F} = i\omega\,\alpha(\omega) \quad (3)$$

$$A(\omega) = \frac{\ddot{x}(t)}{f(t)} = \frac{-\omega^2\bar{X}e^{i\omega t}}{Fe^{i\omega t}} = -\,\omega^2\alpha\,(\omega) \quad (4)$$

It follows that log-log plots of mobility or acceleration will show some differences with respect to the receptance plot, resulting from the fact that mass and stiffness, though still displayed as straight lines, have different slopes. This is shown in Fig. 4 where the mobility and acceleration magnitudes of the system described by Fig. 5 are plotted against frequency (a dB magnitude scale is now used) [6].

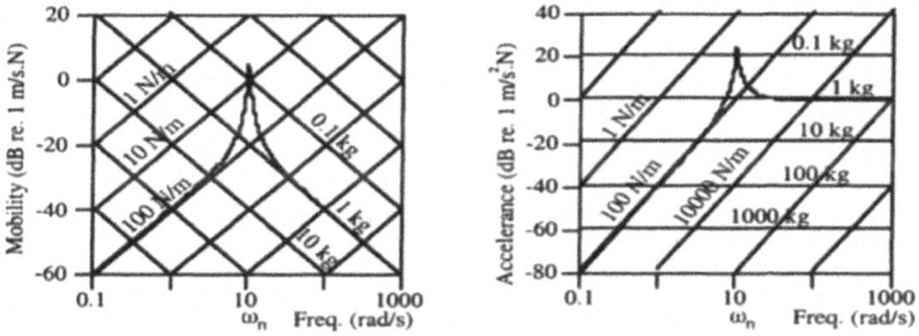

Fig. 4. Annotated log-log plots of the magnitudes of mobility and acceleration versus frequency

4 Conclusions

Modal analysis evolves more in parallel with control theory. Inverse structural dynamic problems such as force identification from measured responses were actively pursued.

Nonlinear dynamic characteristics were studied experimentally.

Modal analysis has entered many fields of engineering and science. Applications range from automotive engineering, aeronautical and astronautical engineering to bioengineering, medicine and science. Numerical (finite element) and experimental modal analysis have become two pillars in structural dynamics.

References

1. Fu, Z.F., Jimin, H.: Modal Analysis, 2nd edn. Elsevier, Amsterdam (2001)
2. https://arxiv.org/pdf/1711.10188.pdf. Accessed 07 May 2019

3. Montalvão E Silva, J.M., Silva, M.M.J., Maia, M.M.N.: Modal Analysis and Testing. Springer, Netherlands (1999)
4. Sujatha, C.: Vibration and Acoustics: Measurement and Signal Analysis. Tata McGraw Hill Education Private Limited, New Delhi (2010)
5. Bilošová, A.: Modal Testing. Ostrava (2011)
6. www.reading.ac.uk/AcaDepts/sp/PPLATO/imp/interactive%20mathematics/loglog2.pdf. Accesed 07 May 2019

Concepts and Mechatronics and Cyber-Mixmechatronics Constructions, Integrated in COBOT Type Technology Platform for Intelligent Industry (4.0)

Gheorghe Gheorghe(✉)

National Institute of Research and Development in Mechatronics and Measurement Technique, Şos. Pantelimon 6-8, Bucharest, Romania
geocefin@yahoo.com

Abstract. The scientific paper focuses on a synthesis, the architecture: why industry 4.0 and the digital enterprise? - for intelligent technological change, intelligent societal change and business paradigm shift, new Intelligent Industry Opportunities (4.0) created by the development of mechatronics and cyber-mixmechatronics - information ensures the satisfaction of man's spiritual needs, computer science adds value to everything, and computer science culture, the vision of INCDMTM - Romania in the intelligent domain research and innovation "Internet of Things - IoT" - integrated in COBOT technology platforms and the INCDMTM - Romania mission for mechatronic and cyber-mixmechronicon concepts and constructions integrated in COBOT technology platforms - for Intelligent Industry (4.0).

The scientific paper concludes the impact of enterprise and industry digitization through mechatronic and cyber-mixmechronic integrated systems that are or are to be implemented in different industrial sectors - automotive, aerospace, agro-industry, medicine, etc. in Romania and their digitization strategy at national and European level.

Keywords: Concepts and mechatronics and Cyber-mixmechatronics constructions · COBOT tehnology platforms · Digital enterprise · Intelligent Industry (4.0)

1 Why Industry 4.0 and Digital Enterprise?

The Industry Initiative 4.0 [1] includes the digitization of production processes based on systems that communicate with each other in an autonomous way along the value chain. This Industry 4.0 considers the potential of the initiative and changes in the business paradigm and the impact of this transformation.

The concept of Industry 4.0 includes, among others:

- technological change;
- societal changes and
- changes in the business paradigm,

G. I. Gheorghe (Ed.): ICOMECYME 2019, LNNS 85, pp. 281–300, 2020.
https://doi.org/10.1007/978-3-030-26991-3_26

as an industrial policy at EU level to support Member States and businesses in the transformation needed to connect and integrate digital technologies with industrial products and services.

Thus, **the three key dimensions** of consistent change for Industry 4.0 are explored through: technological, social and business paradigm.

Essentially, Industry 4.0 describes **the organization of intelligent production processes** based on technologies and products/systems that **communicate autonomously with each other along the value chain: a smart factory model** of the future where computer-driven systems monitor physical processes, create a virtual copy of the physical world and make decentralized decisions based on self-organizing systems.

The Industry 4.0 concept takes into account the digital growth of manufacturing industries where physical objects are perfectly integrated into the information network, allowing for decentralized production and real-time adaptation in the future.

Industry 4.0 has been developed by developed European countries to create a coherent policy framework for maintaining industrial competitiveness and productivity and for integrating Internet objects, Internet services, Industrial Internet, advanced manufacturing, and intelligent manufacturing [2].

For Industry 4.0 to be succeed, some key requirements such as system standardization, platforms, protocols, changes in work organization, reflecting new business models, digital security and know-how protection, availability of suitably qualified workers, research and investment, and a legal framework (EU common) to support the dissemination of Industry 4.0 on the internal and external market.

As the industry's 4.0 Implementation horizon is to have "pilots" in the pipeline and their full implementation, starting in 2025.

In support of Industry 4.0, **the policy approach** [3] is **to develop new core markets into a dual strategy,** where:

- technology and services in Industry 4.0 can be sold, and
- production and other products can be sold more easily due to increased productivity and competitiveness.

For Industry 4.0, the dimensions of change are relevant: technological change, social change, and changes in the business paradigm.

Regarding technological change, digitization is a determinant of changes throughout the value chain and while many enterprises recognize the need for adaptation, much less, especially among SMEs, are prepared for it. There are significant challenges (costs and risks) for businesses in terms of digital security in: the protection of intellectual property, personal data and privacy; system design and operability; environmental protection and health and safety.

Regarding social change, there is little awareness of Industry 4.0. Although there is a gap in skills (as well as a discrepancy in the desire to adapt to the digital single market), qualification requirements to adapt to Industry 4.0 are much higher. The provision of skills and capabilities in Industry 4.0 across the EU is uneven, which may lead to increased concentration and competition between existing centers.

Regarding the change of business paradigm, there are difficulties for SMEs to participate in the Supply Chain industry 4.0 (costs, risks, reduced flexibility and strategic reduction, independence). The public sector can play a role in creating an ecosystem that will help SMEs move to Industry 4.0.

Standardization [4] remains a major challenge in the widespread implementation of Industry 4.0.

The question here is whether Industry 4.0 will consolidate national industries or EU industry, or whether it is more necessary to maintain its position or whether it will inevitably shift to new emerging economies such as China through the international dissemination of technology by multinationals remains a response to it.

Public sector intervention could take different forms but the most promising seems to be supporting research at EU and Member State level and coordinating initiatives across the EU, through a platform and to illustrate the good practices of initiatives in some Member States that others could follow. In order to maximize value added, initiatives should only go beyond technical issues and the production sector and should reflect the differences in the economic structures of the Member States.

In conclusion are recomended:

(a) a review of existing measures for Industry 4.0 and taking into account the most important issues (skills, migration, change of business business models, clusters, cross-border business collaboration programs, IT security and standards, etc.).
(b) adopting new measures to identify gaps at EU and Member State level to monitor the latest developments, fund research and support SMEs, raise public awareness of challenges and opportunities, support the development of a framework including standards and role coordination.

2 The New Oportunities from Intelligent Industry (4.0) Created by the Development of Mechatronics and Cyber-MixMechatronics

Synthesized are presented the new Opportunities in the Intelligent Industry (4.0) created by the development of Mechatronics and Cyber-MixMechatronics, architecture for 2009 ÷ 2018.

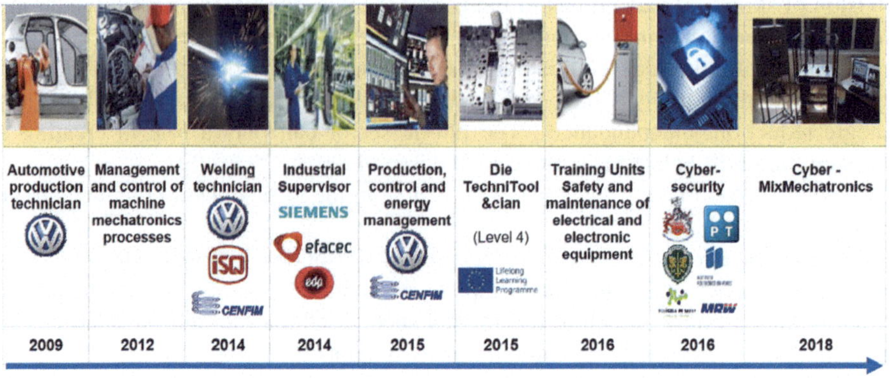

Mechatronics technology puts the focus on information, which is the tonal component in relation to material and energy.

This position of the information is motivated (by the Japanese) by the following arguments:

- information ensures the satisfaction of the spiritual needs of human;
- only computer science increases the added value of all things;
- computer science means culture.

Promoting information links in the technical systems structure ensures flexibility and reconfigurability.

Quantitative and qualitative assessment of information is an essential issue in education, research and production.

Information is equally important in medicine, literature, art, music, sports, etc.

3 The Vision of INCDMTM Romania in Research and Innovation for the Smart Domain "Internet of Things - IoT" and Integrated in COBOT Type Technology Platforms

Research and Innovation domains in Romania are due to the advanced development of information and communication technology, along with integrative mechatronics that generate more and more objects/objects integrated with sensors and communication capability with other objects/things that transform the physical world into a matrix informative-knowledge system [5].

The Internet Challenge of Objects - IoT allows today that things/objects in our environment are active participants who share information with other actors or members of the network, wired/wireless, using the same Internet Protocol-IP protocol that connects the Internet. Thus, objects/things are able to recognize events and changes around them and act and react in a quasi-autonomous way without human intervention.

In this exposed context, challenges in research, development and innovation create an "intelligent planet", where physical, digital and virtual worlds converge to create smart environments that can make energy, transport, cities and many other intelligent domains [6–8].

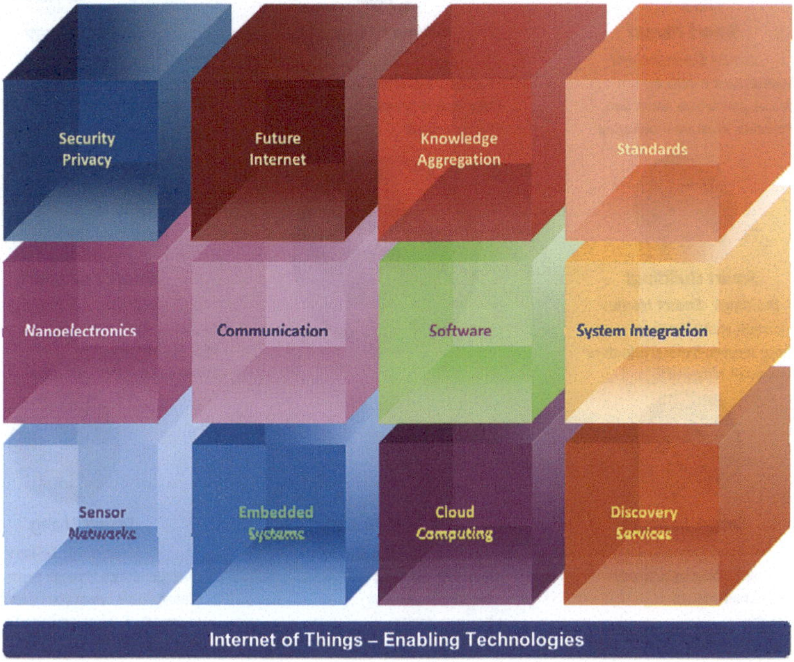

The development of some generic technologies such as **"nano electronics", "communications", "sensors", "smart phones", "embedded systems", "cloud computing and software"** will be among the key to supporting future innovations of IoT and cyber – mechatronics products, which affects many industrial sectors.

In Romania, many projects and initiatives address technologies and knowledge about the Internet of Things and Cyber-Mix Mechatronics [9].

Thus, the integration of knowledge in this context is conceptualized as the process through which disparate, specialized knowledge, located in several projects in Romania (and at the level of the European Union), **are increasingly combined, applied and assimilated. INCDMTM-Romania** is developing research in IoT on the Internet of Things that aims at defining IoT technology and developing research challenges at national (and European) level with a view to the global development of multidisciplinary science. Motivation for the Internet of Things is to address the high potential of IoT-based capabilities in Romania (as well as in Europe) - to coordinate/encourage the convergence of ongoing activities on the most important issues - to build a broad consensus based on modalities IoT in Romania (as in Europe).

The applications of the Internet energy are connected through The Future Internet and "Internet of Things, allowing for smooth and secure interactions of intelligent systems embedded over heterogeneous communication infrastructures.

Thus, developments in the global environment and developments in Romania's policy, science and technology policy and intelligent industrial policy and digitized enterprise converge to the objective of supporting these links at national level (in Romania).

That is why INCDMTM to achieve real coordination and cooperation in Romania [10], as well as at European level, is to support the national link of the scientific groups with other national scientific and innovation networks and research based on the intelligent industrial application research, digitized enterprise.

The creation of IoT national networks will allow better coordination of IoT knowledge-producing projects at national level with cross-country interdependencies and cooperation at European level through integration and interconnection.

This concept of linkage will contribute to the consolidation and replication of the success factors achieved through specific IOT projects and will be an instrument that will help to promote the exchange of ideas, solutions, results and their validation among research projects at national level and European.

Integration of IoT with mechatronic and cyber-mixmechronic systems for telemetry, telecontrol, telemonitoring and teleconfiguration processes have led **to new concepts of intelligent system systems.**

The following are the concepts of **COBOT technology platforms** for different applications in intelligent industrial environments.

In the architecture below, the "**basic principle for intelligent agriculture**" is presented:

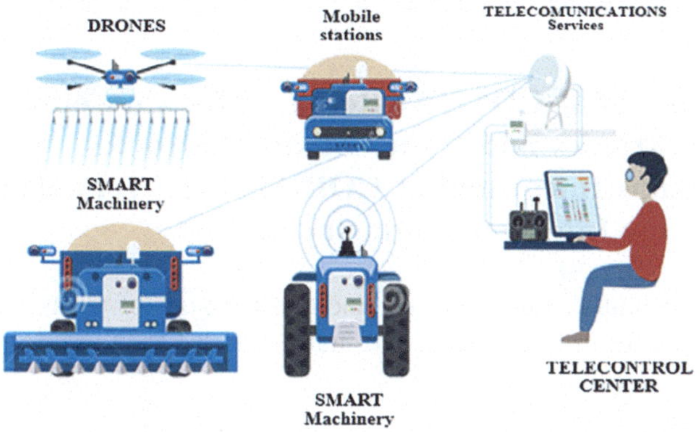

Robotics and more and more integrated technical innovation have boosted the agricultural world for many years. Of these, drones, robots that perform more tasks, such as weeding, milking, sowing, etc. These allow, in addition to suppressing the notion of "hardness" related to certain tasks, to collect accurate and useful data. They promote farm productivity and optimization. These autonomous and digital tools can, for example, be used to reduce the use of agricultural inputs and hence to restore their environmental impact.

Therefore, technological improvements appear for agriculture as a tool and as a key development. And this, as well as environmental aspects such as traceability of products and improvement of operator activity, are defining.

The architecture presented below, represents "**Industry 4.0 as a interconnected reality of industrial production flows**":

Interconnected production and control processes in industrial environments that operate complex machines will determine the future of Industry 4.0 that contributes to the implementation of dynamic, real-time and self-sustaining processes by providing active and passive detection and interconnection solutions [11].

The interconnection of individual production steps means that these steps can be combined in any way they need, with maximum traceability of information and field status. This enables us to create fully observable production lines in real time, efficiency and flexibility being the main features of intelligent factories.

4 The Mission of INCDMTM Romania for Concepts and Mechatronics and Cyber-Mixmechatronics Constructions Integrated in COBOT Type Tehnology Platforms for Intelligent Industry (4.0)

Thus, there are presented some concepts and constructions of technological platforms COBOT type [12], for different fields and industries, designed in original solutions of the author of this scientific paper [13]:

- in Fig. 1: COBOT Technology Platform for Integrated Verification and Control Processes in the Digitized Enterprise and Industry 4.0 in the MixMecatronic field.

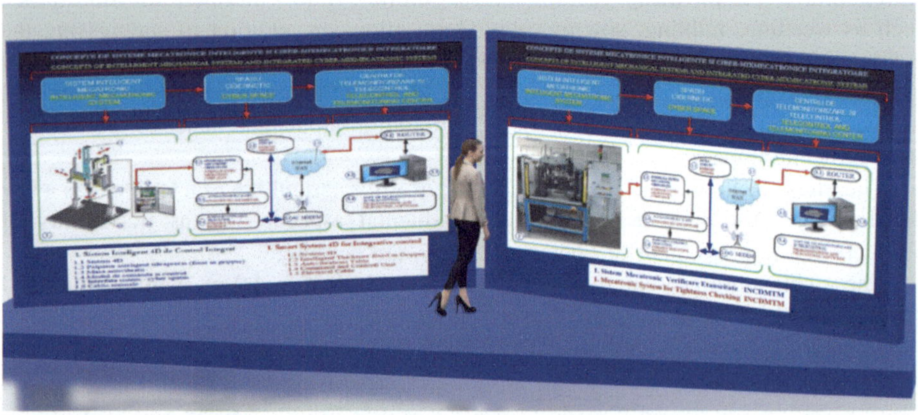

Fig. 1. COBOT technology platform for integrated verification and control processes in the digitized enterprise and Industry 4.0 in the MixMecatronic field.

The COBOT technology platform for integrated verification and control processes is logically structured on the intelligent mecatronic and cyber-mixmecatronic architecture (4D/digital ultra-precision measuring system/control and control system/ pneutronic antivibration system/etc.) + architecture Mechatronic system for leakage of casting parts (casting system/casting/casting system/instrument pneumatic air/plunger system which ensures the closure of all the holes of the casting and thus ensures the space enclosed in the pneumatic/Input/Output Actuator/Mechatronic Mechatronic Metering System/Leakage Mechatronic System Checkpiece (after result display)/Mechatronic Operating and Leak Test System optical signaling system leakage checking process - system operation - system malfunction alarm - total system/ pneumatic system/system power blocking system.

The architecture of the cyberpaths and the architecture of telemonitoring - tele-service - teleconfiguration and telecontrol centers and with a very important role in collaborative activities with intelligent 4D mechatronic systems and leakproofness verification, the collaborative man, highlighting in fact the technological “cobot”.

- in Fig. 2: COBOT technology platform for manufacturing (welding and handling parts) from the digitized enterprise and Industry 4.0 in the field **of machine building.**

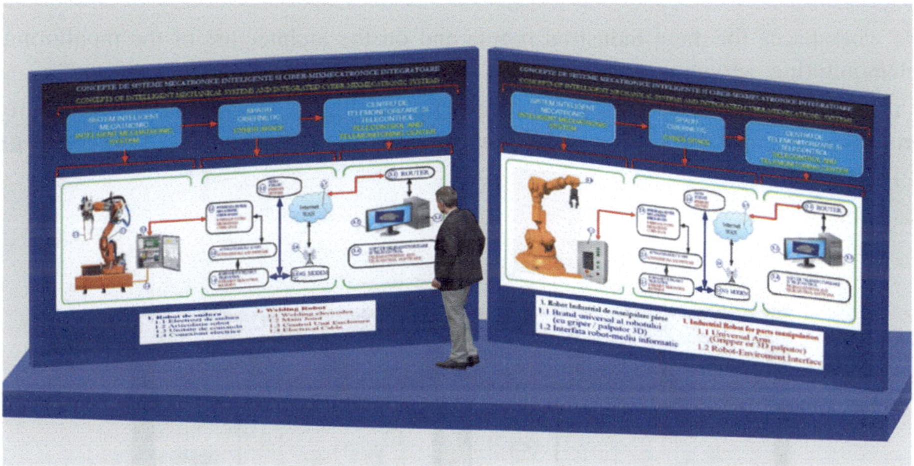

Fig. 2. COBOT-type technological platform for manufacturing (welding and handling parts) of enterprise and Industry 4.0 in machine building

The COBOT-type technological platform for manufacturing processes (welding and handling parts) is logically structured on the welding robot architecture and the parts handling robot (welding robot/robot handling parts/welding robot interface system/robot interface handling parts/mechatronic system for control and control robot welding/mechatronic system for control and control robot handling parts/etc./),

- in Fig. 3: COBOT-type technology platform for CNC Machine-Machine Tooling processes in Enterprise and Industry 4.0 in **Machine Tools-Building.**

The COBOT Technology Platform for CNC Machine Assistance is logically built/structured on the architecture of intelligent mechatronic systems - Assistant Robot Machine Tools (Industrial robots specific to CNC MU assistance/interfaces between industrial robots and cyber-working spaces for industrial robots) on the architecture of

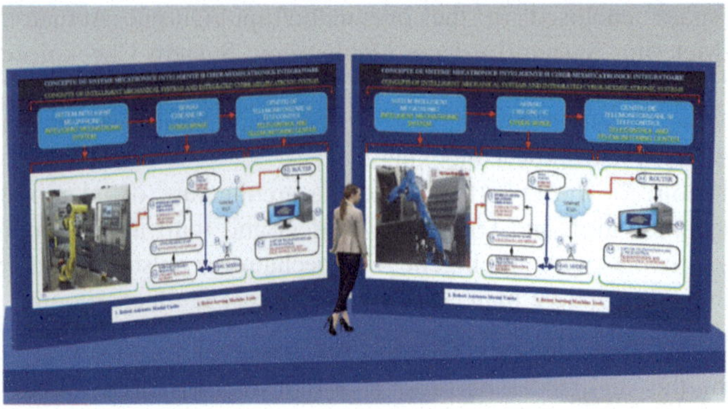

Fig. 3. COBOT technology platform for assistive processes machine tools with CNC enterprise and Industry 4.0, M.U.

the workspaces for these industrial robots and on the architecture of the monitoring, telemonitoring, servicing, telecontrol and telecontrol centers (routers/computers).

• in Fig. 4: COBOT technology platform for intelligent measurement and control processes in the digitized enterprise and Industry 4.0 in the field of **Technology Equipments.**

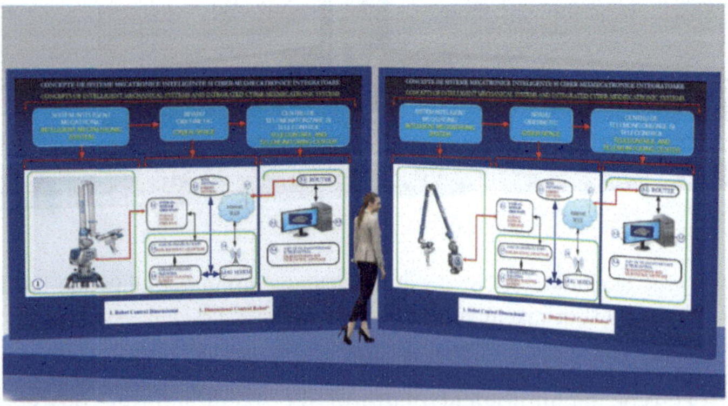

Fig. 4. COBOT technology platform for smart measurement and control processes in enterprise and Industry 4.0, technology equipment

The COBOT technology platform for intelligent dimensional measurement and control is logically structured on the dimensional control mechatronic robot architecture (intelligent mechatronic systems like control robots/control robot interfaces - cybernetic spaces/intelligent mechatronic systems for managing and coordinating processes/support systems for control robots/digital touch probe systems for measurement and control/mechatronic systems for operating and collaborating with the robot operator (the collaborative man), on the architecture of cyberspace working and collaboration with the digitized enterprise and on the architecture of monitoring and telemonitoring centers, teleconfiguration, service - teleservice and control and telecontrol (routers/computers/etc.) and collaboration with collaborative human with control robots (special programs for measuring and controlling processes/on special software for measurement and control/human collaborative, etc.).

- in Fig. 5: COBOT technology platform for surgical operations and laboratory analysis (for diagnosis of medical treatments and services) in the **Intelligent Medical field**

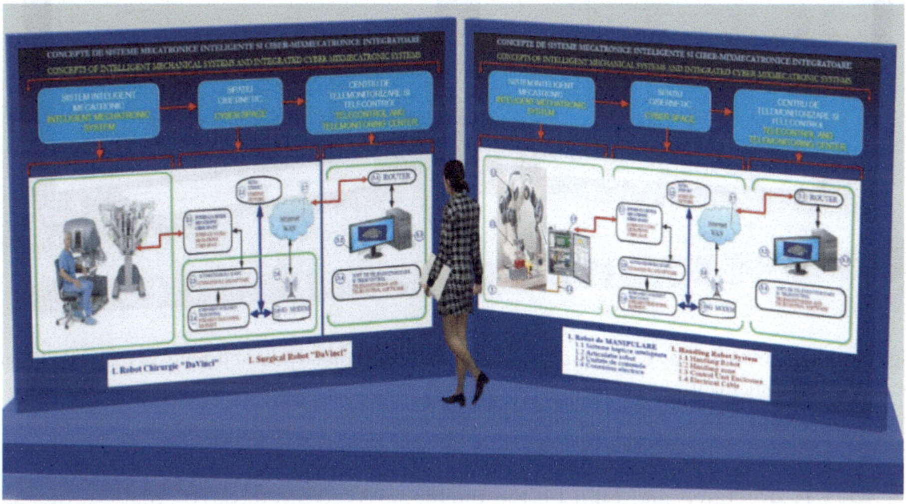

Fig. 5. COBOT technology platform for surgical operations and laboratory analyzes in the field of intelligent medical

The COBOT platform for surgical operations and laboratory analyzes is structured on the architecture of the medical surgical robot (DaVinci), a smart medical robot/medical related instrument/mecatronic interface/mecatronic system of operation and coordination/mecatronic intelligent intelligence system surgical operations/ mechatronic system for precise and precise manipulation of medical instrumentation/ mechatronic intelligent warning and alarm system/mecatronic TV transmission system from "clean space" intended for "surgical operations"/etc.), on the architecture of the robot handling and servicing (robot proper/intelligent haptic systems/control unit/electrical connection systems/space drives/mechatronic auxiliary and adaptive systems to medical devices, etc.), the architecture of the medical robots working spaces (medical interface - cyber space/cyber space for each medical robot/antenna/4G modem/Internet WAN/Intranet/etc.), on the architecture of the monitoring centers - telemonitoring/service - teleservice/control - telecontrol/configuration - teleconfiguration (routers, computers, programs for surgical and medical analyzes, special software, etc.) (the collaborative human of the platform).

- in Fig. 6: COBOT Technology Platform for Patient Analysis and Walking Recovery in **Intelligent Medical and Biomedical field**

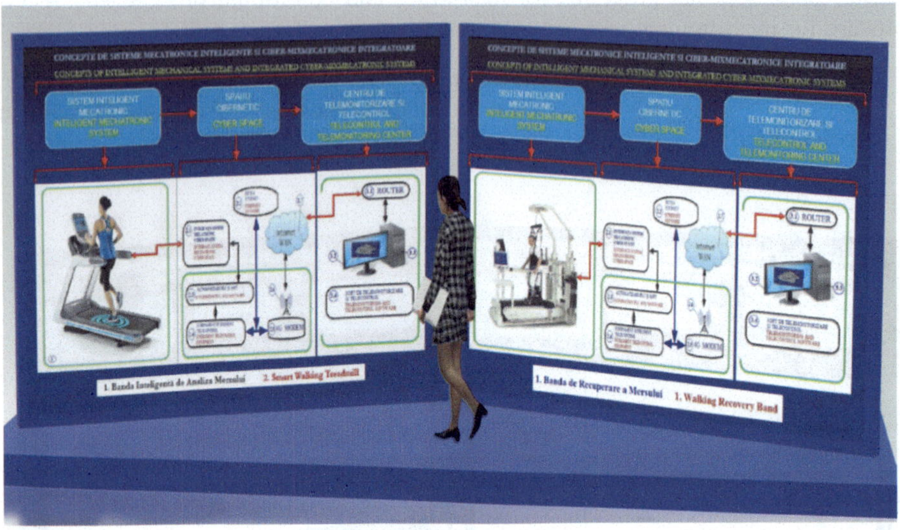

Fig. 6. COBOT technology platform for walking through analysis and recovery in intelligent medical and biomedical

The COBOT Technology Platform for Patient Walking through Analysis and Recovery is structured and built on the Intelligent Walking through Analysis Mechatronic System architecture (the actual stripe/warning and alarm system/stop-lock system/individual operating system/etc.) on the mechatronic track gauge architecture (the actual band/the computerized system for medical recovery parameters/warning and alarm system/stop-lock/individual operating system/etc.) on (4GMODEM/Antenna/Internet WAN/Intranet, etc.), on the architecture of telemonitoring/teleservice/telecontrol/teleconfiguration centers (routers, computers, software and software), the architecture of the cybernetic spaces of the two intelligent mechatronic systems of Analysis and Walkback Recovery Specialized Analysis and Recovery/etc.).

- in Fig. 7: COBOT technology platform for control and automotive manufacturing processes in the Enterprise and Industry 4.0 **Automotive industry.**

The COBOT technology platform for Intelligent Manufacturing in Intelligent Control Processes is built and structured on the Automobile Robot Architecture for Control (Intelligent Operating Mechronic Intelligent Operating System/Intelligent Alarm and Alarm System/Mechatronic Coordinating System and control/cyber-interfacing system/intelligent self-adjusting and self-positioning system/etc.), on the robot architecture of automotive casting control (engine/gearbox/injection pump/etc.) Intelligent control/Mechatronic Intelligent Operating System/Intelligent Alarm and Alarm System/Mechatronic Coordination and Control System/Mechatronic

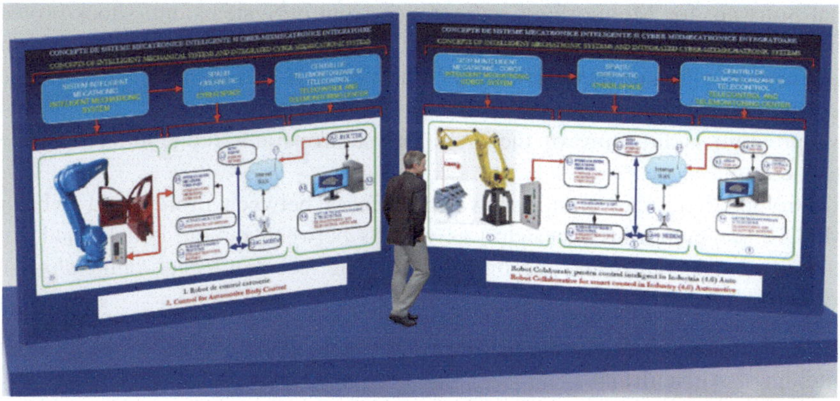

Fig. 7. COBOT technology platform for control processes and manufacturing lines in enterprise and Industry 4.0

Coordinating and Command System with Measuring Metering/etc.), Cyberspace Architecture (Interface between control and external cyberspace/Internet WAN/Intra net/4GMODEM/etc.), on the collaborative program architecture of the IT operator (collaborative man with control robots) and on the architecture of the telemonitoring centers - teleconfiguration - teleservice - telecontrol (routers, computers/measurement/control/analysis/operational decisions).

- in Fig. 8: a COBOT technology platform for social services in the Intelligent City

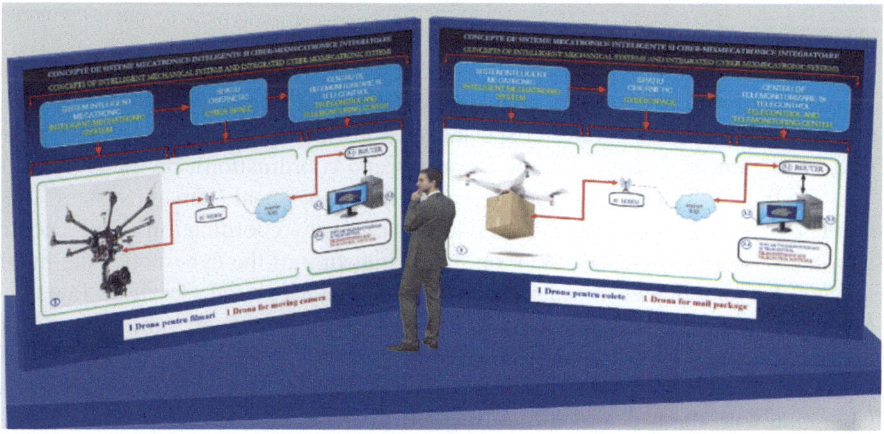

Fig. 8. COBOT platform for social services in the intelligent city

The COBOT social services platform is built and structured on Drone's architecture for filming (Electronic Drone/Fastener Film/Camera/Travel and Transport Controller/

Special Software for Monitoring/Configuration/Control/Systems anti-shock/protection systems for wind turbines/etc.), on the parcels' drone architecture (electronic drones/parcel shelf/camera/controllers for transport and transport/special software for monitoring/anti-shock systems/wind turbine protection systems, etc.) and the collaborative program architecture of the drone operator/operator (collaborative man with Electronic Drone).

Electronic drones are equipped with interfaces between drones and cyberspace (4GMODEM/antennas/interfaces/etc.) as well as between remote control and remote control centers, telecontrol and teleconfiguration (interfaces/routers/computers/programs - specialized software/etc.

- in Fig. 9: COBOT technology platform for agricultural processes **in the field of Intelligent Agriculture**

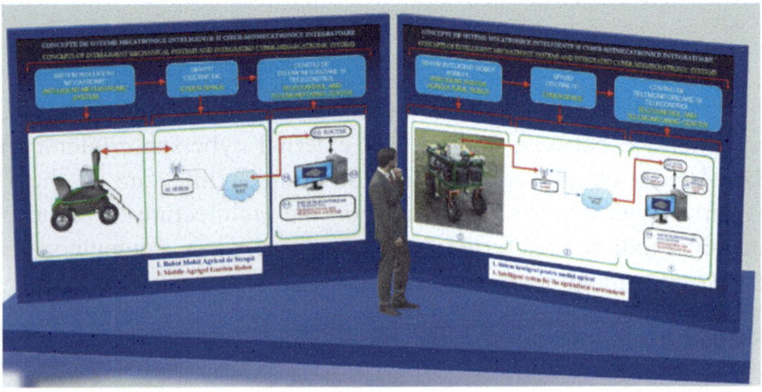

Fig. 9. COBOT technology platform for agricultural processes in intelligent agriculture

The COBOT technology platform for agricultural processes is structured and built on the architecture of the mobile sprinkler mobile robot (the actual robot/agricultural instrument set/monitoring controller/specialized software/drive systems for agricultural services/intelligent mechatronic control systems and coordination), on the architecture of the intelligent system for the agricultural environment (smart system itself/ instrument set for agricultural services/intelligent coordination and control system/interface for cyberspace/etc.), on the architecture of the cybernetics the mobile robot and the intelligent mecatronic system for agricultural services (antennas/ 4GMODEM)/Internet WAN/Intranet/etc.), the architecture of the monitoring/ telemonitoring, configuration/teleconfiguration and control/telecontrol centers (routers, programs - specialized software/etc.) and hand architecture operator geriatric for collaboration with the two agricultural systems (the collaborative man of the COBOT platform).

- in Fig. 10: COBOT-type technological platform for industrial assembly processes from the Digitized Enterprise and Industry 4.0 **in the field of Machine Building**

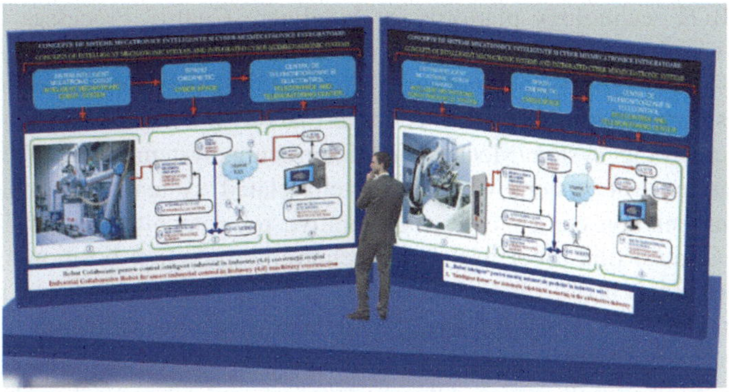

Fig. 10. COBOT technology platform for assembly and industrial control in enterprise and Industry 4.0, machine building

The COBOT technology platform for industrial control processes is structured and built on the collaborative robot architecture for intelligent industrial control of the geometry of the space where another industrial benchmark is to be assembled (the actual robot/packet of prehensive devices, mechatronic operating system and coordination/interfacing with cyberspace/etc.), on the intelligent robot architecture for automotive windscreen assembly (the actual robot/instrumentation and control set/mechatronic system for operation and coordination/interfacing with cyberspace/programs - specialized software/etc.), on the operator's management architecture (through programs and softwares), the collaborative man (with the two control and installation robots), on the architecture of the antenna spaces/4GMODEM)/Internet WAN/Intranet/etc.) and on the architecture of the monitoring centers has/telemonitoring, configuration/teleconfiguration, control/telecontrol, etc. (routers, computers/programs - specialized software/etc.).

- in Fig. 11: COBOT Technology Platform for Intelligent Control Processes in the Digitized Enterprise and Industry 4.0 **in the Electronics and Electrotechnics field.**

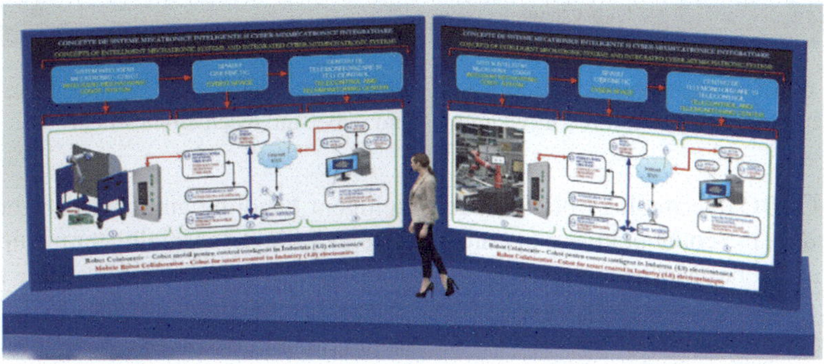

Fig. 11. COBOT technology platform for intelligent control processes in enterprise and Industry 4.0, in electronics and electrotechnics

The COBOT technology platform for Intelligent Control Processes in the Electronics and Electrotechnics field is structured and built on the collaborative robot architecture - the COBOT for intelligent laser control (the robot itself/mechatronic precision positioning system of the electronic marker)/Intelligent operation and coordination mechatronic system/Mechatronic interfacing system with cyberspace/laser related system/control system viewing system/related system with measuring and control instrumentation etc.), on the collaborative robot architecture - COBOT Intelligent Laser Control (the robot itself/the mechatronic precision positioning system of the electronic circuit board (PCB)/intelligent operation and coordination mechatronic system/cybernetic interfacing system/laser related system/visualization of the measurement process (4GMODEM/Antennas/Internet/Intranet/etc.) and on the architecture of the telemonitoring, teleservice, teleconfiguration and telecontrol centers (the collaboration of the robots), the architecture of the cyberspace routers/computers/programs - specialized software/etc.).

- in Fig. 12: **COBOT technology platform for intelligent manipulation and control processes in the Digitized Enterprise and Industry 4.0, in the Mechatronics and Cyber-MixMechatronics**

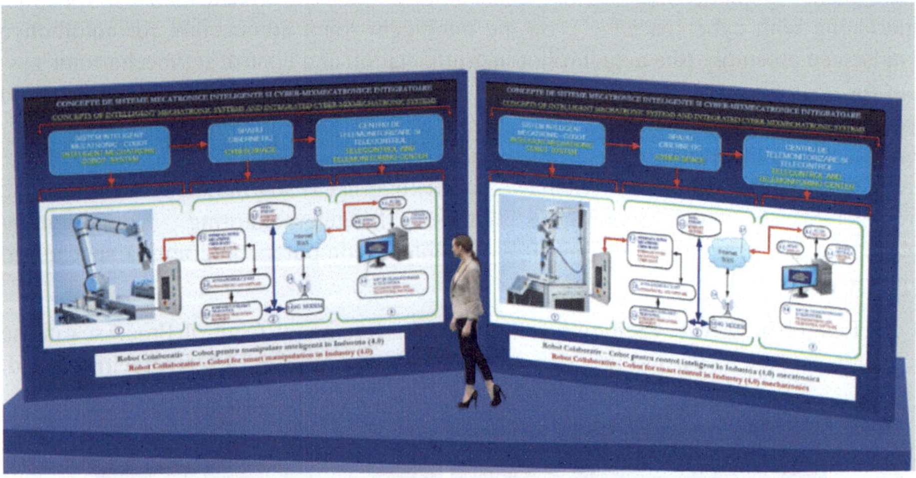

Fig. 12. COBOT technology platform for intelligent handling and control processes in enterprise and Industry 4.0 in the mechatronics and cyber-mixmechatronics field

The COBOT technology platform for intelligent handling and control is built and structured on the collaborative robot architecture - COBOT for intelligent manipulation (the actual robot/Mechatronic control and operation system/mechatronic viewing system/cybernetic interfacing system/the system for handling devices/etc.), on the collaborative robot architecture - COBOT for intelligent control (the robot itself/smart mechatronic interface for cyberspace/command and operation system/intelligent viewing system/system (4GMODEM/antennas/Internet/Intranet/etc.), on the architecture of the monitoring centers, configuration, remote control (routers/computers/software - etc.) and on the collaborative management architecture of the operation (the collaborative man of intelligent robots), without which the COBOT platform can not function, the human-robot collaboration being mandatory and deployed according to an appropriate structure of the COBOT platform software.

- in Fig. 13: COBOT technology platform for industrial control processes from the Digitized Enterprise and Industry 4.0 **in the Aeronautics field**

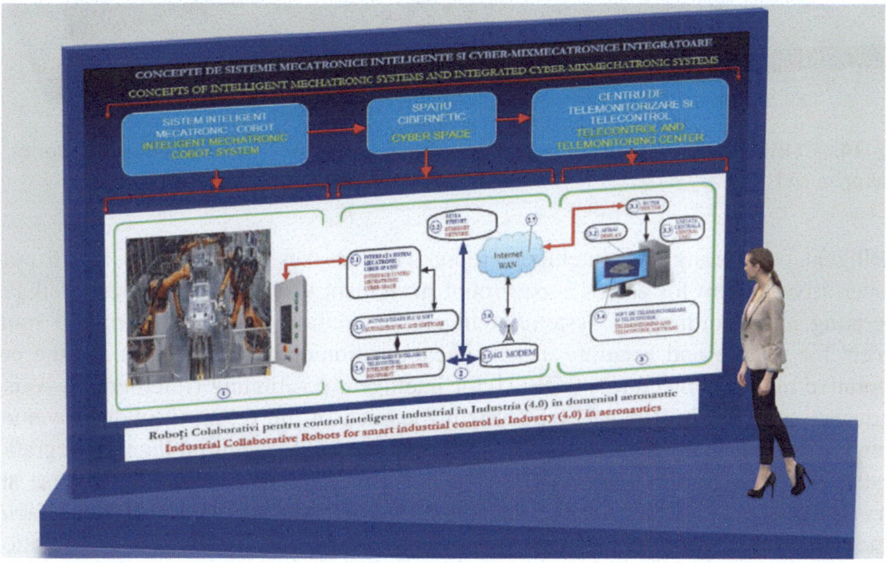

Fig. 13. COBOT technology platform for industrial control processes in enterprise and Industry 4.0 aeronautics

The COBOT technology platform for aeronautical industrial landmark control processes is built and structured on the collaborative control robots architecture of aeronautical industrial landmarks (robots/mechatronic operating and control systems/mechatronic viewing systems/mechatronic interfaces with cyber spaces/etc.), on the architecture of cyberspace (4GMODEM/Antennas/Internet/Intranet/IoT/CPS/CMS/etc.), on the architecture of monitoring/telemonitoring, service/teleservice, control/telecontrol and teleconfiguration (routers/computers/programs - specialized software/etc.) and collaborative management architecture (the collaborative man with the technological platform, where the human-robot collaboration process defines COBOT and the obligation through the technological platform.

- in Fig. 14: COBOT technology platform for positioning processes in Metrology Laboratories in Enterprise and Industry 4.0, **in the field of Intelligent Metrology**

The COBOT technology platform for positioning processes in metrology laboratories is built and structured on the collaborative robot architecture - COBOT for

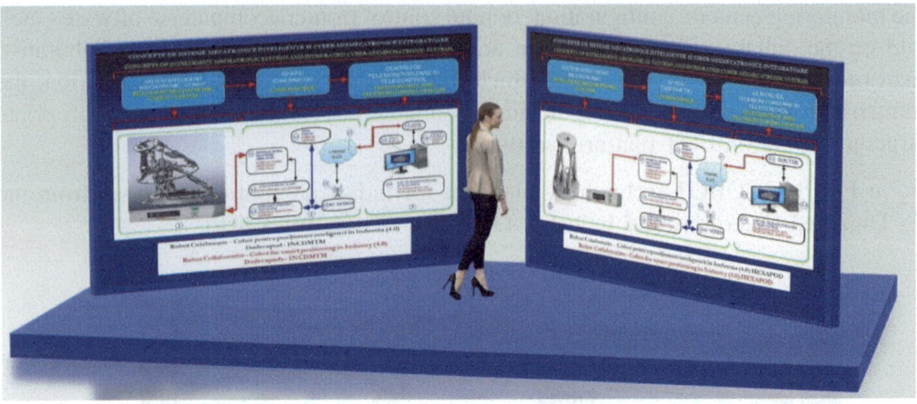

Fig. 14. COBOT technology platform for positioning processes in metrology laboratories in enterprise and Industry 4.0

intelligent positioning (the intelligent robot with 12 axes - 6 Cartesian and 6 polar proper/drive system for each 12 axes/robot movement visualization system/intelligent operating and coordinating system/intelligent integration system sensors and transducers for precise and accurate measurement/positioning processes/etc.), on the collaborative robot architecture - COBOT for intelligent positioning (Intelligent 6-Axis-3 Cartesian and 3-Axle Robot/Drive System for Each of the 6 Axes/Robot Movement Viewing System/Intelligent Operating and Coordinating System/Intelligent Integration System Sensors and Transducers for Measurement/Positioning Processes precise and very precise/etc.), per country (antennas/4GMODEMs/Internet/Intranet/interfacing systems/etc.), on the architecture of monitoring/telemonitoring, configuration/teleconfiguration, control/telecontrol and service/teleservice centers (routers/computers/programs - specialized software/etc.).

5 Conclusions

The scientific paper shows that industrial digitization will impact both horizontally and vertically on the value chain, which implies that on the one hand companies need to integrate and digitize their data flow vertically. to the development of products and purchases to processing and transport logistics, and on the other hand, involves a horizontal collaboration with key suppliers, customers and other partners in the value chain

Companies, businesses and industry in general Industry 4.0 should be involved in the implementation and deployment of complex digital solutions, and all committed staff are fully confident that industrial digitization is the most appropriate and necessary choice for the future.

The scientific work also synthesizes the beginnings of industrial digitization by introducing intelligent concepts and solutions proposed by the author for mechatronic and cyber-mixmechronic integrated systems that are or are to be implemented in different industrial sectors (automobiles, aerospace, agriculture, medicine, etc.) from Romania.

The Enterprise and Industry Digitization Strategy 4.0 is synthesized at national and European level in the following chart.

Pillar 1

Single Digital Market - Free cross-border access to online services and information

Pilllar 2

Interoperability and standards - integration, devices, applications, data and services in the code of social ethics

Pilllar 3

Trust and security - Increase Internet users' trust in electronic services and online transactions through transparency and security

Pillar 4

Fast and ultrafast access to the Internet - aims to invest in infrastructure in broadband equipment

Pillar 5

Research and Innovation in ICT - Stimulates adequate funding for increased competitiveness

Pillar 6

Increasing the digital literacy of skills and inclusion - Creating a bridge to the digital divide

Pillar 7

ICT benefits for EU society - ICT's ability to reduce bureaucracy, support elderly care, improve health services, and deliver public services

Goals to be Achieved by 2020:

- Employment (75% of people between 20 and 65 years should be employed)
- Research/Development (3% of GDP should be allocated to R/D)
- In the field of education (40% of people between 30 and 34 years to complete the third level of education)
- Combating poverty and social exclusion.

References

1. German Engineering Association (www.vdma.org), German association of ICT industry (www.bitkom.org), and German association of electrics and electronics industry (www.zvei.org)
2. Club IT&C – TAG Media – mai 2018: Digital Transformation 2018. ISSN 1583-5111
3. FU şi Acatech: Asigurarea viitorului industriei prelucrătoare din Germania - Recomandări pentru implementarea inițiativei strategice INDUSTRIE 4.0 – Raport Final (2013)

4. Gheorghe, I.G., Constantin, A., Ilie, I.: Mechatronics and cyber mechatronics in intelligent applications from industry and society. In: Proceedings of OPTIROB 2016, Romania (2016)
5. Gheorghe, I.G.: Mecatronica şi Cyber-MixMecatronica în Industria 4.0, Editura CEFIN. ISBN 978-606-8261-28-7, Ed. CEFIN (2018)
6. https://pt.dreamstime.com/ilustra%C3%A7%C3%A3o-stock-agricultura-esperta-image83848448
7. http://digital-agri.fr/agriculture-et-digitalisation/
8. http://electronica-azi.ro/2017/07/07/industrie-4-0-realitatea-interconectata-a-fluxurilor-industriale-de-productie/
9. https://media.hotnews.ro/media_server1/document-2018-10-11-22752549-0-studiu-guvernare.pdf
10. Boston Consulting Group: Industria 4.0: Viitorul productivităţii şi creşterii în industria prelucrătoare (2015)
11. BMWi: Digitizarea industriei - Plattforma Industrie 4.0, Raport de activitate - Aprilie 2016 (2016)
12. Karnouskos, S., Colombo, A.W., Bangemann, T., Manninen, K., Camp, R., Tilly, M., Stluka, P., Jammes, F., Delsing, J., Eliasson, J.: A SOA-based architecture for empowering future collaborative cloud based industrial automation. In: IECON 2012 - 38th Annual Conference on IEEE Industrial Electronics Society, pp. 5766–5772 (2012)
13. Gheorghe, G.: Original concepts and achievements for designing of smart mechatronic and cyber-mixmechatronic systems used in laboratories and in the industry. In: 18th IFAC Conference will take place by support of IFAC (International Federation of Automatic Control), TECIS 2018, 13–15 september, Baku, Azerbaijan, publicata de IFAC-PapersOnLine (2018)

Author Index

G. I. Gheorghe (Ed.): ICOMECYME 2019, LNNS 85, pp. 301–302, 2020.
https://doi.org/10.1007/978-3-030-26991-3

The manufacturer's authorised representative in the EU is Springer Nature Customer Service Centre GmbH, Europaplatz 3, 69115 Heidelberg, Germany. If you have any concerns regarding our products, please contact ProductSafety@springernature.com

Printed and bound by CPI Group (UK) Ltd, Croydon, CR0 4YY
05/10/2026
02228139-0012